Springer Series in
OPTICAL SCIENCES 84

founded by H.K.V. Lotsch

Springer
Berlin
Heidelberg
New York
Barcelona
Hong Kong
London
Milan
Paris
Tokyo

http://www.springer.de/phys/

Springer Series in
OPTICAL SCIENCES

The Springer Series in Optical Sciences, under the leadership of Editor-in-Chief *William T. Rhodes*, Georgia Institute of Technology, USA, and Georgia Tech Lorraine, France, provides an expanding selection of research monographs in all major areas of optics: lasers and quantum optics, ultrafast phenomena, optical spectroscopy techniques, optoelectronics, quantum information, information optics, applied laser technology, industrial applications, and other topics of contemporary interest.
With this broad coverage of topics, the series is of use to all research scientists and engineers who need up-to-date reference books.

The editors encourage prospective authors to correspond with them in advance of submitting a manuscript. Submission of manuscripts should be made to the Editor-in-Chief or one of the Editors. See also http://www.springer.de/phys/books/optical_science/

Satoshi Kawata
Motoichi Ohtsu
Masahiro Irie
(Eds.)

Nano-Optics

With 258 Figures

Springer

Professor Satoshi Kawata
Department of Applied Physics, Osaka University,
Suita, Osaka 565-0871, Japan

Porfessor Motoichi Ohtsu
Tokyo Institute of Technology
4259 Nagatuda, Midori-ku, Yokohama 226-8502, Japan

Professor Masahiro Irie
Department of Chemistry and Biochemistry,
Graduate School of Engineering, Kyushu University, Japan

ISSN 0342-4111

ISBN 3-540-41829-6 Springer-Verlag Berlin Heidelberg New York

Library of Congress Cataloging-in-Publication Data

Nano-optics / Satoshi Kawata, Motoichi Ohtsu, Masahiro Irie (eds.).
p. cm. – (Springer series optical sciences ; v. 74) Includes bibliographical references and index.
ISBN 3540418296 (alk. papaer)
1. Nanostructure materials. 2. Near-field microscopy. 3. Quantum optics. 4. Photonics.
I. Kawata, Satoshi, 1966- II. Ohtsu, Motoichi. III. Irie, Masahiro. IV. Series.
TA418.9.N35 N32 2001
621.36–dc21 2001054264

Springer-Verlag Berlin Heidelberg New York
a member of BertelsmannSpringer Science+Business Media GmbH

http://www.springer.de

Printed in Germany

Camera-ready by authors using a Springer TEX macropackage
Cover concept by eStudio Calamar Steinen using a background picture from The Optics Project. Courtesy of John T. Foley, Professor, Department of Physics and Astronomy, Mississippi State University, USA.
Cover production: *design & production* GmbH, Heidelberg

Printed on acid-free paper SPIN 10798011 57/3141/yu 5 4 3 2 1 0

Preface

Nano-optics is a new optical science dealing with evanescent photons confined in a nanometer scale volume smaller than the spot size determined by classical diffraction theory. Such photons can be manipulated with a nanoprobe located in the optical near field of structures. High-density data storage can be performed with nano-optics for microscope imaging with nanometric resolution, and also for high-resolution photofabrication, single-molecule detection, and local spectral analysis, This book contains the latest results of the development of imaging theories, both electromagnetic and quantum optics, and discussions on instrumentation both aperture probes and scattering ones. Applications of nano-optics to chemistry, quantum device physics, and biological science are also shown. The authors were selected from the members of the Japanese National Project "Near-Field Nano-Optics," funded by the Ministry of Education, Culture, Sports, Science and Technology, Japan from 1997 to 2000 with 70 professors and scientists in Japan.

Osaka,
February 2002

Satoshi Kawata
Motoichi Ohtsu
Masahiro Irie

Contents

List of Contributors

Hidefumi Akiyama
Institute for Solid State Physics
University of Tokyo
Japan

Jongsuck Bae
Tohoku University
Aoba
Sendai 980-8578, Japan

Kikuo Cho
Osaka University
Machikaneyama 1–3
Toyonaka 560-8531, Japan

Patrick Degenaar
Japan Advanced Institute
of Science and Technology
Japan

Masayoshi Esashi
New Industry Creation Hatchery
Center
(NICHE)
Tohoku University
Japan

Masamichi Fujihira
Tokyo Institute of Technology
Nagatsuta, Midori-ku
Yokohama, 226–8501, Japan

Masuo Fukui
Guraduate school of Engineering
Tokushima University
Japan

Makoto Kuwata-Gonokami
Department of Applied Physics
University of Tokyo
Japan

Kazuhiro Hane
Graduate School of Engineering
Tohoku University
Japan

Hirokazu Hori
Yamanashi University
4-3-11 Takeda
Kofu 400–8511, Japan

Yasushi Inouye
Department of Applied Physics
Osaka University
Suita
Osaka 565-0871, Japan

Masahiro Irie
Department of Chemistry and
Biochemistry
Graduate School of Engineering
Kyushu University
Japan

Yoshiharu Ishii
Single Molecule Processes Project
ICORP, JST
2-4-14 Senba-Higashi, Mino
Osaka 562-0035 Japan

Shinzaburo Itoh
Kyoto University
Yoshida
Sakyo–ku
Kyoto 606-8501, Japan

Kotaro Kajikawa
Tokyo Institute of Technology
Yokohama, 226-8502, Japan

Olaf Karthaus
Chitose Institute
Science and Technology
Japan

Toshihiko Kataoka
Department of Precision
Science and Technology
Osaka University
Suita 565-0871, Japan

Satoshi Kawata
Department of Applied Physics
Osaka University
Suita
Osaka 565-0871, Japan

Kazuo Kitahara
International Christian University
3–10–2 Osawa
Mitaka 181-8585, Japan

Akihiro Kusumi
ERATO, JST
Nagoya, 460–0012, Japan
Department of Biological Science
Nagoya University
464-8602, Japan

Gerard Marriott
Department of Physiology
University of Wisconsin–Madison,
USA

Hiroshi Muramatsu
R & D Center
Seiko Instruments
563 Takatsuka-shinden, Japan

Ryusuke Nishitani
Kyushu Institute of Technology
Iizuka
Fukuoka 820, Japan

Shun-ichi Ogawa
Kusumi Membrane Organizer
Project
ERATO, JST
Nagoya, 460-0012, Japan

Satoko Ohta-Iino
Kusumi Membrane Organizer
Project
ERATO, JST
Nagoya, 460-0012, Japan

Motoichi Ohtsu
Tokyo Institute of Technology
4259 Nagatuda
Midori-ku
Yokohama 226-8502, Japan

Takayuki Okamoto
RIKEN
Wako 351-0198, Japan

Satoshi Okazaki
Kyoto University
Yoshida
Sakyo-ku
Kyoto 606-8501, Japan

Takahito Ono
Graduate School of Engineering
Tohoku University
Japan

Masatoshi Osawa
Hokkaido University
Sapporo 060-0811, Japan

Kiwamu Saito
Graduate School of Natural Science and Technology
Kanazawa University
Kanazawa, 920-1192, Japan

Yasushi Sako
Department of Physiology and Biosignaling
Graduate School of Medicine
Osaka University
ROREST, JST
Suita
Osaka 565-0871, Japan

Keiji Sawada
Sinshu University
500 Wakazato
Nagano 380-8553, Japan

Yasuhiro Sugawara
Graduate School of Engineering
Osaka University
Japan

Atsushi Takahara
Institute for Fundamental Research of Organic Chemistry
Kyusyu Univesity
Hakozaki, Higashi-ku
Fukuoka 812-8581, Japan

Junichi Takahara
Graduate School of Engineering Science
Osaka University
Japan

Eiichi Tamiya
Japan Advanced Institute of Science and Technology
Tatsunokushi
Ishikawa, 923-1292, Japan

Kazuo Tanaka
Department of Electronics and computer engineering
Gifu University
Gifu, Gifu 501-1193, Japan

Takuo Tanaka
Osaka University
Toyonaka
Osaka 560-8531, Japan

Hitoshi Tatsumi
Department of Physiology
Graduate School of Medicine
Nagoya University
Nagoya, 466-8550, Japan
CREST, JST

Norihiro Umeda
Tokyo University of Agriculture and Technology
Koganei
Tokyo 184-8588, Japan

Hirofumi Yamada
Graduate School of Engineering
Kyoto University
Japan

Aishi Yamamoto
Tohoku University
Aoba
Sendai 980-8578, Japan

Sadahiko Yamamoto
Osaka University
Toyonaka
Osaka 560-8531, Japan

Toshio Yanagida
Department of Physiology and Biosignaling
Graduate School of Medicine
Osaka University
Suita
Osaka 565-0871, Japan
Single Molecule Processes Project
ICORP, JST
Osaka 562-0035, Japan

1 Quantum Theory for Near-Field Nano-Optics

K. Cho, H. Hori, and K. Kitahara

Near-field optics and related techniques are based on the electromagnetic interactions of matter in the quasi-static regime, where the electromagnetic fields in a mode coupled with matter play the fundamental role. Here, the quasi-static regime implies that the dominant electromagnetic interactions between matter take place across a distance much smaller than the optical wavelength. Recent developments in microfabrication provide a diversity of probe tips with the potential to pick the local fields out of the sea of macroscopic electromagnetic interactions [1]. In the near-field regime, the optical properties of matter and associated near-fields depend strongly on the sizes and shapes of the matter involved. In fact, the optical response of matter is determined by internal electronic processes including the interaction with optical fields consistent with both electronic and electromagnetic boundary conditions. The resulting scattered fields reflect the properties of these internal processes in the illuminated objects, especially when they are observed in an optical near-field. The scattered fields exhibit asymptotic behavior in the far-field limit as propagating optical waves of a retarded nature that carry electromagnetic energy out of the object [2]. In this case, the optical response of matter can be represented in macroscopic quantities, such as dielectric functions, which enable us to reproduce the macroscopic electromagnetic boundary conditions correctly [3]. In contrast, when an observation is carried out in a subwavelength vicinity of objects, the scattered field involves strong near-fields showing steep decay, which reveal the details of optical interactions taking place inside the objects. Theoretical descriptions of optical near-field processes should therefore be based on detailed studies of the electromagnetic interaction of matter at the microscopic level.

It is possible to separate theoretically the near-field relevant components from those of a propagating nature within ordinary scattering problems in the approximation that the observation process exerts a small or neglegible disturbance on the optical response of the object. The basic optical near-field properties can be found in evanescent waves that arise at illuminated material boundaries and decay exponentially in the direction perpendicular to the boundary surface [3]. The evanescent waves mediate optical interactions only in a short range in the direction normal to the surface and have a wave vector along the surface with magnitudes larger than those of optical waves in free space. This makes it possible to drive an optical interaction localized in a spatial extent narrower than the optical wavelength in vacuum which corresponds to the origin of ultrahigh spatial resolution realized in optical near-field techniques. Optical fields exhibit such localized behavior only as a result of interactions with matter, so that optical near-fields can be considered modes of optical fields coupled with matter.

Here, we have three different characteristic scales with respect to the observation processes of scattered optical fields: the size a of the scatterer, the distance r between the observation point and the object, and the optical wavelength λ_0 under consideration. We can observe several different characteristics involved in optical interactions of matter depending on the relationship between these parameters. The definition of the observation point also depends on our techniques of observation and involves a number of important issues discussed with respect to the way far-field detection of signals arises from near-field interactions. Optical near-field techniques are, therefore, based on fabricating probe-tips that enable us to localize the observation point according to the purpose of observation. Keeping this in mind, we will consider the opservation point as localized virtually at a point in space in our general theoretical treatment in this chapter.

In electronic systems on mesoscopic scales, the size and shape of objects are reflected significantly in the optical response since a macroscopic electronic excitation due to collective motion comes into resonance with a cavity mode in the object. Examples are plasmons, excitons, and their polaritons excited in metallic and semiconductor small objects or thin films. The resonant behavior sometimes strongly enhances the optical near-field interaction, which can be used as a probe-tip with high sensitivity to the specific mode of excitation. In the microscopic limit, the optical response depends strongly on the shape of the object whose size is close to electronic de Broglie wavelengths, due to the nonlocal responses of quantum mechanical electronic systems to optical fields. The optical near-field probes in this microscopic region provide us the possibility of observing and controlling the local excitations and observations in nonlocal modes [4]. Extensive studies in this direction will provide us with novel electronic devices of nanometer size in which quantum coherence and mesoscopic transport properties play very important roles.

Besides the local behaviors, optical near-field techniques also involve important issues of the way local modes and local material responses can be excited and observed by using macroscopic apparatus. In fact, for optical near-field obsevation using an optical system, we should first prepare a setup that can illuminate near-field objects or probe-tips by using a macroscopic light source in the far-field. And we should then have a system appropriate for selectively collecting the scattered light from the small object or probe-tip by using a macroscopic photodetector. It is very difficult to consider such a complicated mixture of microscopic and macroscopic phenomena as a whole from both the experimental and theoretical points of view. The problems can, however, be very much simplified if the system under consideration can be separated into several characteristic subsystems, each of which has a different characteristic scale described as an optical mode. The near-field studies, therefore, involve considerations of the separability of the system into characteristic subsystems with clear meaning and also of the relationship between the properties of electromagnetic interactions and the scales of matter and observation. Note that such separation into meaningful subsystems is possible because the optical wavelength is usually quite large compared with the sizes of microscopic objects and probe tips available in our observations. In contrast to this, in electronic near-field apparatus, such as STM, even an atomic-sized probe-tip produces an electron field localized in a space as wide as the electronic wavelength in the bulk materials. This makes it difficult to separate the near-field relevant subsystems, so that one should take all the major scattering events of electrons inside matter into account to evaluate the microscopic object-probe interaction.

In this section, we will investigate the basic features of optical near-field interaction of matter based on quantum mechanical treatment. One of the most important issues is to understand that the properties of interactions depend strongly on the local configurations of electronic systems, which can be found in mesoscopic electronic systems. Mode descriptions of optical near-fields and field quantization are also useful for understanding the radiative problems of atoms, molecules, and other mesoscopic electronic systems in the near-field regime. Evanescent waves that decay exponentially provide the basis for these studies, in which we discuss several interesting properties such as pseudomomentum and pseudoangular momentum conservation that reflect the nature of the optical near-field as electromagnetic fields coupled with matter. These properties enlarge our techniques to control the state and motion of microscopic and mesoscopic objects, which cannot be obtained by using propagating optical fields. Interactive processes in the optical near-field regime also involve profound problems related to transport properties in quantum mechanical systems. Optical near-field interactions can be interpreted as the tunneling problem of photons, simulating electron tunneling effects at a potential barrier. Detailed study in this direction from the viewpoint of nonequilibrium open systems and dissipation or decoherence of quantum states will provide us with a background for exploration into photoelectronic devices on a nanometer scale.

In Sect. 1.2, we will study the most profound feature of the optical near-field interaction through the nonlocal response theory and its application to resonant near-field optics, which will show us a number of important properties involved in near-field interactions of mesoscopic electronic systems.

In Sect. 1.3, we will study the mode description of optical near-field and field quantization in terms of evanescent waves based on the angular-spectrum representation of scattered fields. We will extend the triplet mode description into a more general one involving detector modes and study the resonant interaction of atoms in near-field regime including spontaneous radiation.

In Sect. 1.4, we will briefly investigate several interesting properties found in near-field optical interactions of matter from the quantum mechanical viewpoint. We study the quantities conserved during resonant interaction of matter with an optical near-field. We also investigate the general poperties of electromagnetic signal transport in mesoscopic electronic systems for studies of electronic devices via analogies to optical near-field phenomena which involve a number of fundamental problems of transport in open thermodynamic systems. We also discuss the compatibility between macroscopic and microscopic descriptions and local mode descriptions.

1.1 Resonant Near-Field Optics

The problems of near-field optics may theoretically be divided into two categories according to the ways of describing matter. In most cases, a matter system (sample, substrate, and probe tip) is described in terms of various dielectric constants. Thus the problem is reduced to solving macroscopic Maxwell equations for various geometries, which is technically demanding because of rather complicated geometries, although the scheme is conceptually well understood. A problem in this type of approach is that one cannot answer how far this scheme is valid as sample size becomes smaller or for resonant processes. The value of the dielctric constant may become different from the bulk value as the sample size gets smaller, or the very concept of a local dielectric constant may be nullified, especially for resonant frequencies [5].

In the second category of theoretical approaches, we study the problem from a microscopic point of view. We describe the induced polarization of matter in a quantum mechanical manner explicitly considering the sizes and shapes of the sample and probe-tip [4]. Especially in resonant processes, this approach includes the nonlocal relationship between source field $\boldsymbol{E}$ and induced polarization $\boldsymbol{P}$ such as (in the case of linear response, for example)

$$\boldsymbol{P}(\boldsymbol{r},\omega) = \int \mathrm{d}\boldsymbol{r}\ \chi(\boldsymbol{r},\boldsymbol{r}';\omega)\ \boldsymbol{E}(\boldsymbol{r}',\omega) \tag{1.1}$$

for a given frequency ω. This type of approach looks rather complicated compared with the first one and has not been used widely until now. However, its importance has become gradually realized as a general problem of radiation–matter interaction since the recent popularity of mesoscopic physics. Thus, the study of near-field optics by this approach is certainly a challenging problem for theorists and will also provide new and useful viewpoints for experimentalists.

In this section, we will sketch the outline of the second approach, and show some applications to near-field optics in the linear response regime.

1.1.1 Outline of Microscopic Nonlocal Response Theory

To study the optical response of mesoscopic (or nanoscale) matter, we need to determine the motion of matter and the electromagnetic (EM) field self-consistently [4,7,8]. For this purpose, it is generally sufficient to consider the current density $\boldsymbol{J}$ of matter and the vector potential $\boldsymbol{A}$ (Coulomb gauge: $\nabla \cdot \boldsymbol{A} = 0$) of the EM field. Their motions are related via the fundamental equations of EM theory and quantum mechanics. In quantum electrodynamics (QED), the relationship is between the two operators, and in the semiclassical framework which we take in this article, equations are set up between their expectation values (i.e., c-numbers).

For a linear response at frequency ω, $\boldsymbol{J}$ can be written as

$$\begin{aligned}\boldsymbol{J}(\boldsymbol{r},\omega) \\ = \frac{1}{c}\sum_{\nu}[g_{\nu}(\omega)F_{\nu 0}(\omega) < 0|\boldsymbol{I}(\boldsymbol{r})|\nu > +h_{\nu}(\omega)F_{0\nu}(\omega) < \nu|\boldsymbol{I}(\boldsymbol{r})|0 >] ,\end{aligned} \tag{1.2}$$

where F's are the variables to be determined, i.e.,

$$F_{\mu\nu}(\omega) = \int \mathrm{d}\boldsymbol{r} < \mu|\boldsymbol{I}(\boldsymbol{r})|\nu > \boldsymbol{A}(\boldsymbol{r},\omega) . \tag{1.3}$$

The current density operator $\boldsymbol{I}(\boldsymbol{r})$ and the factors (g_{ν}, h_{ν}) are defined as

$$\boldsymbol{I}(\boldsymbol{r}) = \sum_{j} \frac{e_j}{2m_j}[\boldsymbol{p}_j\delta(\boldsymbol{r}-\boldsymbol{r}_j) + \delta(\boldsymbol{r}-\boldsymbol{r}_j)\boldsymbol{p}_j] , \tag{1.4}$$

$$g_{\nu} = \frac{1}{E_{\nu 0} - \hbar\omega - i0^+} , \quad h_{\nu} = \frac{1}{E_{\nu 0} + \hbar\omega + i0^+} . \tag{1.5}$$

where $e_j, m_j, \boldsymbol{p}_j$ and $\boldsymbol{r}_j$ are the charge, mass, momentum, and coordinates of the jth particle, respectively, and $E_{\nu 0}(= E_{\nu} - E_0)$ is the νth excitation energy of the matter from its ground state. The matter is assumed to be in the ground state initially, i.e., $T = 0\,\mathrm{K}$. In the expressions for g_{ν} and h_{ν}, we have neglected a term proportional to the charge density of the matter's ground state, which plays a minor role in resonant processes.

From the solution of Maxwell's equations, the vector potential $\boldsymbol{A}(\boldsymbol{r},\omega)$ is given as

$$\boldsymbol{A}(\boldsymbol{r},\omega) = \boldsymbol{A}_0(\boldsymbol{r},\omega) + \frac{1}{c}\int \mathrm{d}\boldsymbol{r}' \; \tilde{\mathbf{G}}_q(\boldsymbol{r}-\boldsymbol{r}') \cdot \boldsymbol{J}(\boldsymbol{r}',\omega) , \tag{1.6}$$

where $\boldsymbol{A}_0(\boldsymbol{r},\omega)$ is an incident field and $\tilde{\mathbf{G}}_q(\boldsymbol{r}-\boldsymbol{r}')$ is the radiative Green function ($q = \omega/c$), defined as

$$\tilde{\mathbf{G}}_q(\boldsymbol{r}-\boldsymbol{r}') = G_q(\boldsymbol{r}-\boldsymbol{r}')\mathbf{1} + \frac{1}{q^2}[G_q(\boldsymbol{r}-\boldsymbol{r}') - G_{q=0}(\boldsymbol{r}-\boldsymbol{r}')]\nabla'\nabla' , \tag{1.7}$$

$$G_q(\boldsymbol{r}) = \frac{e^{iq|\boldsymbol{r}|}}{|\boldsymbol{r}|} = \frac{1}{2\pi^2}\int \mathrm{d}\boldsymbol{k} \; \frac{1}{k^2 - (q + i0^+)^2} \; \mathrm{e}^{i\boldsymbol{k}\boldsymbol{r}} . \tag{1.8}$$

The coupled equations (1.2) and (1.6) for $\boldsymbol{J}$ and $\boldsymbol{A}$ can be rewritten as simultaneous linear equations for the variables $[F_{\mu\nu}(\omega)]$ as

$$F_{\mu 0}^{(0)} = \sum_{\nu}[(\delta_{\mu\nu} - \tilde{A}_{\mu 0,0\nu} g_\nu) F_{\nu 0} - \tilde{A}_{\mu 0,\nu 0} h_\nu F_{0\nu}] \ , \tag{1.9}$$

$$F_{0\mu}^{(0)} = \sum_{\nu}[-\tilde{A}_{0\mu,0\nu} g_\nu F_{\nu 0} + (\delta_{\mu\nu} - \tilde{A}_{0\mu,\nu 0} h_\nu) F_{0\nu}] \ , \tag{1.10}$$

where $F_{\mu\nu}^{(0)}$ is defined by (1.3) with $\boldsymbol{A}$ substituted by the incident field $\boldsymbol{A}_0$. The coefficients $\tilde{A}$ represent the retarded interaction (via a transverse field) between two components of the current density, defined as

$$\tilde{A}_{\mu\nu,\tau\sigma} = \frac{1}{c^2} \int \mathrm{d}\boldsymbol{r} \int \mathrm{d}\boldsymbol{r}' < \mu|\boldsymbol{I}(\boldsymbol{r})|\nu > \tilde{\mathbf{G}}_q(\boldsymbol{r} - \boldsymbol{r}') < \tau|\boldsymbol{I}(\boldsymbol{r}')|\sigma > \ . \tag{1.11}$$

Since $\tilde{\mathbf{G}}_q(\boldsymbol{r} - \boldsymbol{r}')$ describes the propagation of a transverse wave, only the transverse components of the current density vectors contribute in (1.11).

The solution of the simultaneous equations (1.9) and (1.10) determines $\boldsymbol{J}$ via (1.2), and from this $\boldsymbol{J}$ we get the (transverse) response field from (1.6). The longitudinal part of the field can be further determined by the induced current density as

$$\boldsymbol{E}_{long}(\boldsymbol{r}, \omega) = \frac{i}{\omega} \nabla\nabla \int \mathrm{d}\boldsymbol{r}' \ \frac{\boldsymbol{J}(\boldsymbol{r}', \omega)}{|\boldsymbol{r} - \boldsymbol{r}'|} \ . \tag{1.12}$$

In this way the initial conditions of matter and a radiative field completely determine the response field and the current density. The extension to nonlinear processes is straightforward [8].

This way of describing the radiation–matter interaction can be summarized as "a generalized Lorentz picture" [9]. In the well-known Lorentz model, matter is modeled by an assembly of electrical (point) oscillators, from which one gets the (frequency) dispersion of a resonant medium. In the generalized version, it is derived from the first principles that these oscillators have spatial extension and contain radiative shift and width. This way of a microscopic semiclassical description of the radiation–matter interaction provides an appropriate bridge between the macroscopic optical response theory and QED.

One of the most remarkable points in this framework, in contrast with the traditional macroscopic local response theory, is that there is no need for the argument about boundary conditions at the matter–surface interface because the boundary conditions are already taken into account in the microscopic description of matter eigenstates. The problem of an additional boundary condition (ABC) for the resonant processes in spatially dispersive media [5] exists only when one sticks to the macroscopic Maxwell equations. In the microscopic response theory for nonlocal (i.e., spatially dispersive) media, it does not exist at all, as explicitly shown above.

The eigenmodes of the coupled radiation–matter system are given, in terms of the coefficient matrix $\mathbf{S}$ of the r.h.s. of (1.9) and (1.10), by $\det(\mathbf{S}) = 0$. They might be called "self-sustaining modes", since they are finite-amplitude solutions in the absence of an incident field. They represent various known concepts such as bulk and surface polaritons, Xrays in a crystal (in a dynamic scattering regime), whispering gallery modes in a dielectric sphere, surface plasma polaritons, etc., for various corresponding situations. The eigenvalues of the self-sustaining modes differ from matter excitatory energies by the radiative corrections, which represent both the shift and width (real and imaginary parts of $\tilde{A}$, respectively) of each matter's excitatory energy.

The radiative width of the νth eigenstate of matter is, in the lowest order, the imaginary part of the diagonal element of the retarded interaction:

$$Im[\tilde{A}_{\nu 0,0\nu}] = \frac{\omega}{4\pi c^3} \int_{|k|=q} \mathrm{d}\Omega_{\boldsymbol{k}} \boldsymbol{I}^{\perp}_{\nu 0}(\boldsymbol{k}) \cdot \boldsymbol{I}^{\perp}_{0\nu}(\boldsymbol{k}) \ , \tag{1.13}$$

where $\boldsymbol{I}^{\perp}_{\mu\nu}(\boldsymbol{k})$ is the perpendicular (to $\boldsymbol{k}$) component of the $\boldsymbol{k}$th Fourier component of $< \mu|\boldsymbol{I}(\boldsymbol{r})|\nu >$, i.e.,

$$\boldsymbol{I}^{\perp}_{\mu\nu}(\boldsymbol{k}) = \int \mathrm{d}\boldsymbol{r}\ \mathrm{e}^{i\boldsymbol{k}\boldsymbol{r}} < \mu|\boldsymbol{I}(\boldsymbol{r})|\nu >^{(\perp)} \ , \tag{1.14}$$

$$= \sum_{i=1}^{2} \frac{e}{m}\ \hat{e}_i(\boldsymbol{k}) < \mu|\hat{e}_i(\boldsymbol{k}) \sum_j \boldsymbol{p}_j \exp[i\boldsymbol{k}\boldsymbol{r}_j]|\nu > \ . \tag{1.15}$$

In (1.15), $\{\hat{e}_i(\boldsymbol{k});\ i = 1, 2\}$ are two mutually orthogonal unit vectors perpendicular to $\boldsymbol{k}$. The above expression for the radiative width agrees with that of the corresponding QED calculation for a two-level atom in vacuum [10]. This equivalence is valid in the long wavelength approximation (LWA) and also for arbitrarily extended states, as shown above.

The index μ of the quantum mechanical states of a matter has generally infinite degrees of freedom, which cause difficulty in any numerical treatment. There is a very popular recipe for this difficulty, i.e., to divide the whole induced polarization (current density) into resonant ($\boldsymbol{P}_{\mathrm{r}}$) and off-resonant components, and to regard the latter as a background component represented by a constant susceptibility. Since the background part has no more dynamic variables to be determined, this part can be renormalized into the EM field. In this way, the problem is reduced to that with a finite number of resonant components of induced polarization and the EM field induced by the resonant and background parts of induced polarization.

The propagation of an EM field in the presence of background polarization ($= \chi_b\ \boldsymbol{E}(\boldsymbol{r}, \omega)$) alone can be described by the Green function, defined as

$$[\nabla \times \nabla \times -q^2 - 4\pi q^2 \chi_b \Theta(\boldsymbol{r})]\tilde{\mathbf{G}}_{\mathrm{b}}(\boldsymbol{r}, \boldsymbol{r}') = 4\pi\delta(\boldsymbol{r} - \boldsymbol{r}')\mathbf{1} \ , \tag{1.16}$$

where the factor Θ is 1 (0) for $\boldsymbol{r}$ inside (outside) the matter described by the local background susceptibility χ_{b}. The Maxwell equations with the source term due both to the resonant and background polarization can be rewritten as

$$[\nabla \times \nabla \times -q^2 - 4\pi q^2 \chi_b \Theta(\boldsymbol{r})]\boldsymbol{E}(\boldsymbol{r},\omega) = 4\pi q^2 \boldsymbol{P}_{\mathrm{r}}(\boldsymbol{r},\omega) \ . \tag{1.17}$$

Its solution is given as

$$\boldsymbol{E}(\boldsymbol{r}) = \boldsymbol{E}_0(\boldsymbol{r}) + q^2 \int \mathrm{d}\boldsymbol{r}' \, \tilde{\mathbf{G}}_b(\boldsymbol{r},\boldsymbol{r}')\boldsymbol{P}_{\mathrm{r}}(\boldsymbol{r}') \ , \tag{1.18}$$

where only the resonant part of polarization appears explicitly as the source term because the nonresonant part, treated as local polarization, is renormalized into Green's function. The free field $\boldsymbol{E}_0$ is the solution of (1.17) in the absence of $\boldsymbol{P}_{\mathrm{r}}$, and thus, it includes the effect of the background local dielectric. The eigenmodes related to χ_{b} (e.g., whispering gallery modes in a dielectric sphere, Fabry–Pérot modes in a slab, etc.) are included in $\boldsymbol{E}_0$, as well as in $\tilde{\mathbf{G}}_{\mathrm{b}}$.

The response field and the induced polarization are determined as before through the self-consistent requirement between the resonant part of polarization $\boldsymbol{P}_{\mathrm{r}}$ and the transverse component of the EM field [6]. However, the forms of the equations to determine the expansion coefficients $\{F_{\mu\nu}\}$ are rather complicated because of the subtraction of the longitudinal component. A more convenient way to obtain a self-consistent response is to use the full Maxwell field in the definitions of $\{F_{\mu\nu}\}$ and Green's function $\tilde{\mathbf{G}}_{\mathrm{b}}$. Then, the equations to determine the expansion coefficients are

$$F^{(0)}_{\mu 0} = \sum_{\nu}{}' (\delta_{\mu\nu} - \tilde{A}_{\mu 0,0\nu} g_\nu) F_{\nu 0} \tag{1.19}$$

where the prime on the summation restricts the index ν only to the resonant states. The use of the full Maxwell field in $\{F_{\mu\nu}\}$ requires that the previously defined "retarded" interaction $\tilde{A}_{\mu 0,0\nu}$ should also be supplied with an "instantaneous" interaction via a longitudinal EM field. This means that the free field Green's function in (1.11) should be replaced with $\tilde{\mathbf{G}}_{\mathrm{b}}$. The energy levels $\{E_{\nu 0}\}$ should also be determined from the "matter Hamiltonian without the Coulomb interaction among the induced polarization charges" [11].

The advantage of this revised version of nonlocal response theory is that the number of unknown variables can be made very small in accordance with the number of the resonant modes in question, and yet the effect of the neglected degrees of freedom is included as that of the background susceptibility. A practical problem is to prepare the renormalized Green function $\tilde{\mathbf{G}}_{\mathrm{b}}$ for the given form of a dielectric. Its analytic form is known for simple geometries, such as a semi-infinite medium, a slab, and a sphere [12,13].

The use of the renormalized Green function $\tilde{\mathbf{G}}_{\mathrm{b}}$ allows correct treatment of the interaction among the components of induced polarization (current density) via a longitudinal and transverse EM field modified by the presence of a background dielectric medium with a given geometry. This includes various effects, such as (a) a change in the radiative lifetime of resonant levels due to the presence of another polarizable matter and (b) "bulk-like" and "image-charge-induced" screening effects of the Coulomb interaction between the components of induced charge density (or polarization). The image charge effect exists also in the evaluation of the energy levels of the matter $\{E_\mu\}$. The image-charge-induced screening in (b) can work both to enhance and reduce the interaction, depending on the difference in the background dielectric constants across a surface/interface. These effects play an essential role in resonant near-field optical processes, some examples of which are given below.

1.1.2 Resonant SNOM

One of the useful applications of the nonlocal response scheme is scanning near-field optical microscopy (SNOM) in a resonant condition. The use of a resonant condition has a twofold meaning: One is the possibility of higher signal intensity than in the off-resonant condition, and the other is to look for the characteristic spatial structure of resonant polarization. The latter feature is not yet very popular in SNOM studies but contains an unexplored region of fundamental problems in SNOM. As scanning tunneling spectroscopy (STS) explores the shape (microscopic structure) and also the electronic structure of a sample, resonant SNOM is a tool for investigating the spatial structure of induced polarization at each resonance. In this sense, its is more than a microscope to see the shape of a sample [14].

When the light frequency in use is close to a resonant level of the matter system, i.e., sample, probe-tip, and substrate, the induced polarization and hence the signal intensity will be large. Considering that the resonance level depends on the relative positions of the sample, probe-tip and substrate, the signal intensity varies quite sensitively with the position of the probe-tip for a fixed frequency of light. Figure 1.1 shows an example of a contour map of signal intensity for a sample consisting of an assembly of four small spheres of radius a_0 placed on a plane $(x, y, z = 0)$ [15]. The probe-tip, which is also modeled by a sphere, is scanned on a plane $(x, y, z = d)$. Each sphere has a resonant level, and they interact via dipole–dipole and photon-mediated coupling. The light frequency is fixed to one of the resonant energies of the matter system for the position of the probe-tip at $(0, 0, d)$. For this particular frequency, the contour map of signal intensity shows very sharp peaks that are spatially much narrower than the size of a sphere. This is due to the effect of "configuration resonance," which was first proposed as a mechanism to increase the resolution based on a simplified model of a sample consisting of point dipoles [16]. For finite sized samples, however, this effect tends to distort the shape of a sample rather than to increase the resolution. Therefore, this effect should be used, not just for looking at the shape of a sample, but

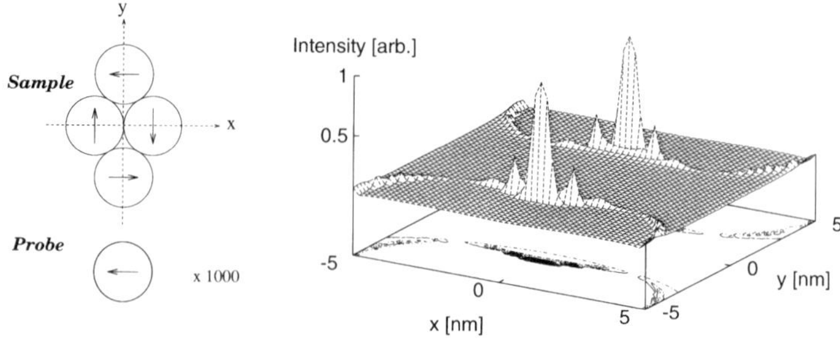

Fig. 1.1. Signal intensity of a particular resonant mode, whose polarization pattern is indicated on the l.h.s.

for studying the details of the spatial structure of induced polarization for each resonant frequency.

Another use of resonant SNOM would be to excite and detect the non-dipolar pattern of induced polarization of a sample. The operating mode that allows this for a sample smaller than the resonant light wavelength is the (internal) reflection mode, where a tip is used to send probe light and collect signal light. Since the excitation and detection are done locally, the tip excites or detects the polarization sufficiently, if the tip is located at the large amplitude of a relevant polarization pattern. And, this is not affected by the dipole or multipole character of the polarization. If the local amplitude is large enough, any modes can contribute to the signal.

Let us see an example [17]. We consider a chain of 10 microspheres of CuCl (radius 3 nm) with a spacing of 5 nm. Each of the spheres is assumed to have a common resonant level. Due to the dipole–dipole coupling of these resonant levels, they form Frenkel excitons extending over the whole chain. The polarization patterns of two typical modes are indicated in Fig. 1.2. Mode 1 is dipole allowed, and mode 2 is dipole forbidden. The former can be excited by a plane wave and can be detected at the far-field, but the latter cannot. If we use the reflection mode SNOM, however, the latter can be excited and detected effectively. As a model of this SNOM, we consider an additional sphere as a SNOM tip, from which the incident field emerges. From the self-consistent solution of induced polarization in every part of the matter system, we pick up the component on the tip as the source polarization for signal light. Through this definition of signal intensity, we get the spectral intensity and its dependence on the tip position for the two resonant modes of the Frenkel exciton, as shown in Fig. 1.3. This example demonstrates that the intensities of the two modes are comparable, i.e., the dipole selection rule no longer works in this circumstance.

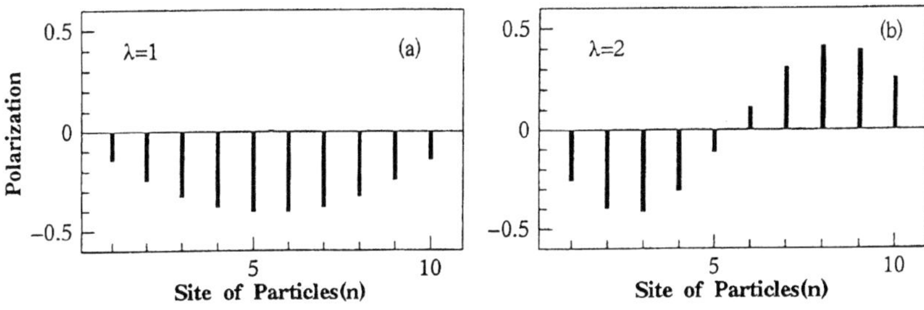

Fig. 1.2. Polarization patterns of dipole-active ($\lambda = 1$) and dipole-forbidden ($\lambda = 2$) modes. The spacing of the spheres is 50 Å

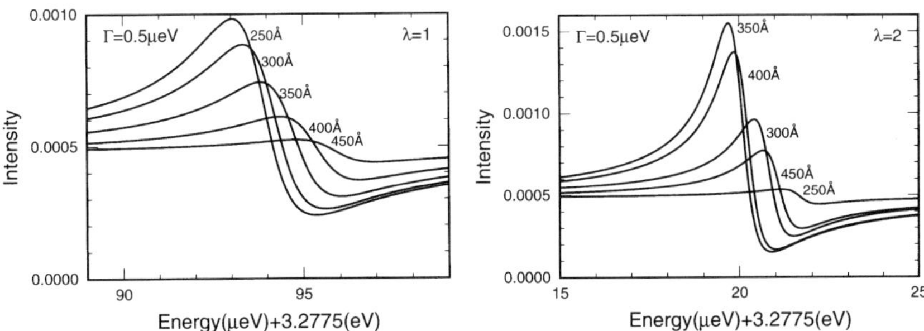

Fig. 1.3. Spectra of signal intensity at various positions of the probe-tip along the chain. The number (in Å) on each curve indicates the tip position. Note the similar intensities of the two cases. A nonradiative decay width of $\Gamma = 0.5\,\mu$eV is considered

1.1.3 Coupling of Cavity Modes and Matter Excitation

Cavity QED is now a popular subject [18] because the radiative lifetime can be severely affected in a cavity, and the interaction between an atomic excitation with a particular cavity mode can be studied with high precision. Since the mode structure of a cavity is common to both classical and quantum mechanical descriptions, we may apply our semiclassical framework to this kind of problem. Let us consider a small resonant matter, such as an atom, a molecule, or a quantum dot (an "atom," hereafter), in the EM field of a microcavity, to see the effect of cavity modes on the resonant frequency and radiative width of the "atom." We may treat the "atom" as a two-level system (1 and 2) corresponding to the relevant transition to observe the radiative width, renormalizing all of the cavity effect in Green's function of the EM field $\tilde{\mathbf{G}}_{\mathrm{b}}$ defined above. Then, the retarded interaction $\tilde{A}$ reduces to a 1×1 matrix $\tilde{A}_{21,12}$. The equation to determine the eigenmodes of the system is

$$E_{21} - \hbar\omega - \tilde{A}_{21,12} = 0\,, \tag{1.20}$$

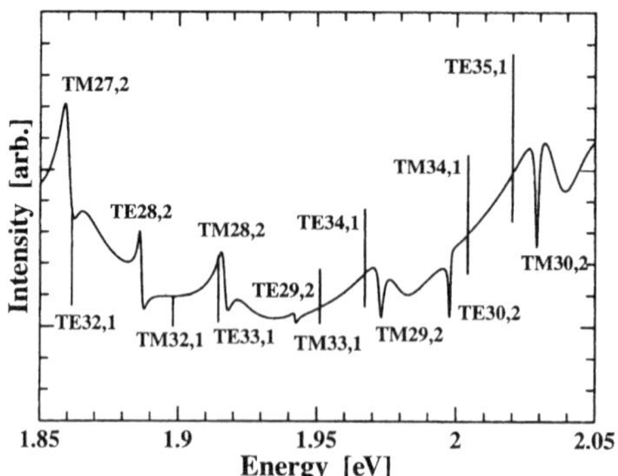

Fig. 1.4. Spectrum of TE and TM WG modes. The two numbers correspond to ℓ and s. Only the modes with $s = 1, 2$ are explicitly assigned

where the matrix element $\tilde{A}_{21,12}$ is generally a function of ω. If its ω-dependence is weak, we may replace the ω by $E_{21}/\hbar$. Then, the complex resonant frequency is

$$\hbar\omega = E_{21} - \tilde{A}_{21,12}(\omega = E_{21}/\hbar) \ . \tag{1.21}$$

The real and imaginary parts of $\tilde{A}$ give the peak shift and width of the response spectrum.

The peculiar feature of a cavity is the existence of resonance modes, which are reflected on the resonant structure of Green's function $\tilde{\mathbf{G}}_{\mathrm{b}}$. If some of the cavity resonances are close to $E_{21}/\hbar$, we must solve the above equation with the ω-dependence of $\tilde{A}$ kept as it is. The solution of such cases leads to the emergence of the coupled modes of "atomic" excitation and cavity modes. Through the coupling, the positions of the original modes are shifted, and they also exchange their widths.

As an example, we show the coupled modes of an "atomic" excitation and the whispering gallery (WG) modes of a spherical cavity of radius a_0 [19]. Figure 1.4 shows the cross section of the sphere alone, with the assignments of various WG modes (in terms of TE, TM, angular momentum ℓ, and order number s). When E_{21} is close to the TM(33, $s = 1$) WG mode, the cross section of the sphere containing an "atom" (at position $r = 0.9a_0$) shows a resonant structure due to the new coupled modes (Fig. 1.5). Note that the sum of the radiative widths is almost conserved among the two coupled modes. Due to the axial symmetry of the system, only the states with the same azimuthal quantum number M couple with one another. This means that there are WG modes unaffected by the "atomic" excitations that remain at the same position in the spectrum. The coupling depends on the field amplitude of the WG mode at the position of the "atom." Therefore, as we change the position of the "atom," the curves in Fig. 1.5 change. When two WG modes couple with the "atomic" excitation, there arises a more complex

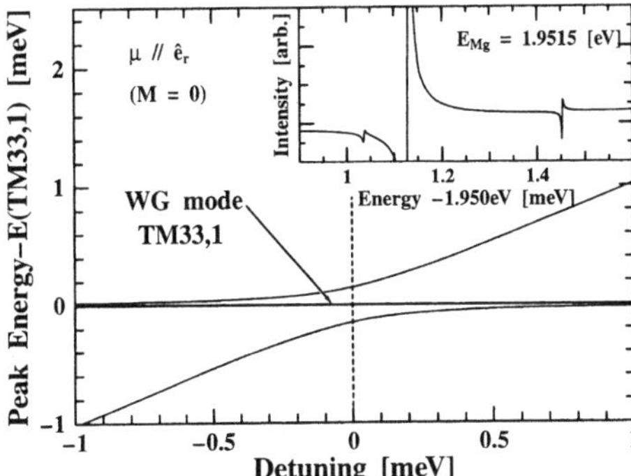

Fig. 1.5. Peak energies of coupled modes as functions of detuning (i.e., the difference between atomic excitatory energy and the eigenenergy of the WG mode.) The horizontal solid line is the energy of the noninteracting WG modes. The inset is the signal spectrum for an atom with a particular detuning energy

exchange of radiative shifts and widths, and, similarly as in the previous example, the sum of the radiative widths is almost conserved among the three modes [19].

The field of near-field nano-optics in the resonant condition has many fundamental problems which have not been studied until now because the theoretical scheme for solids has long remained macroscopic. The microscopic nonlocal scheme outlined in this note is useful for the study of micro-, meso-, and macroscopic systems from a unified point of view. It has been applied to various mesoscopic systems [9], and the examples given here already show some unique results with their general character in describing various optical processes from first principles. A further attempt to broaden the class of problems is now considered.

1.2 Quantization of Evanescent Waves and Optical Near-Rield Interaction of Atoms

Mode descriptions of optical near-fields are useful for understanding the basic properties such as locality and range of interaction as well as for field quantization [1]. In this section, we will show that near-field interactions can be described in terms of evanescent waves as the angular spectrum representation by assuming a virtual planar boundary. The range and locality of interactions are expressed as the width and peak of the angular spectrum. The detector-mode triplet is introduced as the basis for the second quantization convenient for radiative problems, which enables also a direct comparison of quantum and classical radiative processes. Rotational group expression is used to include the vector nature of the fields and to evaluate multipole radiative processes. Modulation of the spontaneous decay rate is demonstrated in the optical near-field regime. The origin of these cavity-QED effects is clarified based on the classical-to-quantum correspondence.

1.2.1 State of Vector Fields

Vector plane waves can be represented in terms of the rotational transform of standard plane waves directed in z with polarization vectors in x and y, as

$$\hat{\boldsymbol{s}}_\mu e^{iK\hat{\boldsymbol{s}}\cdot\boldsymbol{r}} \quad (\mu = \mathrm{TE}, \mathrm{TM}), \quad \hat{\boldsymbol{s}} = (\sin\alpha\ \cos\beta,\ \sin\alpha\ \sin\beta,\ \cos\alpha), \tag{1.22}$$

with the wave number K, its directional angle (α, β), and the polarization vectors $\hat{\boldsymbol{s}}_\mu$ for TE (transverse electric) and TM (transverse magnetic) modes as

$$\hat{\boldsymbol{s}}_{\mathrm{TE}} = (-\sin\beta,\ \cos\beta,\ 0), \quad \hat{\boldsymbol{s}}_{\mathrm{TM}} = (\cos\alpha\ \cos\beta,\ \cos\alpha\ \sin\beta,\ -\sin\alpha). \tag{1.23}$$

Extending the rotational of angle α into the complex region, we can obtain the descriptions of inhomogeneous modes, i.e., evanescent waves [3], including their polarization states. We consider the contour C for the analytic continuation of α as shown in Fig. 1.6 with $-\pi \le \beta < +\pi$.

Vector spherical waves are labeled by wave number K, total angular momentum J, its z-projection m (magnetic quantum number), and polarization state μ (electric; E or magnetic+ M). The so-called cavity mode functions corresponding to incoming waves are written as the helicity representation:

$$\boldsymbol{A}^{\mathrm{M}}_{K,J,m}(\boldsymbol{r}) = i^J j_J(\rho)\boldsymbol{Y}_{J,J,m}(\hat{\boldsymbol{r}}), \tag{1.24}$$

$$\boldsymbol{A}^{\mathrm{E}}_{K,J,m}(\boldsymbol{r}) = i^{J+1}\sqrt{\frac{J}{2J+1}}j_{J+1}(\rho)\boldsymbol{Y}_{J,J+1,m}(\hat{\boldsymbol{r}})$$

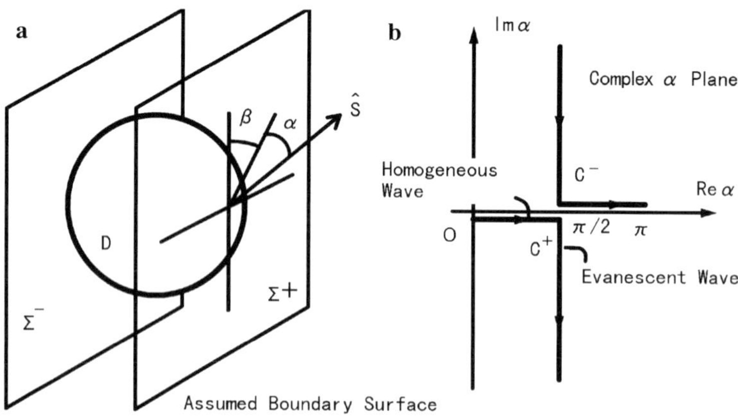

Fig. 1.6. Half-space configuration of the contour of analytic continuation for α as the basis of angular spectrum representation

$$+i^{J-1}\sqrt{\frac{J+1}{2J+1}}j_{J-1}(\rho)\boldsymbol{Y}_{J,J-1,m}(\hat{\boldsymbol{r}}) , \tag{1.25}$$

where $\hat{\boldsymbol{r}}$ is the unit vector of $\boldsymbol{r}$, $\rho = K\mathrm{r}$, $j_\ell(\rho)$ the spherical Bessel function, and $\boldsymbol{Y}_{J,\ell,m}(\hat{\boldsymbol{r}})$ the vector spherical harmonics [2] defined by

$$\boldsymbol{Y}_{J,\ell,m}(\hat{\boldsymbol{r}}) = \sum_{\mu=-1}^{+1} < \ell, 1; m-\mu, \mu | J, m > \mathrm{Y}_\ell^{m-\mu}(\hat{\boldsymbol{r}})\hat{\boldsymbol{s}}_\mu , \tag{1.26}$$

and $< \ell, 1; m-\mu, \mu | J, m >$ stands for the Clebsch-Gordan coefficients. The so-called radiative mode functions corresponding to outgoing waves can be obtained by replacing the spherical Bessel function in the cavity mode functions by the spherical Hankel function of the first kind, $h^{(1)}(\rho)$

$$\boldsymbol{A}_{K,J,m}^{\lambda\ (\mathrm{RA})}(\boldsymbol{r}) = \boldsymbol{A}_{K,J,m}^{\lambda}(\boldsymbol{r}) \left[j(\rho) \to h^{(1)}(\rho) \right] \quad (\lambda = \mathrm{E}, \mathrm{M}) . \tag{1.27}$$

It is convenient to employ the transverse electric, magnetic, and the longitudinal spherical vector functions as follows:

$$\boldsymbol{Y}_{J,m}^{\mathrm{E}}(\hat{\boldsymbol{s}}) = \sqrt{\frac{J}{2J+1}}\boldsymbol{Y}_{J,J+1,m}(\hat{\boldsymbol{s}}) + \sqrt{\frac{J+1}{2J+1}}\boldsymbol{Y}_{J,J-1,m}(\hat{\boldsymbol{s}}) , \tag{1.28}$$

$$\boldsymbol{Y}_{J,m}^{\mathrm{M}}(\hat{\boldsymbol{s}}) = \boldsymbol{Y}_{J,J,m}(\hat{\boldsymbol{s}}) , \tag{1.29}$$

$$\boldsymbol{Y}_{J,m}^{\mathrm{L}}(\hat{\boldsymbol{s}}) = -\sqrt{\frac{J+1}{2J+1}}\boldsymbol{Y}_{J,J+1,m}(\hat{\boldsymbol{s}}) + \sqrt{\frac{J}{2J+1}}\boldsymbol{Y}_{J,J-1,m}(\hat{\boldsymbol{s}}) . \tag{1.30}$$

Using circular polarization vectors with well defined helicity $+1, 0$ and -1,

$$\hat{\boldsymbol{s}}_{\pm 1} = \mp\frac{1}{\sqrt{2}}(\hat{\boldsymbol{s}}_{\mathrm{TM}} \pm \mathrm{i}\hat{\boldsymbol{s}}_{\mathrm{TE}}), \quad \hat{\boldsymbol{s}}_0 = \hat{\boldsymbol{s}}, \tag{1.31}$$

the vector spherical harmonics are expanded by

$$\boldsymbol{Y}_{J,m}^{\lambda}(\hat{\boldsymbol{s}}) = \sum_{\mu\,=\,-1}^{+1} (-1)^\mu [\hat{\boldsymbol{s}}_{-\mu}\boldsymbol{Y}_{J,m}^{\lambda}(\hat{\boldsymbol{s}})]\hat{\boldsymbol{s}}_\mu \quad (\lambda = \mathrm{E}, \mathrm{M}, \mathrm{L}), \tag{1.32}$$

where the nonvanishing components are as follows:

$$\hat{\boldsymbol{s}}_{\pm 1}\boldsymbol{Y}_{J,m}^{\mathrm{E}}(\hat{\boldsymbol{s}}) = \pm\hat{\boldsymbol{s}}_{\pm 1}\boldsymbol{Y}_{J,m}^{\mathrm{M}}(\hat{\boldsymbol{s}}) = -\sqrt{\frac{2J+1}{8\pi}}D_{m,\mp 1}^{(J)\ *}(\hat{\boldsymbol{s}}), \tag{1.33}$$

$$\hat{\boldsymbol{s}}_0\boldsymbol{Y}_{J,m}^{\mathrm{L}}(\hat{\boldsymbol{s}}) = +\sqrt{\frac{2J+1}{4\pi}}D_{m,0}^{(J)\ *}(\hat{\boldsymbol{s}}), \tag{1.34}$$

where $D_{m,m'}^{(J)}(\hat{\mathrm{s}}) =< J, m|\hat{D}(\hat{\mathrm{s}})|J, m' >$ are the matrix elements of the rotational operators $\hat{D}(\hat{\boldsymbol{s}}) = \hat{D}_z(\beta)\hat{D}_y(\alpha) = \mathrm{e}^{-i\hat{J}_z\beta}\mathrm{e}^{-i\hat{J}_y\alpha}$. The transform formulas between spherical and planar mode functions are given by [20]

$$\boldsymbol{A}_{K,J,m}^{\lambda(\mathrm{RA})}(\boldsymbol{r}) = \frac{1}{2\pi}\sum_{\mu}^{\mathrm{TE,TM}} \int_{\mathrm{C}}\int_{-\pi}^{+\pi} \sin\alpha \mathrm{d}\alpha \mathrm{d}\beta(\mu, \alpha, \beta|\lambda, J, m)\hat{\boldsymbol{s}}_\mu \mathrm{e}^{iK\hat{\boldsymbol{s}}\cdot\boldsymbol{r}}, \tag{1.35}$$

$$\hat{\mathbf{s}}_\mu e^{iK\hat{\mathbf{s}}\cdot\boldsymbol{r}} = 4\pi \sum_{\lambda}^{\mathrm{M,E}} \sum_{J=1}^{\infty} \sum_{m=-J}^{+J} (\lambda, J, m|\mu, \alpha, \beta) \boldsymbol{A}^{\lambda}_{K,J,m}(\boldsymbol{r}) , \tag{1.36}$$

where K, J and m are the wave number, total angular momentum of fields, and its z projection, respectively. The expansion coefficients are given by

$$(\mathrm{E/M}, J, m\,|\mathrm{TE}, \alpha, \beta)$$
$$= \sqrt{\frac{2J+1}{8\pi}} \left(\pm\frac{i}{\sqrt{2}}\right) [D^{(J)}_{m,+1}(\hat{\mathbf{s}}) \pm D^{(J)}_{m,-1}(\hat{\mathbf{s}})] , \tag{1.37}$$

$$(\mathrm{E/M}, J, m|\mathrm{TM}, \alpha, \beta) = -\mathrm{i}(\mathrm{M/E}, J, m|\mathrm{TE}, \alpha, \beta) . \tag{1.38}$$

The reversal formulas can be obtained simply by $(\psi|\Phi) = (\Phi|\psi)^*$. As a result, we can obtain the angular spectrum representation [20,21] as

$$\boldsymbol{A}^{\lambda(\mathrm{RA})}_{K,J,m}(\mathbf{r}) = \frac{1}{2\pi} \int_C \int_{-\pi}^{+\pi} d\Omega_{\hat{\mathbf{s}}} \boldsymbol{Y}^{\lambda}_{J,m}(\hat{\mathbf{s}})\, e^{iK\hat{\mathbf{s}}\boldsymbol{r}} ,$$
$$d\Omega_{\hat{\mathbf{s}}} = \sin\alpha \mathrm{d}\alpha \mathrm{d}\beta . \tag{1.39}$$

One of the advantages of our representation is that the interaction can be divided into homogeneous and evanescent components. This provides a clear way to understand the range and locality of interaction in the near-field regime, that is, the peak of the angular spectrum shows us the dominant spatial frequency involved in the interaction propagator at a given interaction distance. The locality of the interaction is also found as the width of the angular spectrum. For example, we will consider the electric dipole propagator

$$\mathrm{G}_{m,\mu}(\boldsymbol{r}) \equiv (-1)^m \sqrt{6\pi} \hat{\boldsymbol{E}}_\mu A^{\mathrm{E(RA)}}_{K,1,-m}(\boldsymbol{r}) , \tag{1.40}$$

which can be described in terms of the angular spectrum. With $\boldsymbol{r} = Z\hat{\boldsymbol{E}}_z$, this becomes for $\mu = m = 0$ [22],

$$\mathrm{G}_{0,0}(Z\hat{\boldsymbol{E}}_z) = \int_0^1 \mathrm{d}t \frac{3}{2}(1-t^2)\mathrm{e}^{i\rho t} + \int_0^\infty \mathrm{d}\xi \left[-i\frac{3}{2}(1+\xi^2)\mathrm{e}^{-\rho\xi}\right] , \tag{1.41}$$

where the first term describes the homogeneous part and the second term the evanescent one. The latter is dominant for the near-field regime, $\rho < 1$. The numerical result is shown in Fig. 1.7 for the evanescent part as a function of the distance between two interacting electric dipoles. We find that the dominant components are those with spatial frequencies around the inverse of the distance Z and the spectral width indicating the lateral locality of interaction also depends on the inverse of the distance. These results show us the origin of the ultrahigh resolution of near-field optical microscopes.

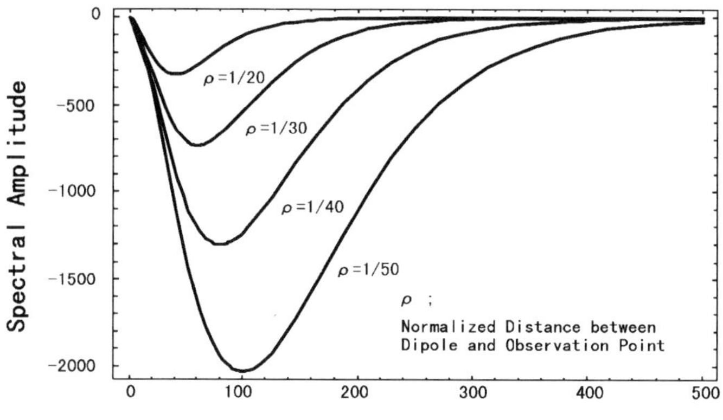

Fig. 1.7. The angular spectrum representation for the electric dipole propagator

1.2.2 Radiative Fields Near a Planar Dielectric Surface

We will now study the reflected and transmitted multipole fields at planar dielectric surfaces. The evaluation of the planar boundary value problem is straightforward by using the angular spectrum representation. Consider the radiative field of multipoles in the lower half-space $z_0 < 0$ which is incident on the dielectric surface. We indicate this component by the superscript (I) and the reflected and transmitted components by (R) and (T), respectively. We will use the coordinate system: the origin is on the dielectric surface, $\boldsymbol{r} = (x, y, z)$ is the general position vector, $\boldsymbol{r} = (x_\mathrm{R}, y_\mathrm{R}, z_\mathrm{R})$ is the position of the multipole, $\boldsymbol{r}_0 = \boldsymbol{r} - \boldsymbol{r}$, $\hat{\mathrm{s}}^{(\mathrm{I})} = \hat{\mathrm{s}}(\alpha, \beta)$, and $\hat{\mathrm{s}}_\mu^{(\mathrm{I})} = \hat{\mathrm{s}}_\mu(\alpha, \beta)$. The vector potential for the fields incident on the dielectric surface is expressed in terms of the angular spectrum with the contour C_- for $0 \le z < z_\mathrm{R}$, and the incident electric multipole field is then given by

$$\boldsymbol{E}(\boldsymbol{r}, t) = -(\partial/\partial t)\mathbf{A}(\boldsymbol{r}, t) = iK\boldsymbol{A}(\boldsymbol{r}, t)\,, \tag{1.42}$$

where we assume a monochromatic radiative field with the temporal evolution e^{-iKt}. With the expansion coefficients defined by

$$f_{\lambda,J,m}(\mu, \alpha, \beta) = (\mu, \alpha, \beta|\lambda, J, m)\mathrm{e}^{-iK\hat{\boldsymbol{s}}(\alpha,\beta)\boldsymbol{R}}\,, \tag{1.43}$$

the incident, reflected,and transmitted electric multipole fields are described as [23]

$$\boldsymbol{E}^{\lambda(\mathrm{I})}_{K,J,m}(\boldsymbol{r}) = \frac{iK}{2\pi}\sum_{\mu}\int_{C_-}\int_{-\pi}^{+\pi}\mathrm{d}\Omega_{\hat{\boldsymbol{s}}}f_{\lambda,J,m}(\mu, \alpha, \beta)\hat{\boldsymbol{s}}_\mu^{(\mathrm{I})}\mathrm{e}^{iK\hat{\boldsymbol{s}}^{(\mathrm{I})}\boldsymbol{r}}\,,$$
$$(0 \le z < z_\mathrm{R}) \tag{1.44}$$

$$\boldsymbol{E}^{\lambda(\mathrm{R})}_{K,J,m}(\boldsymbol{r}) = \frac{iK}{2\pi}\sum_{\mu}\int_{C_-}\int_{-\pi}^{+\pi} \mathrm{d}\Omega_{\hat{\boldsymbol{s}}} f_{\lambda,J,m}(\mu,\alpha,\beta)\mathcal{R}_\mu \hat{\boldsymbol{s}}^{(\mathrm{R})}_\mu \mathrm{e}^{iK\hat{\boldsymbol{s}}^{(\mathrm{R})}\boldsymbol{r}}\,,$$
$$(z \geq 0) \tag{1.45}$$

$$\boldsymbol{E}^{\lambda(\mathrm{T})}_{K,J,m}(\boldsymbol{r}) = \frac{iK}{2\pi}\sum_{\mu}\int_{C_-}\int_{-\pi}^{+\pi} \mathrm{d}\Omega_{\hat{\boldsymbol{s}}} f_{\lambda,J,m}(\mu,\alpha,\beta)\mathcal{T}_\mu \hat{\boldsymbol{s}}^{(\mathrm{T})}_\mu \mathrm{e}^{inK\hat{\boldsymbol{s}}^{(\mathrm{T})}\boldsymbol{r}}\,,$$
$$(z < 0)\,, \tag{1.46}$$

where the mode functions vanish out of the region specified above. Here, $\hat{\boldsymbol{s}}^{(\mathrm{R})}\hat{\boldsymbol{s}}(\alpha^{(\mathrm{R})},\beta) =$ and $\hat{\boldsymbol{s}}^{(\mathrm{R})}_\mu = \hat{\boldsymbol{s}}_\mu(\alpha^{(\mathrm{R})},\beta)$. The angle of reflection is given by $\alpha^{(\mathrm{R})} = \pi - \alpha$. $\hat{\boldsymbol{s}}^{(\mathrm{T})} = \hat{\boldsymbol{s}}(\alpha^{(\mathrm{T})},\beta)$, and $\hat{\boldsymbol{s}}^{(\mathrm{T})}_\mu = \hat{\boldsymbol{s}}_\mu(\alpha^{(\mathrm{T})},\beta)$. The angle of refraction is given by $\sin\alpha = n\sin\alpha^{(\mathrm{T})}$, where $\alpha^{(\mathrm{T})}$ takes the contour C_-, and confirms the convergence of integration for $z < 0$. Here, the reflection $\mathcal{R}_\mu$ and transmission $\mathcal{T}_\mu$ coefficients are given by the Fresnel relations as [3]

$$\begin{aligned}\mathcal{R}_\mu &= \mathcal{R}_\mu(\alpha,\alpha^{(\mathrm{T})}) \\ &= \frac{\cos\alpha - n\cos\alpha^{(\mathrm{T})}}{\cos\alpha + n\cos\alpha^{(\mathrm{T})}}\delta_{\mu,\mathrm{TE}} + \frac{n\cos\alpha - \cos\alpha^{(\mathrm{T})}}{n\cos\alpha + \cos\alpha^{(\mathrm{T})}}\delta_{\mu,\mathrm{TM}}\,,\end{aligned} \tag{1.47}$$

$$\begin{aligned}\mathcal{T}_\mu &= \mathcal{T}_\mu(\alpha,\alpha^{(\mathrm{T})}) \\ &= \frac{2\cos\alpha}{\cos\alpha + n\cos\alpha^{(\mathrm{T})}}\delta_{\mu,\mathrm{TE}} + \frac{2\cos\alpha}{n\cos\alpha + \cos\alpha^{(\mathrm{T})}}\delta_{\mu,\mathrm{TM}}\,.\end{aligned} \tag{1.48}$$

In addition, the electric field corresponding to the multipole radiation directly into the upper half-space $z_0 > 0$ can be derived as

$$\boldsymbol{E}^{\lambda\,(\mathrm{D})}_{K,J,m}(\boldsymbol{r}) = \frac{iK}{2\pi}\sum_{\mu}\int_{C_+}\int_{-\pi}^{+\pi} \mathrm{d}\Omega_{\hat{\boldsymbol{s}}} f_{\lambda,J,m}(\mu,\alpha,\beta)\hat{\boldsymbol{s}}^{(\mathrm{D})}_\mu \mathrm{e}^{iK\hat{\boldsymbol{s}}^{(\mathrm{D})}\boldsymbol{r}}\,,$$
$$(z_{\mathrm{R}} < z)\,, \tag{1.49}$$

where $\hat{\boldsymbol{s}}^{(\mathrm{D})} = \hat{\boldsymbol{s}}(\alpha,\beta)$ and $\hat{\boldsymbol{s}}^{(\mathrm{D})}_\mu = \hat{\boldsymbol{s}}_\mu(\alpha,\beta)$ and the field vanishes for $z < z_{\mathrm{R}}$. The total electric field radiated by the multipole plus surface system is obtained simply by adding field components as

$$\boldsymbol{E}(\boldsymbol{r}) = \boldsymbol{E}^{\lambda\,(\mathrm{D})}_{K,J,m}(\boldsymbol{r}) + \boldsymbol{E}^{\lambda\,(\mathrm{I})}_{K,J,m}(\boldsymbol{r}) + \boldsymbol{E}^{\lambda\,(\mathrm{R})}_{K,J,m}(\boldsymbol{r}) + \boldsymbol{E}^{\lambda\,(\mathrm{T})}_{K,J,m}(\boldsymbol{r})\,. \tag{1.50}$$

The associated magnetic field is obtained from Maxwell's equations.

Investigating the angular spectrum representation, we find that the scattering at the planar dielectric surface involves the following three characteristic processes [23] shown in Fig. 1.8:

1. For homogeneous incident and reflected waves, $(\pi/2) < \alpha \leq \pi$, the transmitted field is homogeneous, $\alpha^{(\mathrm{T})}_{\mathrm{c}} < \alpha^{(\mathrm{T})} \leq \pi$, where $\alpha^{(\mathrm{T})}_{\mathrm{c}}$ is the critical angle defined by $\sin\alpha^{(\mathrm{T})}_{\mathrm{c}} = (1/n)$

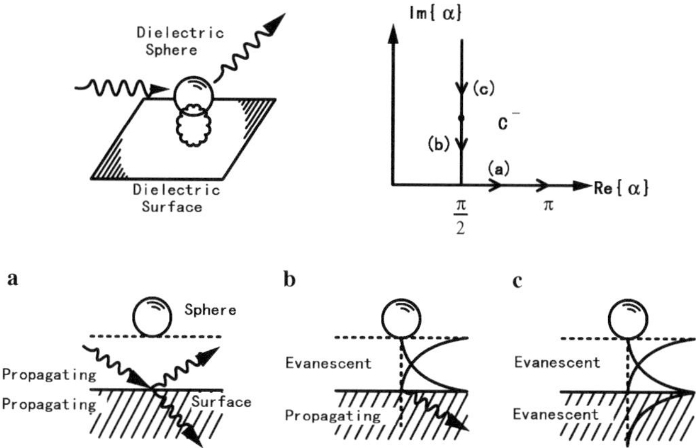

Fig. 1.8. Three characteristics processes involved in multipole radiation near a surface

2. For evanescent incident and reflected waves, i.e., $\alpha = (\pi/2) + i\gamma,\ 0 \leq \gamma < \gamma_c$, with the critical angle $(\pi/2 + i\gamma_c)$ related by $\sin(\pi/2 + i\gamma_c) = n$, the transmitted field is homogeneous, $\alpha_c^{(T)} \geq \alpha^{(T)} > (\pi/2)$.
3. For evanescent incident waves, $\alpha = (\pi/2) + i\gamma,\ \gamma_c \leq \gamma < \infty$, the reflected field is evanescent, $\alpha^{(R)} = (\pi/2) - i\gamma,\ \gamma_c < \gamma < \infty$, and the transmitted field is totally evanescent, i.e., $\alpha = (\pi/2) + i\gamma^{(T)},\ 0 \leq \gamma^{(T)} < \infty$.

Here, case 2 plays the important role in the cavity-QED effect [19,24]. Case 3 is particularly interesting in the confined interactions via localized mode on the surface, such as surface plasmons and varieties of polaritons. This corresponds, therefore, to multiple scattering between the emitter and the surface.

1.2.3 Detector-Mode Functions and Field Quantization

We consider a half-space problem: the half is filled with a homogeneous, isotropic, dielectric medium of refractive index n (left half-space, $z < 0$), and another half is vacuum (right half-space, $z \geq 0$). We consider the Fourier component $\boldsymbol{E}(\boldsymbol{r}, t) = \boldsymbol{E}(\boldsymbol{r}) \exp(-iKt)$ with frequency K $(c = 1)$. $\mathbf{E}(\boldsymbol{r})$ satisfies the Helmholtz equation

$$\nabla^2 \boldsymbol{E}(\boldsymbol{r}) + K^2 n^2(\mathbf{r}) \boldsymbol{E}(\boldsymbol{r}) = \mathbf{0}\,;$$
$$n(\boldsymbol{r}) = n(z < 0)\,,$$
$$n(\boldsymbol{r}) = 1(z \geq 0)\,, \tag{1.51}$$

where n is assumed real. Regarding the electromagnetic boundary conditions, the electromagnetic modefunctions should be composed either of the triplet of a single incident plus two outgoing plane waves or the triplet of two incident plus a single outgoing plane wave. The former is the so-called triplet mode [25], and the two outgoing waves correspond to the reflected and transmitted plane waves useful for the scattering problem. For radiative problems, we should consider the correlation between two outgoing waves. The latter is the so-called detector mode [22,26] involving only a single outgoing wave resulting from interference between the reflected and transmitted waves for the incidence of the phase-correlated waves. The detector-mode description is useful for evaluating radiative problems near the surface in quantum mechanics because the final state density of a single outgoing wave can be evaluated simply and the correspondence between quantum and classical processes is straightforward [22].

It is convenient to define the wave vectors of incoming fields from the left of the boundary as $\boldsymbol{k} = (k_x, k_y, k_z)$, and those of outgoing fields to the right of the boundary as $\boldsymbol{k}^{(D)} = (K_x, K_y, -K_z)$, which satisfy the relations

$$k = nK, \quad k_z = +\sqrt{(k^2 - k_x^2 - k_y^2)}, \tag{1.52}$$

$$K_x = k_x, \quad K_y = k_y, \quad K_z = -\sqrt{(K^2 - K_x^2 - K_y^2)}\,, \tag{1.53}$$

where $\boldsymbol{k}$ is always real, but $\boldsymbol{k}^{(D)}$ becomes complex when $K_x^2 + K_y^2 > K^2$. The incoming wave vector from the right is $\boldsymbol{k} = (K_x, K_y, K_z)$, and the outgoing wave vector to the left is $\boldsymbol{k}^{(D)} = (k_x, k_y, -k_z)$. Their projections onto the plane $z = 0$ are $\boldsymbol{k}_\parallel = (k_x, k_y, 0)$ and $\boldsymbol{k}_\parallel = (K_x, K_y, 0)$, where the relations $\boldsymbol{k}_\parallel = \boldsymbol{k}_\parallel = \boldsymbol{k}_\parallel^{(D)} = \boldsymbol{k}_\parallel^{(D)}$ are due to the phase continuity or invariance under the parallel displacement of the system with respect to the planar boundary. This implies pseudomomentum conservation along the boundary surface [27,28]. We can define the electric fields of the R-detector mode as

$$\begin{aligned}\boldsymbol{\mathcal{E}}_{\mathrm{DR}}(\boldsymbol{k}^{(D)}, \mu, \boldsymbol{r}) = \boldsymbol{\mathcal{E}}_{\mathrm{DR}}^{(I)}(\mathbf{K}^{(D)}, \mu, \boldsymbol{r}) \\ + \boldsymbol{\mathcal{E}}_{\mathrm{DR}}^{(R)}(\boldsymbol{k}, \mu, \boldsymbol{r}) + \boldsymbol{\mathcal{E}}_{\mathrm{DR}}^{(T)}(\boldsymbol{k}, \mu, \boldsymbol{r})\,,\end{aligned} \tag{1.54}$$

where μ indicates polarization. With the reflection and transmission coefficients, the components are defined by $\mathcal{R}_{R,\mu}$ and $\mathcal{T}_{R,\mu}$, as follows:

$$\begin{aligned}\boldsymbol{\mathcal{E}}_{\mathrm{DR}}^{(I)}(\boldsymbol{k}^{(D)}, \mu, \boldsymbol{r}) &= \frac{1}{\sqrt{2}} \hat{\mathbf{s}}_\mu(\alpha_1, \beta_1) \exp[iK\hat{\mathbf{s}}(\alpha_1, \beta_1)\boldsymbol{r}]\,, \\ (z \geq 0) &= 0\,, \quad (z < 0)\,,\end{aligned} \tag{1.55}$$

$$\begin{aligned}\boldsymbol{\mathcal{E}}_{\mathrm{DR}}^{(R)}(\boldsymbol{k}, \mu, \boldsymbol{r}) &= \frac{1}{\sqrt{2}} \hat{\mathbf{s}}_\mu(\alpha_1', \beta_1')\mathcal{R}_{R,\mu} \exp[iK\hat{\mathbf{s}}(\alpha_1', \beta_1')\boldsymbol{r}]\,, \\ (z \geq 0) &= 0\,, \quad (z < 0)\,,\end{aligned} \tag{1.56}$$

$$\begin{aligned}\boldsymbol{\mathcal{E}}_{\mathrm{DR}}^{(T)}(\boldsymbol{k}, \mu, \boldsymbol{r}) &= \frac{1}{\sqrt{2}} \hat{\mathbf{s}}_\mu(\alpha_2, \beta_2)\mathcal{T}_{R,\mu} \exp[inK\hat{\mathbf{s}}(\alpha_2, \beta_2)\boldsymbol{r}]\,, \\ (z < 0) &= 0\,, \quad (z \geq 0)\,.\end{aligned} \tag{1.57}$$

The associated magnetic fields are obtained by using Maxwell's equations. Here, $\hat{\mathbf{s}}(\alpha_1, \beta_1) = \boldsymbol{k}^{(D)}/K$, $\hat{\mathbf{s}}(\alpha_1', \beta_1') = \boldsymbol{k}/K$, and $\hat{\mathbf{s}}(\alpha_2, \beta_2) = \boldsymbol{k}/nK$). Snell's law, $\sin\alpha_1 = n\sin\alpha_2$, requires that $\beta_1 = \beta_2$ and $\alpha_1' = \pi - \alpha_1$. We consider in a practical setup that the component $\boldsymbol{\mathcal{E}}^{(I)}_{\mathrm{DR}}$ is coupled to a single photodetector located in the far-field, so that $\boldsymbol{\mathcal{E}}^{(I)}_{\mathrm{DR}}$ is a homogeneous wave having real polar angles $\alpha_1 (0 \le \alpha_1 < \pi/2)$, $\beta_1 (0 \le \beta_1 < 2\pi)$, α_2, and β_2 $(0 \le \alpha_2 < \alpha_{2\mathrm{C}}$ with $n\sin\alpha_{2\mathrm{C}} = 1$, $0 \le \beta_2 < 2\pi)$.

The L-detector mode functions are represented by

$$\begin{aligned}\boldsymbol{\mathcal{E}}_{\mathrm{DL}}(\boldsymbol{k}^{(D)}, \mu, \boldsymbol{r}) &= \boldsymbol{\mathcal{E}}^{(I)}_{\mathrm{DL}}(\boldsymbol{k}^{(D)}, \mu, \boldsymbol{r}) \\ &\quad + \boldsymbol{\mathcal{E}}^{(R)}_{\mathrm{DL}}(\boldsymbol{k}, \mu, \boldsymbol{r}) + \boldsymbol{\mathcal{E}}^{(T)}_{\mathrm{DL}}(\boldsymbol{k}^*, \mu, \boldsymbol{r}), \end{aligned} \tag{1.58}$$

where $\boldsymbol{k}^{(D)}$ indicates the wave vector of the single outgoing wave. Here,

$$\begin{aligned}\boldsymbol{\mathcal{E}}^{(I)}_{\mathrm{DL}}(\boldsymbol{k}^{(D)}, \mu, \boldsymbol{r}) &= \frac{1}{\sqrt{2}}\hat{\mathbf{s}}_\mu(\alpha_2', \beta_2')\exp[inK\hat{\mathbf{s}}(\alpha_2', \beta_2')\boldsymbol{r}], \\ &\quad (z<0) = 0, \quad (z \ge 0),\end{aligned} \tag{1.59}$$

$$\begin{aligned}\boldsymbol{\mathcal{E}}^{(R)}_{\mathrm{DL}}(\boldsymbol{k}, \mu, \boldsymbol{r}) &= \frac{1}{\sqrt{2}}\hat{\mathbf{s}}_\mu(\alpha_2, \beta_2)\mathcal{R}^*_{L,\mu}\exp[inK\hat{\mathbf{s}}(\alpha_2, \beta_2)\boldsymbol{r}], \\ &\quad (z<0) = 0, \quad (z \ge 0),\end{aligned} \tag{1.60}$$

$$\begin{aligned}\boldsymbol{\mathcal{E}}^{(T)}_{\mathrm{DL}}(\boldsymbol{k}^*, \mu, \boldsymbol{r}) &= \frac{1}{\sqrt{2}}\hat{\mathbf{s}}_\mu(\alpha_1'^*, \beta_1')\mathcal{T}^*_{L,\mu}\exp[iK\hat{\mathbf{s}}(\alpha_1'^*, \beta_1')\boldsymbol{r}], \\ &\quad (z \ge 0) = 0, \quad (z < 0),\end{aligned} \tag{1.61}$$

with $\hat{\mathbf{s}}(\alpha_2', \beta_2') = \boldsymbol{k}^{(D)}/(nK)$ $\alpha_2' = \pi - \alpha_2$. The asterisk denotes a complex conjugate. We consider $\boldsymbol{\mathcal{E}}^{(I)}_{\mathrm{DL}}$ also a homogeneous wave, so that (α_2') and β_2 are real $[(\pi/2) < (\alpha_2') \le \pi$, $0 \le \beta_2 < 2\pi]$. Snell's law implies that α_1' can be a complex value, so that $\boldsymbol{\mathcal{E}}^{(T)}_{\mathrm{DL}}$ represents the homogeneous waves for $(\pi - \alpha_{2\mathrm{C}}) < \alpha_2' \le \pi$ and the evanescent waves for $(\pi/2) < \alpha_2' \le (\pi - \alpha_{2\mathrm{C}})$, where $\sin[(\pi/2) + i\gamma_{1\mathrm{C}}] = n$. The reflection and transmission coefficients have the relationship $\mathcal{R}_{L,\mu} = -\mathcal{R}_{R,\mu}$, $\mathcal{T}_{L,\mu} = n(\cos\alpha_2/\cos\alpha_1)\mathcal{T}_{R,\mu}$. The L- and R-detector modes correspond to the eigenstates of the pseudomomentum operator $-i\hbar\nabla_\parallel$ as

$$(-i\hbar\nabla_\parallel)\boldsymbol{\mathcal{E}}_{\mathrm{DL}}(\boldsymbol{k}^{(D)}, \mu, \boldsymbol{r}) = (\hbar\mathbf{k}_\parallel)\boldsymbol{\mathcal{E}}_{\mathrm{DL}}(\boldsymbol{k}^{(D)}, \mu, \boldsymbol{r}), \tag{1.62}$$

$$(-i\hbar\nabla_\parallel)\boldsymbol{\mathcal{E}}_{\mathrm{DR}}(\boldsymbol{k}^{(D)}, \mu, \boldsymbol{r}) = (\hbar\mathbf{k}_\parallel)\boldsymbol{\mathcal{E}}_{\mathrm{DR}}(\boldsymbol{k}^{(D)}, \mu, \boldsymbol{r}), \tag{1.63}$$

where$-i\hbar\nabla_\parallel$ operates on each component of $\boldsymbol{\mathcal{E}}$. The field components are orthogonal to each other with respect to μ, $\boldsymbol{k}^{(D)}$, and $\boldsymbol{k}^{(D)}$.

The quantized electric field operators are then defined as follows:

$$\hat{\boldsymbol{E}}(\boldsymbol{r}, t) = \frac{1}{(2\pi)^{3/2}}\sum_{\mu=\mathrm{TE}}^{\mathrm{TM}}\left(\frac{\hbar K}{\epsilon_0}\right)^{1/2}$$

$$\times \left\{ \int_{(-k_z)<0} \mathrm{d}^3 k^{(D)} [\hat{a}_{\mathrm{DL}}(\boldsymbol{k}^{(D)}, \mu) \boldsymbol{\mathcal{E}}_{\mathrm{DL}}(\boldsymbol{k}^{(D)}, \mu, \boldsymbol{r}) \mathrm{e}^{-iKt} + \mathrm{H.c.}] \right.$$

$$\left. + \int_{(-K_z)>0} \mathrm{d}^3 K^{(D)} [\hat{a}_{\mathrm{DR}}(\boldsymbol{k}^{(D)}, \mu) \boldsymbol{\mathcal{E}}_{\mathrm{DR}}(\mathbf{K}^{(D)}, \mu, \boldsymbol{r}) \mathrm{e}^{-iKt} + \mathrm{H.c.}] \right\} \quad (1.64)$$

where

$$\int_{(-k_z)<0} \mathrm{d}^3 k^{(D)} = \int_{-\infty}^{\infty} \int_{-\infty}^{\infty} \int_{-\infty}^{0} \mathrm{d}k_x \mathrm{d}k_y \mathrm{d}(-k_z) = \int_{k_z>0} \mathrm{d}^3 k,$$

$$\int_{(-K_z)>0} \mathrm{d}^3 K^{(D)} = \int_{-\infty}^{\infty} \int_{-\infty}^{\infty} \int_{0}^{\infty} \mathrm{d}K_x \mathrm{d}K_y \mathrm{d}(-K_z) = \int_{K_z<0} \mathrm{d}^3 K .$$

The operators $\hat{a}_{\mathrm{DL}}(\boldsymbol{k}^{(D)}, \mu)$, $\hat{a}^{\dagger}_{\mathrm{DL}}(\boldsymbol{k}^{(D)}, \mu)$, and $\hat{a}_{\mathrm{DR}}(\boldsymbol{k}^{(D)}, \mu)$, $\hat{a}^{\dagger}_{\mathrm{DR}}(\boldsymbol{k}^{(D)}, \mu)$ are, respectively,the annihilation and creation operators of photons with the polarization μ and the wave vectors $\boldsymbol{k}^{(D)}$ and $\boldsymbol{k}^{(D)}$ of outgoing waves, with the commutative relations:

$$\left[\hat{a}_{\mathrm{DL}}(\boldsymbol{k}^{(D)}, \mu), \hat{a}^{\dagger}_{\mathrm{DL}}(\boldsymbol{k}^{'(D)}, \mu^{'})\right] = \delta_{\mu,\mu'} \delta(\boldsymbol{k}^{(D)} - \boldsymbol{k}^{'(D)}), \quad (1.65)$$

$$\left[\hat{a}_{\mathrm{DR}}(\boldsymbol{k}^{(D)}, \mu), \hat{a}^{\dagger}_{\mathrm{DR}}(\boldsymbol{k}^{'(D)}, \mu^{'})\right] = \delta_{\mu,\mu'} \delta(\boldsymbol{k}^{(D)} - \boldsymbol{k}^{'(D)}) . \quad (1.66)$$

The orthgonal relations for the detector modes and the consistency with the usual triplet modes [25] are verified in terms of the time-reversal and spatial rotational transforms between them [22].

The Hamiltonian $\hat{\mathcal{H}}$, the number operator $\hat{\mathcal{N}}$, and the pseudomomentum operator $\hat{\mathcal{P}}_{\|}$ of the quantized electromagnetic field are defined as follows:

$$\hat{\mathcal{H}} = \int_{(-k_z)<0} \mathrm{d}^3 k^{(D)} \sum_{\mu=\mathrm{TE}}^{\mathrm{TM}} \hbar K \hat{a}^{\dagger}_{\mathrm{DL}}(\boldsymbol{k}^{(D)}, \mu) \hat{a}_{\mathrm{DL}}(\boldsymbol{k}^{(D)}, \mu)$$

$$+ \int_{(-K_z)>0} \mathrm{d}^3 K^{(D)} \sum_{\mu=\mathrm{TE}}^{\mathrm{TM}} \hbar K \hat{a}^{\dagger}_{\mathrm{DR}}(\boldsymbol{k}^{(D)}, \mu) \hat{a}_{\mathrm{DR}}(\boldsymbol{k}^{(D)}, \mu), \quad (1.67)$$

$$\hat{\mathcal{N}} = \int_{(-k_z)<0} \mathrm{d}^3 k^{(D)} \sum_{\mu=\mathrm{TE}}^{\mathrm{TM}} \hat{a}^{\dagger}_{\mathrm{DL}}(\boldsymbol{k}^{(D)}, \mu) \hat{a}_{\mathrm{DL}}(\boldsymbol{k}^{(D)}, \mu)$$

$$+ \int_{(-K_z)>0} \mathrm{d}^3 K^{(D)} \sum_{\mu=\mathrm{TE}}^{\mathrm{TM}} \hat{a}^{\dagger}_{\mathrm{DR}}(\boldsymbol{k}^{(D)}, \mu) \hat{a}_{\mathrm{DR}}(\boldsymbol{k}^{(D)}, \mu), \quad (1.68)$$

$$\hat{\boldsymbol{\mathcal{P}}}_{\|} = \int_{(-k_z)<0} \mathrm{d}^3 k^{(D)} \sum_{\mu=\mathrm{TE}}^{\mathrm{TM}} \hbar \mathbf{k}_{\|}^{(D)} \hat{a}^{\dagger}_{\mathrm{DL}}(\boldsymbol{k}^{(D)}, \mu) \hat{a}_{\mathrm{DL}}(\boldsymbol{k}^{(D)}, \mu)$$

$$+ \int_{(-K_z)>0} \mathrm{d}^3 K^{(D)} \sum_{\mu=\mathrm{TE}}^{\mathrm{TM}} \hbar \boldsymbol{k}_{\|}^{(D)} \hat{a}^{\dagger}_{\mathrm{DR}}(\boldsymbol{k}^{(D)}, \mu) \hat{a}_{\mathrm{DR}}(\boldsymbol{k}^{(D)}, \mu). \quad (1.69)$$

Here, $\boldsymbol{k}_{\parallel}^{(D)} = \boldsymbol{k}_{\parallel}$. The eigenstates of $\hat{\mathcal{H}}$, $\hat{\mathcal{N}}$, and $\hat{\boldsymbol{\mathcal{P}}}_{\parallel}$ are produced by operating $\hat{a}_{\mathrm{DL}}^{\dagger}(\boldsymbol{k}^{(D)}, \mu)$ and $\hat{a}_{\mathrm{DR}}^{\dagger}(\boldsymbol{k}^{(D)}, \mu)$ in the vacuum state $|0\rangle$. For example, the single photon eigenstates with the energy eigenvalue $\hbar K$ are given by

$$|D, 1(\boldsymbol{k}^{(D)}, \mu)\rangle = \hat{a}_{\mathrm{DL}}^{\dagger}(\boldsymbol{k}^{(D)}, \mu)|0\rangle, \quad |D, 1(\boldsymbol{k}^{(D)}, \mu)\rangle = \hat{a}_{\mathrm{DR}}^{\dagger}(\boldsymbol{k}^{(D)}, \mu)|0\rangle. \tag{1.70}$$

1.2.4 Multipole Radiation near a Dielectric Surface

We will now study the spontaneous emission of excited atoms near planar dielectric boundaries. The atom–field interaction operator is given by

$$\hat{V}(t) = -(e/m)\hat{\boldsymbol{p}}\hat{\boldsymbol{A}}(\boldsymbol{r}_0 + \boldsymbol{r}, t), \tag{1.71}$$

where $\hat{\boldsymbol{p}}$ is the electron momentum, e and m, respectively, the electron charge and mass, $\boldsymbol{r}_0$ the relative position vector of electron from the nucleus, $\boldsymbol{r} = (X, Y, Z)$ the nuclear position vector, and $\hat{\boldsymbol{A}}$ the vector potential given by

$$\begin{aligned}\hat{\boldsymbol{A}}(\boldsymbol{r}, t) = {} & \frac{-i}{(2\pi)^{3/2}} \sum_{\mu=\mathrm{TE}}^{\mathrm{TM}} \left(\frac{\hbar}{K\epsilon_0}\right)^{1/2} \\ & \times \left\{ \int_{K_z^{(D)}>0} d^3K^{(D)} [\hat{a}_{\mathrm{DR}}(\boldsymbol{k}^{(D)}, \mu)\boldsymbol{\mathcal{E}}_{\mathrm{DR}}(\boldsymbol{k}^{(D)}, \mu, \boldsymbol{r})\mathrm{e}^{-iKt} - \mathrm{H.c.}] \right. \\ & \left. + \int_{k_z^{(D)}<0} d^3k^{(D)} [\hat{a}_{\mathrm{DL}}(\boldsymbol{k}^{(D)}, \mu)\boldsymbol{\mathcal{E}}_{\mathrm{DL}}(\boldsymbol{k}^{(D)}, \mu, \boldsymbol{r})\mathrm{e}^{-iKt} - \mathrm{H.c.}] \right\}.\end{aligned} \tag{1.72}$$

The negative-frequency part of $\hat{\boldsymbol{A}}$ represents the single photon emission within the first-order perturbation. The matrix elements for spontaneous emission are, therefore, given by

$$\langle D, 1(\boldsymbol{k}^{(D)}, \mu)|\hat{a}_{\mathrm{DR}}^{\dagger}(\boldsymbol{k}'^{(D)}, \mu')|0\rangle = \delta_{\mu,\mu'}\delta(\boldsymbol{k}^{(D)} - \boldsymbol{k}'^{(D)}), \tag{1.73}$$

$$\langle D, 1(\boldsymbol{k}^{(D)}, \mu)|\hat{a}_{\mathrm{DL}}^{\dagger}(\boldsymbol{k}'^{(D)}, \mu')|0\rangle = \delta_{\mu,\mu'}\delta(\boldsymbol{k}^{(D)} - \boldsymbol{k}'^{(D)}). \tag{1.74}$$

We can introduce the time-independent matrix element as $V_{fi}(t) = V_{fi}\exp[-i(\omega_0 - K)t]$, where ω_0 is the resonant frequency of the atomic two-level system. The emission of a photon in the mode with $\boldsymbol{k}^{(D)}, \mu$ for $Z > 0$ is given in terms of the detector-mode formalism by

$$\begin{aligned}V_{fi} = {} & \frac{e}{im}\left[\frac{\hbar}{(2\pi)^3 K\epsilon_0}\right]^{1/2} \Big[\langle\psi_f|\hat{\boldsymbol{p}}[\boldsymbol{\mathcal{E}}_{\mathrm{DR}}^{(I)}(\boldsymbol{k}^{(D)}, \mu, \boldsymbol{r}_0)]^*|\psi_i\rangle\exp(-i\boldsymbol{k}^{(D)}\boldsymbol{r}) \\ & + \langle\psi_f|\hat{\boldsymbol{p}}[\boldsymbol{\mathcal{E}}_{\mathrm{DR}}^{(R)}(\boldsymbol{k}^{(D)}, \mu, \boldsymbol{r}_0)]^*|\psi_i\rangle\exp(-i\boldsymbol{k}\boldsymbol{r})\Big],\end{aligned} \tag{1.75}$$

where $|\psi_i\rangle$ and $|\psi_f\rangle$ are, respectively, the excited and the ground states of the atom. For the photonic mode $\boldsymbol{k}^{(D)}$ we can also derive μ for $Z > 0$

$$V_{fi} = \frac{e}{im}\left[\frac{\hbar}{(2\pi)^3 K\epsilon_0}\right]^{1/2} \langle\psi_f|\hat{\boldsymbol{p}}[\boldsymbol{\mathcal{E}}_{\mathrm{DL}}^{(T)}(\boldsymbol{k}^{(D)}, \mu, \boldsymbol{r}_0)]^*|\psi_i\rangle\exp(-i\boldsymbol{k}\boldsymbol{r}). \tag{1.76}$$

The Fermi golden rule provides us with the transition probability dP for the $i \rightarrow f$ atomic transition with a single photon emission as

$$\mathrm{d}\Gamma = \frac{2\pi}{\hbar^2} \left| V_{fi} \right|^2 \delta(\omega_0 - K)\mathrm{d}\rho(K), \tag{1.77}$$

where $\mathrm{d}\rho(K)$ is the mode density for the photonic final states. The detector modes have a single outgoing plane wave and allow us a straightforward evaluation of the final-state mode density as

$$\mathrm{d}\rho(K) = \mathrm{d}^3\boldsymbol{k}^{(D)} = K^2\mathrm{d}K\mathrm{d}\Omega(\alpha_1, \beta_1), \quad \text{(in vacuum)} \tag{1.78}$$

$$\mathrm{d}\rho(k) = \mathrm{d}^3\boldsymbol{k}^{(D)} = n^3K^2\mathrm{d}K\mathrm{d}\Omega(\alpha_2', \beta_2), \quad \text{(in a medium)}. \tag{1.79}$$

Integrating over K, we obtain

$$\mathrm{d}\Gamma_R(\mu, \alpha_1, \beta_1) = \frac{2\pi K^2}{\hbar^2} \left| V_{fi} \right|^2 \mathrm{d}\Omega(\alpha_1, \beta_1), \tag{1.80}$$

$$\mathrm{d}\Gamma_L(\mu, \alpha_2', \beta_2) = \frac{2\pi n^3 K^2}{\hbar^2} \left| V_{fi} \right|^2 \mathrm{d}\Omega(\alpha_2', \beta_2)\,. \tag{1.81}$$

Substituting the matrix element, we obtain the explicit forms:

$$\begin{aligned}\mathrm{d}\Gamma_R(\mu, \alpha_1, \beta_1) = &\frac{K}{\epsilon_0\hbar}\left(\frac{e}{2\pi m}\right)^2 \\ &\times\Big|\left\langle\psi_f\middle|\hat{\mathbf{p}}\big[\boldsymbol{\mathcal{E}}^{(I)}_{\mathrm{DR}}(\boldsymbol{k}^{(D)}, \mu, \boldsymbol{r}_0)\big]^*\middle|\psi_i\right\rangle \exp(-i\rho\cos\alpha_1) \\ &+ \left\langle\psi_f\middle|\hat{\mathbf{p}}\big[\boldsymbol{\mathcal{E}}^{(R)}_{\mathrm{DR}}(\boldsymbol{k}^{(D)}, \mu, \boldsymbol{r}_0)\big]^*\middle|\psi_i\right\rangle \exp(i\rho\cos\alpha_1)\Big|^2 \\ &\times \mathrm{d}\Omega(\alpha_1, \beta_1)\,,\end{aligned} \tag{1.82}$$

$$\begin{aligned}\mathrm{d}\Gamma_L(\mu, \alpha_2', \beta_2) = &\frac{n^3K}{\epsilon_0\hbar}\left(\frac{e}{2\pi m}\right)^2 \\ &\times\Big|\left\langle\psi_f\middle|\hat{\mathbf{p}}\big[\boldsymbol{\mathcal{E}}^{(T)}_{\mathrm{DL}}(\boldsymbol{k}^{(D)}, \mu, \boldsymbol{r}_0)\big]^*\middle|\psi_i\right\rangle \exp(i\rho\cos\alpha_1)\Big|^2 \\ &\mathrm{d}\Omega(\alpha_2', \beta_2)\,.\end{aligned} \tag{1.83}$$

Here, $\rho = KZ$, and $\sin\alpha_1$ is real, but $\cos\alpha_1$ is possibly complex.

Based on these formulas, we can evaluate the angular distribution of the spontaneous atomic radiation and lifetime. We can describe the spinless atomic excited state as the superposition of the states with orbital angular momentum ℓ, magnetic quantum number m, and the other quantum number ν, as

$$\varphi_i = \sum_m (-1)^m C_{\ell,-m}\varphi_{\nu,\ell,m}, \qquad \left(\sum_m \left| C_{\ell,m} \right|^2 = 1 \right), \tag{1.84}$$

where $C_{\ell,m}$ is the spherical tensor of rank ℓ with magnetic quantum number m and the atomic ground state corresponds to $\varphi_f = \varphi_{\nu',0,0}$. With respect to spontaneous radiation, the most interesting term is that involving atomic interaction with evanescent waves. For example, we can derive the angular distribution of dipole radiation into the medium side as

$$\mathrm{d}\Gamma(\boldsymbol{\kappa}^{(D)},\mu) = \left(\frac{2\pi n^3 K^2}{\hbar^2}\right) \times \left|\boldsymbol{d}_{fi}^{(T)}(\mu)\boldsymbol{E}_M(\boldsymbol{k}^{(D)},\mu,\boldsymbol{r}^{(T)})\right|^2 \mathrm{d}\Omega(\boldsymbol{\kappa}^{(D)}) . \quad (1.85)$$

Here, it can be shown that the radiative process viewed from the medium side is equivalent to that from an electric dipole $\boldsymbol{d}_{fi}^{(T)}(\mu) = \langle\varphi_f|\boldsymbol{d}^{(T)}(\mu)|\varphi_i\rangle$ placed at $\boldsymbol{r}^{(T)} = (X, Y, \xi Z)$ in the entirely filled medium, which can be described in terms of the transmission coefficient as

$$\begin{aligned}\boldsymbol{d}^{(T)}(TE) &= \frac{1}{\xi}\left(\frac{2K_z}{K_z - k_z}\right)\boldsymbol{d} ,\\ \boldsymbol{d}^{(T)}(TM) &= \frac{1}{\xi}\left(\frac{2nK_z}{n^2K_z - k_z}\right)\boldsymbol{d}^{(T)} ,\end{aligned} \quad (1.86)$$

with $\boldsymbol{d}^{(T)} = ne(\xi x_0, \xi y_0, z_0)$. Note that such a direct correspondence between the quantum and classical radiative processes is available only for the detector-mode descriptions. Numerical results are in good agreement with both the results of classical calculations [23,29] and experimental results obtained from excitation spectroscopy of the $\mathrm{CsD_2}$ line at 852.1 nm [30].

1.2.5 Spontaneous Radiative Lifetime in an Optical Near-Field

Taking advantage of the detector-mode description, it is straightforward to evaluate the spontaneous radiative rate of an atomic multipole near a planar dielectric surface. The atomic radiative rate is obtained by integrating the radiative probability over all of the possible electromagnetic final states as

$$\Gamma = \Gamma_{\mathrm{R}}^{(I)} + \Gamma_{\mathrm{R}}^{(C)} + \Gamma_{\mathrm{L}}^{(H)} + \Gamma_{\mathrm{L}}^{(E)} , \quad (1.87)$$

where the first term, $\Gamma_{\mathrm{R}}^{(I)}$, corresponds to the interaction of the atomic multipole with the field component $\boldsymbol{\mathcal{E}}_{\mathrm{DR}}^{(I)}(\mu,\alpha_1,\beta_1)$ of the R-detector mode and with the field component $\boldsymbol{\mathcal{E}}_{\mathrm{DR}}^{(R)}(\mu,\pi-\alpha_1,\beta_1)$ of the R-detector mode; the second term, $\Gamma_{\mathrm{R}}^{(C)}$, corresponds to the interference of these fields due to the phase difference $2\rho\cos\alpha_1$ and the phase shifts obtained from the symmetry and the refraction coefficient, and the third and fourth terms, $\Gamma_{\mathrm{L}}^{(H)}$ and $\Gamma_{\mathrm{L}}^{(E)}$, correspond to the interactions of the atomic multipole with photons in the homogeneous waves and the evanescent waves involved in the L-detector mode, respectively. It is obvious that the sum of the components $\Gamma_{\mathrm{R}}^{(I)} + \Gamma_{\mathrm{L}}^{(H)}$ give rise to an intrinsic spontaneous radiative rate, therefore the spontaneous lifetime of the atomic excited state. The modulation of the spontaneous emission rate, the cavity-QED effect [31], arises from the interference term $\Gamma_{\mathrm{R}}^{(C)}$

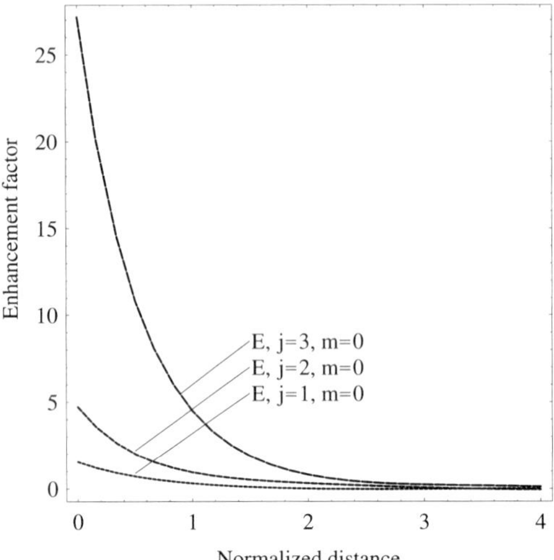

Fig. 1.9. Decay rate of the atomic excited state in the near field of a planar dielectric surface

between the direct and reflected waves into the vacuum side and the interaction with evanescent waves $\Gamma_{\mathrm{L}}^{(E)}$ giving rise to the outgoing waves into the dielectric side beyond the critical angle of total internal reflection. The former suppresses spontaneous emission by decreasing the electromagnetic final state density due to interference, and the latter enhances it by providing an additional electromagnetic final state via optical near-field interactions. The explicit forms are as follows [22]:

$$\begin{aligned}\Gamma_{\mathrm{R}}^{(C)} = \Gamma_0\left(\frac{2J+1}{8}\right)\sum_{\mu=\mathrm{TE}}^{\mathrm{TM}}(-1)^{J+m+\lambda+\mu+1}\int_0^{\pi/2}\sin\alpha_1\mathrm{d}\alpha_1 \\ \times\left[d_{m,+1}^{(J)}(\alpha_1)+(-1)^{\lambda+\mu+1}d_{m,-1}^{(J)}(\alpha_1)\right]^2 2\mathcal{R}_{\mathrm{R},\mu}\cos(2\rho\cos\alpha_1),\end{aligned} \tag{1.88}$$

$$\begin{aligned}\Gamma_{\mathrm{L}}^{(E)} = \Gamma_0\left(\frac{2J+1}{8}\right)\Re\mathrm{e}\left\{\sum_{\mu=\mathrm{TE}}^{\mathrm{TM}}(-1)^{J+m+\lambda+\mu+1}\int_{\pi/2}^{(\pi/2)-i\gamma_{1\mathrm{C}}}\sin\alpha_1\mathrm{d}\alpha_1\right. \\ \left.\times\left[d_{m,+1}^{(J)}(\alpha)+(-1)^{\lambda+\mu+1}d_{m,-1}^{(J)}(\alpha_1)\right]^2 2\mathcal{R}_{\mathrm{R},\mu}\exp(i2\rho\cos\alpha_1)\right\},\end{aligned} \tag{1.89}$$

where $\Re\mathrm{e}\{\ \}$ denotes the real part. Numerical results are shown in Fig. 1.9 for atomic multipole radiation of several orders.

1.3 Quantum Mechanical Aspects of Optical Near-Field Problems

1.3.1 Properties of Near-Field Optical Interactions

We investigate briefly the conserved quantities involved in optical near-field interactions of matter, which exhibit interesting properties arising from the properties of near-fields that depend strongly on the shapes and sizes of the system under consideration. Such modes of optical fields coupled with matter provide us with a number of interesting characteristics useful for innovations of the functional devices in nanometer size. Here, we focus on the pseudomomentum and pseudoangular momentum transferred from optical near-fields to matter through resonant absorption processes. Here, we will consider atomic two-level systems for example.

It is expected that the pseudomomentum $\hbar \boldsymbol{k}_{\parallel}$ is transferred from optical fields to atoms during resonant interactions via evanescent waves. The magnitude of the momentum transferred is expected to be larger than that of free photons with the same frequency, according to the dispersion relation of photons in triplet modes with a complex wave number:

$$K^2 = \left(\frac{\omega}{c}\right)^2 = \left|\boldsymbol{k}_{\parallel}\right|^2 - \left|k_z\right|^2 . \tag{1.90}$$

This is expected because the interacting atom–photon systems near a planar boundary surface show invariance under spatial parallel displacement only in the direction along the surface whereas the imaginary wavenumber implies merely the exponential decay of the strength of the interaction in the direction normal to the surfaces. The conserved quantities with respect to the limited degrees of freedom or symmetries according to the system configuration are referred generally to as pseudoquantities. One of the experimental evidences is the Doppler shift of atomic resonance frequencies, $\Delta\omega_D = \boldsymbol{k}_{\parallel}\boldsymbol{v}$, observed by laser spectroscopy using evanescent waves at a planar dielectric surface, where $\boldsymbol{v}$ is the velocity of atoms moving near the surface. In fact, we can produce evanescent waves of the same frequency with different magnitudes of pseudomomentum by changing the incident angles for total internal reflection. This has been demonstrated by one of the authors using the high-resolution nonlinear laser spectroscopy of Cs atoms at the D_2 resonance line [28], in which two counterpropagating evanescent waves from a single laser are incident at different angles under total internal reflection conditions, produce hole-burning at two different frequencies in the Doppler-broadened absorption profile of hyperfine atomic spectra. This can be observed as the shift of the so-called crossover resonance via the technique of saturated absorption spectroscopy. Note that the short penetration depth of the evanescent waves and the photon emission properties into the evanescent waves have also been demonstrated for the same atom–field configurations [30]. These results show us the consistency between the quantities describing evanescent waves, the penetration depth, the pseudomomentum, and also with detector-mode-based second quantization.

Another interesting quantity involved in optical near-field interactions is the pseudoangular momentum transferred from optical near-fields to atoms through resonant absorption processes. This quantity manifests itself as a conserved quantity under the local rotational symmetry of the atom plus the surface system interacting with photons via evanescent waves. The remarkable property is that the pseudoangular momentum is expected to move normally to the surface and therefore perpendicularly to the direction of wave propagation. This exibits a significant contrast to optical fields in vacuum, in which axes of angular momenta must move parallel to wave vectors due to global symmetry under spatial rotation in the plane perpendicular to the direction of propagation at the speed of light. One interesting setup has been proposed by authors [33]: two TE-polarized evanescent waves closs-propagating at a right angle with equal amplitude E and penetration depth $1/\kappa$ produce a local circularly rotating electric field similar to circular polarization in the direction normal to the surface along the line on which the phase difference between two evanescent waves corresponds exactly to $\pi/2$. Within the dipole approximation, the superposed electric fields serve as a ladder operator $x \pm iy$ for the orbital angular momentum of the atoms, where x and y lie in the plane parallel to the surface. In fact, the resonant dipole transition probability of an atom moving parallel to the surface exactly on this $\Delta\phi = \pm\pi/2$ line at a distance z_0 above the surface is given by [34]

$$T_{fi}\left(\pm\frac{\pi}{2}\right) = \frac{2\pi}{\hbar^2}\left|E\right|^2 e^{-2\kappa z_0} \left|< f|d_x \pm id_y|i >\right|^2 \delta(\omega_0 - \boldsymbol{k}_{\|} \cdot \boldsymbol{v} - Kc), \quad (1.91)$$

where ω_0 is the atomic resonance frequency corresponding to the $i \to f$ transition, $\boldsymbol{v}$ the atomic velocity, d the atomic dipole moments, $\boldsymbol{k}_{\|}\boldsymbol{v}$ corresponds, therefore, to the Doppler shift of the atomic resonance frequency, and Kc the frequency of the cross-propagating evanescent waves. If an atom has spin in its ground state, the absorption of a photon from this field can optically pump the atom by repeated cycles of the excitation with $\Delta\ell = \pm 1$, resulting in electronic spin polarization in the z direction via spin-orbit (LS) coupling and the succeeding spontaneous emission of photons with $\Delta\ell = 0$, on average. Experimental demonstrations are under way by one of the authors [35]. A cylindrical-mode representation of the near-field is useful for describing such processes, which can be easily obtained with the angular-spectrum representation discussed in the previous section [20]. These studies of pseudoangular momentum in the near-field regime will be especially useful in relation to the magneto-optical effect in the near-field regime and the novel techniques of optical diagnosis for electronic spin states at material surfaces or in a two-dimensional electronic system in condensed matter.

1.3.2 Observations and Transport Properties in the Near-Field

The evanescent waves in optical half-space problems exhibit characteristics similar to the tunneling effects of electrons into a potential barrier [32], when the optical system is viewed as a one-dimensional system obtained by projection onto the direction normal to the planar boundary surface. The analogy with respect to the boundary conditions is to TE-polarized electromagnetic waves. It is therefore meaningful to study quantum mechanical transport properties in terms of optical analogs, such as the problem of traversal time of electron wave packets through potential barriers, which often involve pulse-profile transport at superluminal velocities. The studies in this direction are also useful for understanding the basic properties of the observation processes in optical near-field problems and for considering their application to novel optoelectronic devices on the nanometer scale.

Evanescent waves, however, involve a curiosity in their interpretations and physical meanings. It is worth investigating this issue for deeper understanding of the observation process of optical near-fields. Here, we will consider evanescent waves when the half space corresponds to free space and another half to the barrier region with a potential height larger than the energy of the incidental waves considered. It can be shown that evanescent waves in the barrier region that exhibit exponential decay in wave functions have the momentum of a pure imaginary value. This corresponds to the fact that there is no probability current in the barrier region of infinite thickness. Such pure imaginary momenta of evanescent waves, however, bring us some difficulty in our consideration of their physical reality and implications. Here, we will consider how we can carry out an observation of an evanescent wave in the barrier region. As a method of observing a real particle which penetrates into the barrier region to a depth scaled by $1/\kappa$ before the observation, it is natural to consider detection of the particle by observing scattering due to a probe particle impinging on the barrier region. This leads to the situation that a final state of observation is prepared as a real state in which the observed particle is propagating in a narrow hole of the potential bored in the barrier region. It can be shown that, when we introduce a narrow hole in a potential barrier in which the transmitted wave finds a state with a real propagative vector inside the hole, the forward and backward propagating waves in the hole region have the absolute values of amplitudes, $\exp(-\kappa a)$, for the normalized wave incidence at the barrier in the limit of the infinitesimally narrow width of the assumed hole region placed at the distance a from the barrier front, that is, the probability amplitude of observing the particle in the assumed narrow hole in the potential barrier exhibits an exponential decay scaled by the thickness of the barrier wall. This provides us with one of the interpretations for the observation of evanescent waves. With respect to scanning probe microscopies, the narrow hole bored in the barrier region corresponds, therefore, to a probe-tip which scatters a particle penetrating into the barrier and results in the final state of the observed particle propagating to a macroscopic detector placed in the far-field. The calculation shows that in the limit of the infinitesimally narrow probe region, the probability of the scattering of a particle into the final state of a propagating property is just proportional to the amplitude of the evanescent wave at that position without the probing hole. The preparation of probe-tips is, therefore, of primary importance for any near-field observations.

We can now remind ourselves of the basic process of quantum measurements, including tunneling problems through potential barriers. Any quantum system is an isolated system in a rigorous sense. An observation of the quantum state is carried out only by making the quantum system interact with an observation system of classical nature which forces the quantum state to exert a transition into one of the eigenstates with a well-defined value of the physical quantity in the classical observation system. The quantum coherence before the observation is, therefore, destroyed during the interaction with the observation system, which corresponds to the dissipation process or the transport of information from the quantum system into the classical system. This implies that the observation process of quantum systems shows the nature of a nonequilibrium open system with respect to physical information.

An another important point should be noted here that, to determine the final state of the scattered particle by a localized probe, we should force the particle in the forward propagating state out of the probing region. The transition process is otherwise reversed coherently since the backward propagating state has amplitude equal to the forward state and produces a reflecting particle at the front of the potential barrier. This corresponds to the dissipation processes necessary for the observation systems to work as nonequilibrium open systems with unidirectional properties for particle transport. Here, we just note that a number of serious problems exist with respect to the transport properties of particles in such nonequilibrium open systems, which are usually discussed in terms of the linear response theory under an external bias field. Optical near-field techniques have potential as a useful tool in challenging these problems.

1.3.3 Local Mode Descriptions and Compatibility with Macroscopic Descriptions

With respect to global descriptions of near-field optic systems, we now investigate further how we can make the microscopic descriptions of electromagnetic fields compatible with macroscopic descriptions. In dielectric systems, for example, macroscopic optical processes, such as total internal reflection and evanescent waves, are considered the results of spatial distributions of dielectric functions, whereas interactions of molecular constituents belong to our microscopic point of view. The microscopic processes correspond, therefore, to local fluctuations in the macroscopic optical system. Here, we will discuss how local fluctuations are incorporated in electromagnetic equations.

We can find the contributions from such fluctuations in molecular polarization $\boldsymbol{P}$ in the electric displacement vector $\boldsymbol{d} = \varepsilon_0 \boldsymbol{E} + \boldsymbol{P}$. In the dielectric case with macroscopic dielectric function ε, without any isolated charge or conduction current, Maxwell's equation for electric field $\boldsymbol{E}$ for Fourier components with temporal behavior $\exp(-ik_0 ct)$ is written as

$$\nabla^2 \boldsymbol{E} + {k_0}^2 \boldsymbol{E} - \nabla(\nabla \boldsymbol{E}) = -\frac{k_0^2}{\varepsilon_0} \boldsymbol{P} . \tag{1.92}$$

Since the macroscopic average is described by the relation $\langle \boldsymbol{P} \rangle = \langle \boldsymbol{d} \rangle - \varepsilon \langle \boldsymbol{E} \rangle = (\varepsilon - \varepsilon_0) \langle \boldsymbol{E} \rangle$, it is instructive to derive a formulation in which the electrogmagnetic waves are driven by the local fluctuation of polarization $\boldsymbol{P} - (\varepsilon - \varepsilon_0) \boldsymbol{E}$ in volume $\mathcal{V}$ containing the microscopic molecular constituent under consideration. Introducing the refractive index n as $n^2 = \varepsilon / \varepsilon_0$, and therefore, the wave numbers of optical waves propagating in the medium as $k = nk_0$, we obtain the electromagnetic equation

$$\nabla^2 \boldsymbol{E} + k^2 \boldsymbol{E} - \nabla(\nabla \boldsymbol{E}) = -\frac{k_0^2}{\varepsilon_0} \left[\boldsymbol{P} - (\varepsilon - \varepsilon_0) \boldsymbol{E} \right] . \tag{1.93}$$

Further, by introducing the Hertz vector $\boldsymbol{\Pi}$ as $\boldsymbol{E} = \nabla(\nabla \boldsymbol{E}) + k^2 \boldsymbol{\Pi}$ [3], we obtain a wave equation of the Helmholtz type

$$\nabla^2 \boldsymbol{\Pi} + k^2 \boldsymbol{\Pi} = -\frac{1}{\varepsilon} \left[\boldsymbol{P} - (\varepsilon - \varepsilon_0) \boldsymbol{E} \right] , \tag{1.94}$$

where the right-hand side represents the deviation of the local polarization from the macroscopic average in a dielectric medium. This applies for a point dipole source when we consider vacuum. In this way, we can derive the relation between the macroscopic and microscopic regimes of electromagnetic equations. Here, it is also noted that a convenient theoretical description is given for dielectric problems in the near-field regime by introducing the vector potential $\boldsymbol{C} = \nabla \times \boldsymbol{\Pi}$ based on the divergence-free properties of the electric displacement as $\boldsymbol{d} = \nabla \times \boldsymbol{C}$, which is driven by the source $\boldsymbol{j}_M = \nabla \times \boldsymbol{P}$ usually referred to as the magnetic current density [36].

In contrast to the above, it is sometimes desirable to establish localized-mode descriptions of optical near-fields, especially in their quantum mechanical treatments. This issue is related to field quantization in configuration space [37], in which one should assume a volume with the linear scale at least comparable with the wavelength of the localized mode under consideration. In fact, we have introduced triplet-mode formulations involving extended waves in the description of evanescent waves in half-space. One possible approach is the quasi-particle description of electromagnetic fields coupled with material excitations [38], such as plasmons and excitons. These localized-mode descriptions allow us to consider clearly the local properties of near-field subsystems and their connections to macroscopic environments, such as waveguides and so on. Several approaches are under extensive study in terms of polariton-mode descriptions [39,40], which have potential as a useful basis in research aimed at novel applications of quantum optical effects in the near-field regime, such as manipulation and trapping of atomic particles and nanometer-sized polariton devices. These approaches will be spotlighted again in the problems of transport properties in mesoscopic regimes.

Acknowledgments. The author is grateful to Professor Y. Ohfuti and Dr. J. Ushida for the presentation of the unpublished figures, Fig. 1.3 and Fig. 1.1, respectively.

References

1. M. Ohtsu and H. Hori: *Near-Field Nano-Optics* (Kluwer Academic/Plenum, New York 1999)
2. J.D. Jackson: *Classical Electrodynamics*, 2nd ed. (John Wiley & Sons, New York 1975)
3. M. Born and E. Wolf: *Principles of Optics*, 3rd ed. (Pergamon Press, Oxford 1965)
4. K. Cho: Prog. Theor. Phys. Suppl. **106**, 225 (1991)
5. J.L. Birman: *Excitons*, E.I. Rashba and M.D. Sturge, eds. (North Holland, Amsterdam 1982) p.72; P. Halevi, *Spatial Dispersion in Solids and Plasmas*, P. Halevi, ed. (Elsevier, 1992)
6. K. Cho and J. Ushida: *Elementary Processes in Excitations and Reactions on Solid Surfaces*, Springer Series in Solid State Sciences **121**, 193 (1996)
7. K. Cho: J. Phys. Soc. Jpn. **66**, 2496 (1997)
8. H. Ishihara and K. Cho: Phys. Rev. B **42**, 1724 (1990); B**48**, 7960 (1993); B**53**, 15823 (1996)
9. K. Cho: J. Luminescence **87–89**, (2000) 7; Mol. Cryst. Liq. Cryst. **314**, 179 (1998)
10. e.g., Haken: *Laser Theory* (Springer, 1984) Sec. V
11. K. Cho: J. Phys. Soc. Jpn. **68**, 683 (1999)
12. W. C. Chew: *Waves and Fields in Inhomogeneous Media* (IEEE Press, New York 1995)
13. A.A. Maradudin and D.L. Mills: Phys. Rev. **11**, 1392 (1974)
14. K. Cho, Y. Ohfuti, and K. Arima: Jpn. J. Appl. Phys. Suppl. **34**, 267 (1994)
15. J. Ushida and K. Cho: J. Luminescence **66&67**, 94 (1996); J. Ushida: PhD Thesis (Osaka Univ., 1999)
16. O. Keller, M. Xiao, and S. Vozhevolnyi: Surf. Sci. **20**, 217 (1993)
17. K. Cho, Y. Ohfuti, and K. Arima: Surf. Sci. **363**, 378 (1996)
18. S. Haroche and D. Kleppner: Phys. Today **1**, 24 (1989)
19. J. Ushida, T. Ohta, and K. Cho: J. Phys. Soc. Jpn. **68**, 2439 (1999)
20. T. Inoue and H. Hori: Opt. Rev. **3**, 458 (1996)
21. E. Wolf and M. Niet-Vesperinas: J. Opt. Soc. Am. **2**, 886 (1985)
22. T. Inoue and H. Hori: Phys. Rev. A **63**, 063805-1-16 (2001)
23. T. Inoue, I. Banno, and H. Hori: Opt. Rev. **5**, 295 (1998)
24. P. Berman, ed.: *Cavity Quantum Electrodynamics* (Academic Press, San Diego 1994)
25. C.K. Carniglia and L. Mandel: Phys. Rev. D **3**, 280 (1971)

26. J.M. Vigoureux and R. Payen: J. Phys. (Paris) **36**, 1327 (1975)
27. D.F. Nelson: Phys. Rev. A **44**, 3985 (1991)
28. T. Matsudo, T. Takahara, H. Hori, and T. Sakurai: Opt. Commun. **145**, 64 (1998)
29. W. Lukosz and R.E. Kunz: J. Opt. Soc. Am. **67**, 1607; 1615 (1977)
30. T. Matsudo, H. Hori, T. Inoue, H. Iwata, Y. Inoue, and T. Sakurai: Phys. Rev. A **55**, 2406 (1997)
31. G.S. Agarwal: Phys. Rev. A **11**, 230; 243; 253 (1977)
32. C. Cohen–Tannoudji, B. Diu, and F. Laloë: *Principles of Optics* (John Wiley & Sons, New York 1997)
33. H. Hori, K. Kitahara, and M. Ohtsu: *The 1st Asia Pacific Workshop on Near-Field Optics*, Seoul, 1966, p. 49
34. H. Hori, K. Kitahara, I. Banno, and M. Ohtsu: (in press)
35. H. Hori, Y. Ohdaira, K. Kijima, Y. Nakamura, T. Sakurai, and K. Kitahara: *Tech. Digest 5th Int. Conf. Near-Field Opt. and Related Techniques*, Shirahama, 1988, pp. 266–267
36. I. Banno, H. Hori, and T. Inoue: Opt. Rev. **3**, 454 (1996)
37. L. Mandel: Phys. Rev. **144**, 1071 (1966)
38. J.J. Hopfield: Phys. Rev. **3**, 1555 (1958)
39. H. Hori and M. Ohtsu: in *Quantum Control and Measurement*, H. Ezawa and Y. Murayama, eds. (North-Holland, Amsterdam 1993)
40. K. Kobayashi and M. Ohtsu: J. Microsc. **194**, 249 (1999)

2 Electromagnetism Theory and Analysis for Near-Field Nano-Optics

S. Kawata, K. Tanaka, and N. Takahashi

Compared with classical microscopy, either conventional light microscopy with a uniform lamp illumination or advanced laser-scanning microscopy such as confocal microscopy and multi-photon microscopy, near-field microscopy is distinctive in terms of the principle of image formation.

Classical microscopy is described based on the paraxial approximation, mostly by Fresnel's diffraction theory or more precisely Kirchoff's diffraction theory. To accommodate the polarization characteristics at the boundary of lens, a vectorial treatment is also used with the geometrical theory.

However, the paraxial approximation is not appropriate for near-field diffraction, because the size of the probe is much smaller than the wavelength of light, and the structure of the sample and the distance between the probe and the sample are also both smaller than the wavelength of light. There is no simplified or unified theory or approximation for near-field microscope imaging, except that the near-field distribution at a one-dimensional edge or at a slit in a screen has been rigorously solved by Sommerfeld [1] and Bouwkamp [2], respectively.

Another distinction of near-field microscope imaging is a strong interaction or multiple scattering/reflection between the probe and the sample in the near field. Compared with other probe microscopies, the near-field optical interaction between the probe and the sample is the most difficult to analyze because they do not have a directional force as a polar set.

As a result, imaging studies for near-field microscopy have been done using numerical analyses [3–6]. In this chapter, we show the three-dimensional modeling and analysis of near-field scanning microscope imaging for an aperture probe. The method for reconstructing the structure from an image obtained with a near-field microscope is discussed for dielectric strips and metallic strips in a dielectric substrate as samples. Lastly, the radiation force exerted on a small particle in the near field of a subwavelength aperture was also numerically calculated.

2.1 Finite-Difference Time-Domain Analysis of a Near-Field Microscope System

2.1.1 Near-Field Microscope as a Multiple Scattering System

We analyze how the interaction between the probe and the field near the sample surface affects the characteristics of the images produced by a near-field scanning optical microscope [3]. The numerical calculation is carried out using the finite-difference time-domain method, and a three-dimensional model that includes a probe of arbitrary shape is considered. The electrical field distribution formed near and in the sample is very sensitive to the optical properties of the sample, such as its geometrical structure, refractive index and absorptivity distribution. The existence of the probe near the sample heavily disturbs the original electric field distribution diffracted by the sample structure under the illumination.

When analyzing near-field scanning optical microscope (NSOM) images, the effects of probe scattering must be taken into account, especially in the case when its complex dielectric constant is comparable to or even greater than that of the sample itself. This occurs when either a metal-coated dielectric probe or a metallic probe is employed. And as both the probe size and the sample structure are nearly comparable to the wavelength dimensions, the scattered light exhibits the characteristics of Mie scattering. As a result, the surface morphology could be different from the observed NSOM image. The characteristics of an NSOM image of the surface is quite sensitive to the manner in which the probe is scanned over the sample surface.

A complete description of the NSOM imaging process is quite difficult to accomplish, because it requires an understanding of not only how the geometric shape and optical constants of the sample manifest themselves in an NSOM image but also how their effects vary as a function of the scanning mode utilized, the variations in the probe–sample distance, shape and material of the probe, and the polarization of the incident light [7].

The complexity of the NSOM imaging process is greatly reduced if we assume that the near-field generated is produced by the sample alone and is not disturbed by the introduction of a probe [8,9]. In practice, however, the original near-field generated by the sample is altered during the insertion of the probe so that the electric field being detected already includes the interactions made by the photons scattered by the probe.

A description of the NSOM imaging process that includes the effects of the interactions of the probe with the surface field has already been done by Girard and Courjan using the dipole approximation method [10]. In their method, the probe–sample system is modeled by the microscopic polarizability constant of a dipole, so that it is reliable only when dealing with very small structures such as a small number of atomic arrays. When applied to describe results produced by probes of practical sizes and shapes, calculational difficulties are encountered because of the large computer memory required and the high number of operations involved. In describing the images of real structures formed by an NSOM, a more relevant macroscopic permittivity constant needs to be considered in the solution of the Maxwell equations (or equivalently the Helmholtz equation).

The corresponding Helmholtz equation that describes the formation of an image by an NSOM has already been solved using the boundary element method (BEM) [11] and the multiple multipole method (MMP) [12]. In the analysis of a two-dimensional (2-D) NSOM model using the MMP, Novotny et al. [13] used a large amount of computer memory that was proportional to the square of the number of discrete modeling points. The memory demand becomes even more serious in both BEM and MMP when more realistic 3-D models are to be analyzed. For example, BEM requires at least 80 GBytes for 10-wavelength-long cube including four different media with 100 discrete points for each cube length. The finite element method (FEM) [14] can be optimized by solving the band matrix. Even if we use such an optimization, a memory space of least 2 GBytes is still required.

The finite-difference time-domain method (FDTD) [15,16] reduces the memory amount required to only about 0.3 GBytes for the same 3-D model. The memory capacity of current supercomputers and even high-end workstations is already within the 0.5–2.0 GByte range. To analyze a probe–sample interaction for NSOM imaging, we used this FDTD algorithm [3].

2.1.2 Finite-Difference Time-Domain Algorithm for NSOM Imaging

The FD–TD method treats Maxwell equations as a set of finite difference equations in both time and space. The model space considered includes both the probe and the sample surface and consists of an aggregation of cubic cells with each cell having its own complex dielectric constant. The finite difference equations corresponding to Ampere's law: curl $H = \tilde{\varepsilon}\frac{\mathrm{d}E}{\mathrm{d}t}$, can be written as:

$$\frac{H_{z(t;x,y+\Delta y,z)} - H_{z(t;x,y-\Delta y,z)}}{2\Delta y} - \frac{H_{y(t;x,y,z+\Delta z)} - H_{y(t;x,y,z-\Delta z)}}{2\Delta z} = \tilde{\varepsilon}_{(x,y,z)} \frac{E_{x(t+\Delta t;x,y,z)} - E_{x(t+\Delta t;x,y,z)}}{\Delta t}, \tag{2.1}$$

$$\frac{H_{x(t;x,y,z+\Delta z)} - H_{x(t;x,y,z-\Delta z)}}{2\Delta z} - \frac{H_{z(t;x+\Delta x,y,z)} - H_{z(t;x-\Delta x,y,z)}}{2\Delta x} = \tilde{\varepsilon}_{(x,y,z)} \frac{E_{y(t+\Delta t;x,y,z)} - E_{y(t+\Delta t;x,y,z)}}{\Delta t}, \tag{2.2}$$

$$\frac{H_{y(t;x+\Delta x,y,z)} - H_{y(t;x-\Delta x,y,z)}}{2\Delta x} - \frac{H_{x(t;x,y+\Delta y,z)} - H_{x(t;x,y-\Delta y,z)}}{2\Delta y} = \tilde{\varepsilon}_{(x,y,z)} \frac{E_{z(t+\Delta t;x,y,z)} - E_{z(t+\Delta t;x,y,z)}}{\Delta t}, \tag{2.3}$$

where $E = E(E_x, E_y, E_z)$ and $H = H(H_x, H_y, H_z)$ are the electric field and the magnetic induction vectors, respectively, and $2\Delta x$, $2\Delta y$, $2\Delta z$ are increments along the three coordinate directions respectively, Δt is the unit time increment, and $\tilde{\varepsilon}(x, y, z)$ is the complex dielectric constant of the medium at that point. Equations (2.1)–(2.3) are simultaneously solved to determine the component values at the time $t + \Delta t$.

The magnetic induction elements H_x, H_y, H_z are calculated using Faraday's law of induction: $\operatorname{curl} E = -\mu \frac{\mathrm{d}H}{\mathrm{d}t}$:

$$\frac{E_{z(t;x,y+\Delta y,z)} - E_{z(t;x,y-\Delta y,z)}}{2\Delta y} - \frac{E_{y(t;x,y,z+\Delta z)} - E_{y(t;x,y,z-\Delta z)}}{2\Delta z} = -\mu_{(x,y,z)} \frac{H_{x(t+\Delta t;x,y,z)} - H_{x(t+\Delta t;x,y,z)}}{\Delta t} , \tag{2.4}$$

$$\frac{E_{x(t;x,y,z+\Delta z)} - E_{x(t;x,y,z-\Delta z)}}{2\Delta z} - \frac{E_{z(t;x+\Delta x,y,z)} - E_{z(t;x-\Delta x,y,z)}}{2\Delta x} = -\mu_{(x,y,z)} \frac{H_{y(t+\Delta t;x,y,z)} - H_{y(t+\Delta t;x,y,z)}}{\Delta t} , \tag{2.5}$$

$$\frac{E_{y(t;x+\Delta x,y,z)} - E_{y(t;x-\Delta x,y,z)}}{2\Delta x} - \frac{E_{x(t;x,y+\Delta y,z)} - E_{x(t;x,y-\Delta y,z)}}{2\Delta y} = -\mu_{(x,y,z)} \frac{H_{z(t+\Delta t;x,y,z)} - H_{z(t+\Delta t;x,y,z)}}{\Delta t} , \tag{2.6}$$

where $\mu(x, y, z)$ is the permeability of the medium.

The two sets of equations described by (2.1)–(2.3) and (2.4)–(2.6) are alternately evaluated to simulate the light propagation. Figure 2.1 shows a schematic diagram of the 3-D NSOM model to be analyzed. The dielectric probe tip exhibits the profile of a cone. The tip is coated with a thin gold film with a film thickness of $2\lambda/70$ and has an aperture opening of $\lambda/10$. The value of $\lambda = 6.0$ µm is considered because it is the illumination wavelength we used in our recent NSOM application. At λ=6.0 µm, the complex refractive index of gold is 0.27+i 7.1.

The surface profile of the dielectric sample (refractive index = 1.5) consists of a flat surface with a single cylindrical protrusion whose height and diameter are $6\lambda/70$ and $\lambda/5$, respectively. The dimensions of each cell are $2\Delta x = \lambda/70$, $2\Delta y = \lambda/70$, $2\Delta z = 3\lambda/70$, and the total space volume considered measures $150 \times 150 \times 80$ cells. The sample is illuminated normally from below and the light intensity on the surface facing the probe is detected in the far-field region (collection-type NSOM) [17].

The effects of light polarization on the generated near-field images are studied for both parallel and perpendicular polarization conditions. When the sample structures possess a preferred orientation, the characteristics of the light scattered depend on the polarization direction (either parallel or perpendicular) relative to the structure orientation. In this paper, we define

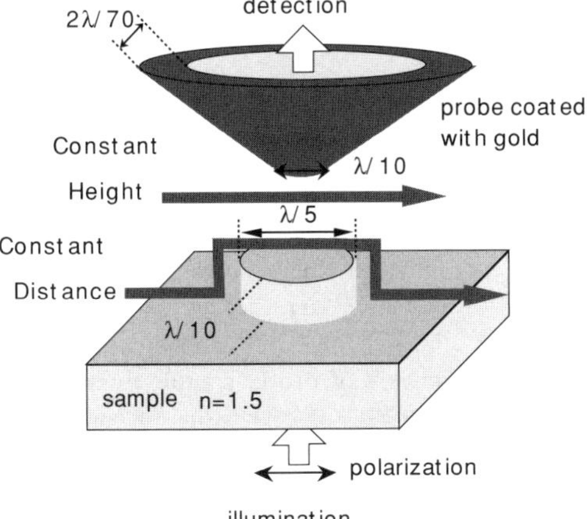

Fig. 2.1. Geometry of three-dimensional NSOM model

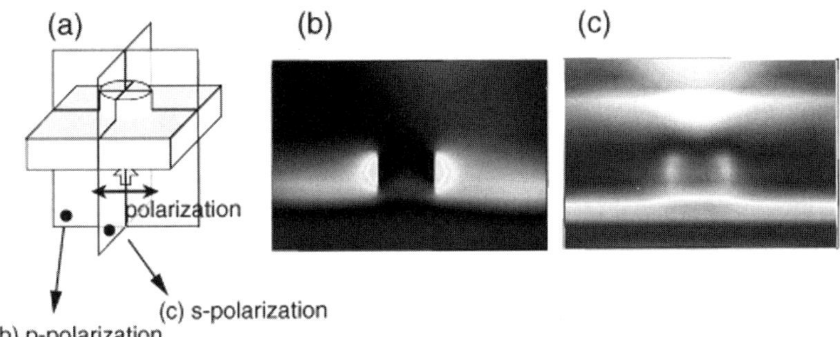

Fig. 2.2. a Convention used when considering the effects of two orthogonal polarization states (s and p) on the calculated electric intensity distribution. Electric field distribution on the sample surface when the effects of probe–sample interactions are not taken into account: **b** p-polarized illumination, and **c** s-polarized illumination

that incident electric field as p-polarized when the electric field vector oscillates in a sectional plane that is perpendicular to the interface of the edge structure (see Fig. 2.2a). On the other hand, s-polarization occurs when the field oscillates in a direction normal to a sectional plane.

2.1.3 NSOM Image Without Effects of Probe–Sample Interaction

We first calculated with the FDTD the electromagnetic field near the sample surface without a probe. Figure 2.2b,c show the calculated results for p-polarization and s-polarization illumination, respectively. Although this model is a rotationally symmetric system, the electric field distributions

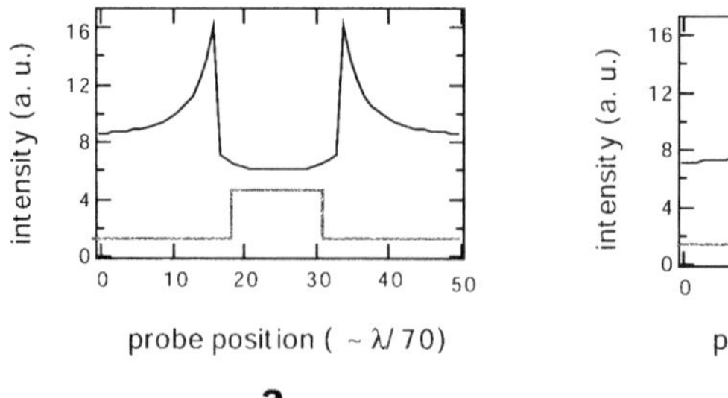

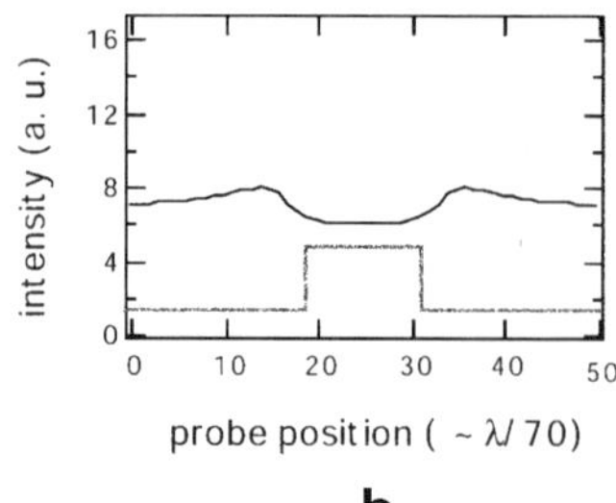

Fig. 2.3. NSOM images generated when the effects of probe–sample interactions are not taken into account (p-polarized illumination): **a** constant-distance-scanning mode, and **b** constant-height-scanning mode

formed in the two orthogonal cross-sections are different from each other due to the difference between the boundary conditions in the edge interface.

In the sectional image of the sample illuminated by p-polarization shown in Fig. 2.2b, the intensity becomes stronger near the edges. At the edge interface the electric flux density is conserved so that the electric field amplitude is greater on the outside of the medium where the permittivity is smaller.

Figure 2.2c shows the same as Fig. 2.2b except that the illumination is s-polarized. Note the decrease (about three times) in the image contrast relative to the image shown in Fig. 2.2b. The electric field distribution is continuous near the edges, because the applied boundary condition used requires the electric field itself to remain equal across the interface. Near the edge, the electric intensity is weaker than in other regions of the sample.

The results correspond to NSOM images obtained with a very weakly scattering probe (Rayleigh particle), in which the interaction of the probe with the electromagnetic field near the sample surface is approximately negligible.

In Fig. 2.3a,b are the NSOM images that were calculated when a weak scattering probe is scanned along the sample surface. These are the cross-sections of the images including the center of the disc protrusion using the constant-distance-scanning mode, and the constant-height-scanning mode for Fig. 2.3a and 2.3b, respectively. In the constant-distance mode, the probe tip is adjusted during scanning such that the probe-surface distance is always kept constant at $d = 3\lambda/70$. On the other hand, in the constant-height mode the probe is always kept at a constant height from a flat surface in the sample regardless of the probe position. The probe height as measured from the highest point in the sample surface to the probe tip is $h = 3\lambda/70$. The images in Fig. 2.3 are produced using p-polarized illumination.

It is clear from these figures that the type of scanning mode utilized affects the characteristic of the image. A higher contrast is produced using the constant-distance-scanning mode. Note that the strongest intensity values are

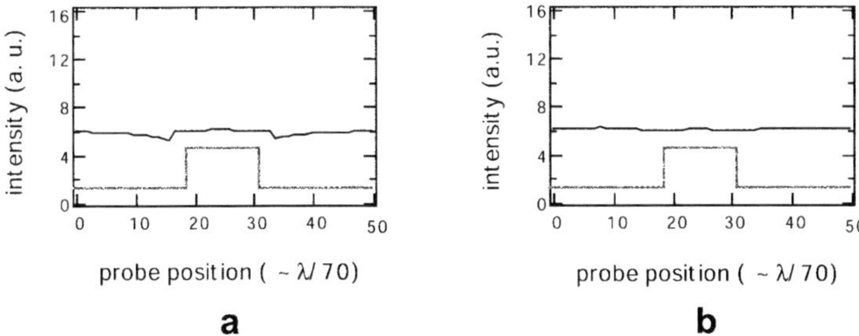

Fig. 2.4. NSOM images generated when probe–sample interaction is not taken into account (s-polarized illumination): **a** constant-distance-scanning mode, and **b** constant-height-scanning mode

always found at the edge interfaces (and not at the center of the cylindrical protrusion) regardless of the scanning mode utilized. It is also important to note that neither constant-height nor constant-distance mode yields the correct geometrical profile of the sample surface.

Figure 2.4a,b are the same as Fig. 2.3a,b, except that the illumination is s-polarized. The contrasts of these images are much lower than those shown in Fig. 2.3.

2.1.4 NSOM Image When the Probe-Sample Interaction is Included

When the permittivity of the probe either becomes comparable to or greater than that of the sample, the multiple scatterings that occur at the probe tip and the sample structure will disturb the field distribution originally formed near the surface. In this section, we show the results of calculated intensity images obtained when the probe–sample interaction is taken into account. Figure 2.5a,b show the NSOM images calculated with the existence of a strongly scattering probe under p-polarized illumination. In both parts, for constant-height and constant-distance mode, the intensity becomes strongest at the center of the sample and the image generated agrees reasonably with the geometrical profile of the sample. Recall that under the weak scatterer approximation the strongest intensities are found near the edges (see Figs. 2.3 and 2.4).

Figure 2.6a,b are the same as Fig. 2.5a,b except that p-polarized illumination is used. The two images look similar to Fig. 2.5a,and 2.5b, showing that the profiles that are consistent with the geometrical profile of the sample surface. The inclusion of the probe–sample interactions in the calculation results in a marked improvement of the image contrast.

For both polarization states, the constant-distance mode produces images (Figs. 2.5a and 2.6a) that are characterized by stronger intensity values even at probe locations far from the cylindrical protrusion compared to images

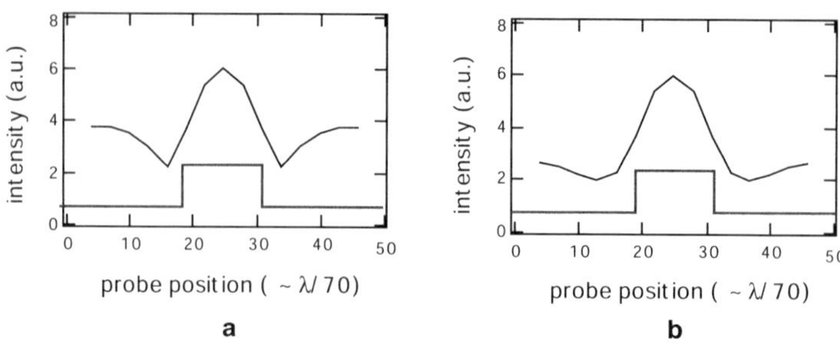

Fig. 2.5. NSOM images generated when probe–sample interaction is included (p-polarized illumination): **a** constant-distance-scanning mode, and **b** constant-height-scanning mode

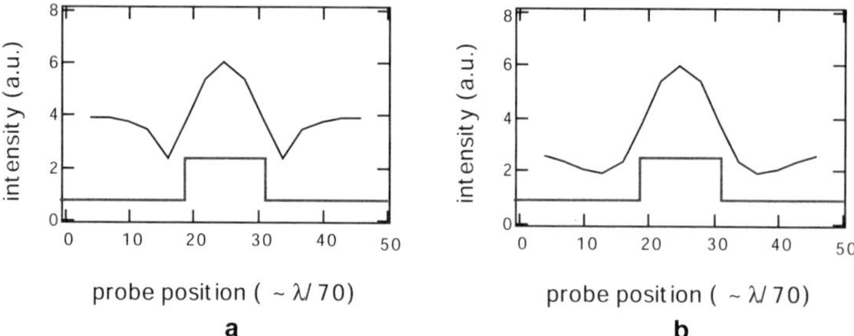

Fig. 2.6. NSOM images generated when probe–sample interaction is considered (s-polarized illumination): **a** constant-distance-scanning mode, and **b** constant-height-scanning mode

(Figs. 2.5b and 2.6b) generated under the constant-height mode. This is because in the constant-distance mode the distance of the surface–probe separation is always maintained at a constant value regardless of the probe location in the entire sample surface. The images obtained using the constant-height mode exhibit better resemblance to the geometrical surface profile than those obtained under the constant-distance mode.

The series of electric field intensity distributions during the scanning over the protrusion (constant-height mode) are shown in Fig. 2.7a–e. The figures are arranged in order that the probe is passing above the sample structure. Note the existence of a localized electric field distribution at the edge of the sample (Fig. 2.7a, also illustrated in Fig. 2.2b). When the probe is at the edge of the sample, the localized field distribution at the edge is decreased because a part of it is transmitted inside the probe. When the probe is above the surface center, the strong electric fields localized in the edges are reduced, and photons are transmitted effectively through the probe aperture for detection. A localized electric field also exists around the probe because of the boundary

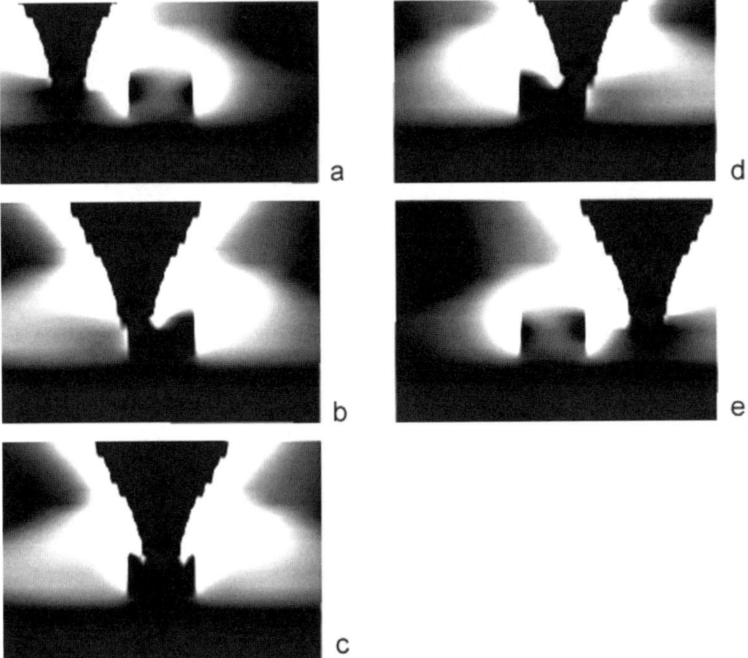

Fig. 2.7. Near-field electric intensity distribution as a function of the probe position relative to the sample surface. **a** far from the edge, **b** nearest to the first edge, **c** right above the sample center, **d** nearest to the second edge, and **e** after passing through the protrusion

condition which supports the existence of a p-polarized electric field on the metallic surface of the probe.

Figure 2.8a–e illustrate the electric field intensity distribution within the gold-coated dielectric probe in corresponding sequence to the images of Fig. 2.7a-e. In the figures, the image contrast is saturated to render the weak intensity values visible. The images indicate that transmission of the field into the probe aperture is most effective when the probe is right above the sample center where a deeper penetration of the field is possible. This effect arises from the excitation of the surface plasmon in the metallic film surface by the p-polarized light. Figure 2.9 shows the orthogonal distributions of the electric field intensity inside the probe when the probe is located right above the center of the sample. Note that the surface plasmon is excited under only p-polarized illumination.

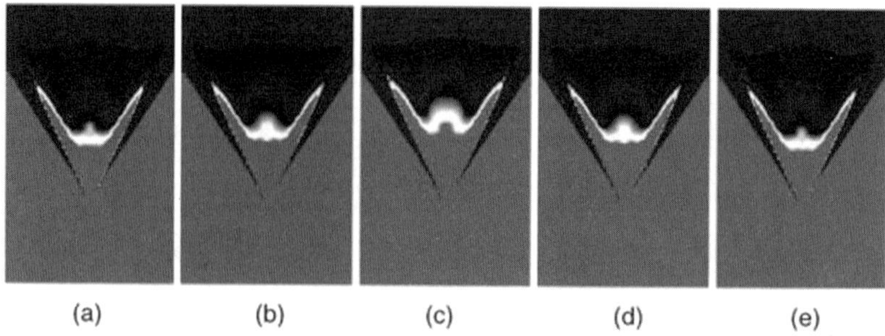

Fig. 2.8. Electric intensity distribution inside the apertured NSOM probe when it is **a** far from the edge, **b** nearest to the edge, **c** right above the sample center, **d** nearest to the second edge, and **e** after passing through the protrusion. The dynamic range of color presentation is saturated to render weak intensity values visible

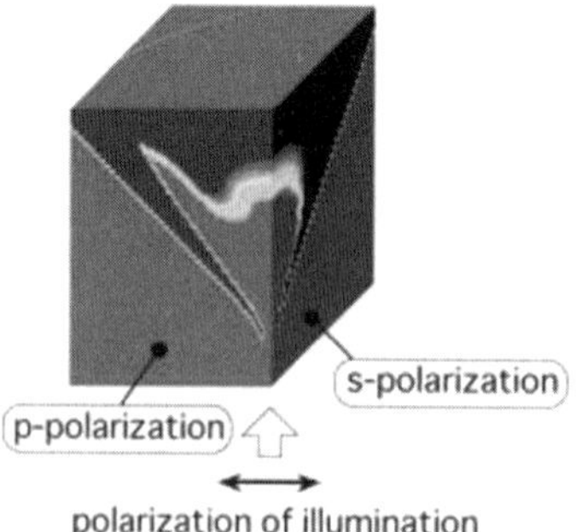

Fig. 2.9. Comparison of the orthogonal electric intensity distributions inside the probe when it is right above the sample center. Note that the surface plasmons are excited when using p-polarized illumination

2.1.5 Effect of the Probe–Sample Distance on the Generated NSOM Images

We have also investigated the behavior of the generated image as a function of the distance (gap) d between the probe tip and the sample surface. Presented in Fig. 2.10a are the resultant images for three different distances for the case of the constant-distance mode with p-polarized illumination. The intensity decreases as the d value increases, or the probe moves away from the sample surface. The generated NSOM intensity image exhibits strong values at both the center of the protrusion and the flat surface, except at the edges when the separation d is small. In the edge structure, multiple scattering occurs between the probe and the edge, but is not coupled with the propagation mode of the probe transmission.

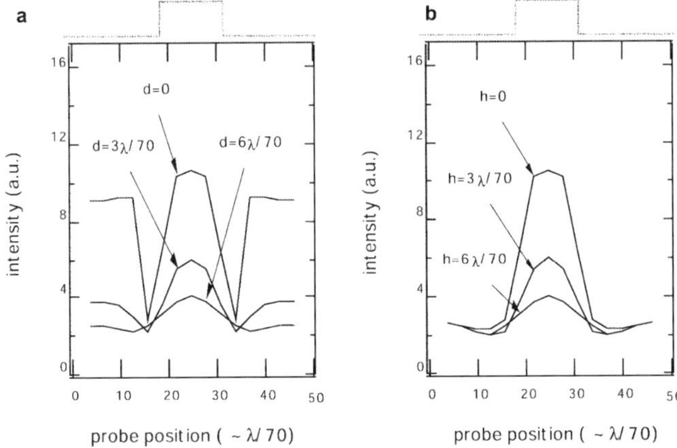

Fig. 2.10. NSOM images (p-polarized illumination) of the same surface as a function of the probe–sample distance when using the **a** constant-distance-scanning mode, and **b** constant-height-scanning mode. The illumination wavelength used is $\lambda = 6.0\ \mu\text{m}$

The results under constant-height mode are presented in Fig. 2.10b. The images shown in Fig. 2.10a,b look similar when the gap distance is as large as $d = h = 6\lambda/70$. This is because for large probe–sample distances the effect of multiple scattering on the image formation process becomes negligible and the image is generated primarily by the coupling effect of the near-field to the waveguide structure of the probe.

The results shown in both Fig. 2.10a and 2.10b indicate that the spatial resolution does not deteriorate with increasing d values in spite of the attendant decrease in the intensity values.

2.1.6 Dependence of NSOM Image on the Spatial Frequency Content of Sample Surface

As for the characteristics of the probe, it is also important to understand the filtering effect of the spatial frequency of the sample structure [18] as well as the imaging resolution. As is seen in Figs. 2.11 and 2.12, the detected intensity is strongest when the sample size is nearly equal to the external diameter of the metal-coated apertured probe.

Figure 2.11 show the generated NSOM images for the samples with protrusions of different sizes. These images were calculated under the constant-height mode. In the calculation, the external diameter of the coated metal is $11\lambda/70$.

When the cylindrical structure of the sample is about twice as big as the external diameter of the probe (Fig. 2.11a), the NSOM image produced is generally consistent with the geometrical profile of the sample, although the

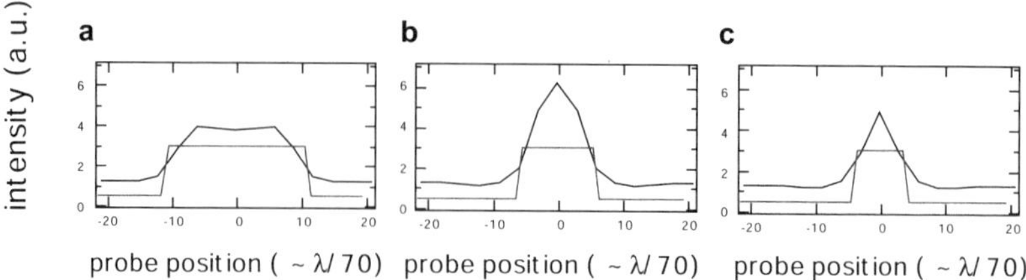

Fig. 2.11. NSOM images as a function of the diameter of the cylindrical protrusion: **a** diameter $= 22\lambda/70$, **b** $11\lambda/70$, **c** $7\lambda/70$. The true surface geometry is given by the *dashed* curve

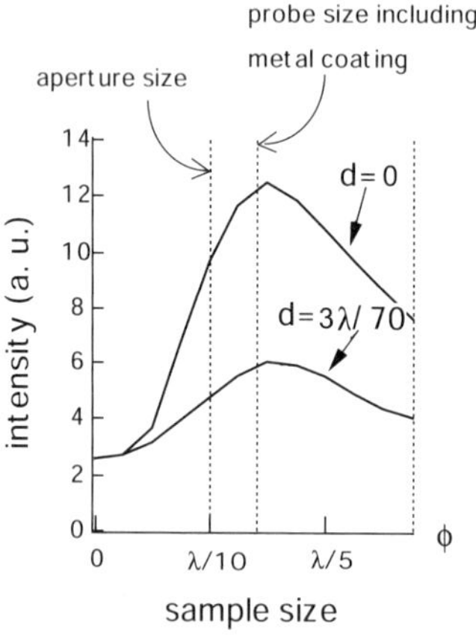

Fig. 2.12. Transfer function of the apertured probe. The intensity vs. sample size curve is calculated for an aperture that is located right above the sample center

intensity is strongest near the edges rather than at the sample center. If the probe is scanned starting from the left side, the detected intensity decreases gradually as the edge location is approached, and then increases steeply right after. As the probe leaves the edge location and approaches the sample center, the intensity again decreases towards a minimum value at the surface center.

However, in the case of smaller diameter values of the protrusion, the intensity is strongest at the surface center (see Fig. 2.11b,c). A relation between the diameter value and the detected signal at the center of the cylindrical protrusion is shown in Fig. 2.12. It can be seen that the detected intensity is strongest when the sample size is nearly equal to the external diameter of the probe. Relative to this value, the intensity decreases monotonously for other diameter values. This indicates that the filtering characteristics of the probe aperture vary with the dimensions of the sample structure relative to the external diameter of the probe.

2.2 Reconstruction of an Optical Image from NSOM Data

2.2.1 Necessity for Numerical Inversion of the NSOM System

The aim of near-field scanning optical microscopes (NSOM) is to obtain an optical image or the spatial distribution of the absorption and refractive index in the sample with nanometer resolution. However, the observed image using NSOMs is often different to the optical image of the sample due to the strong interaction between sample structure and the probe, or the artifact caused by the topography [19]. As a result, an NSOM image varies depending on the polarization of the illumination [7], the distance between the probe and the sample [20], and the shape of the probe [3]. To invert the imaging system of the NSOM it is necessary to reconstruct the optical image from an observed NSOM image.

Here again we use the FD–TD method to simulate the electromagnetic near field in three dimensions with a realistic probe structure with a nanometric sample structure, and reconstruct the optical image using a deconvolution method with a non-negativity constraint and a nonlinear optimization method [4].

2.2.2 NSOM Image of Dielectric Strips

Figure 2.13 shows the three-dimensional model of NSOM imaging for our study. Two narrow strips, each 15 nm wide and 45 nm deep, are embedded in a dielectric substrate (refractive index = 1.5) with a flat surface. This

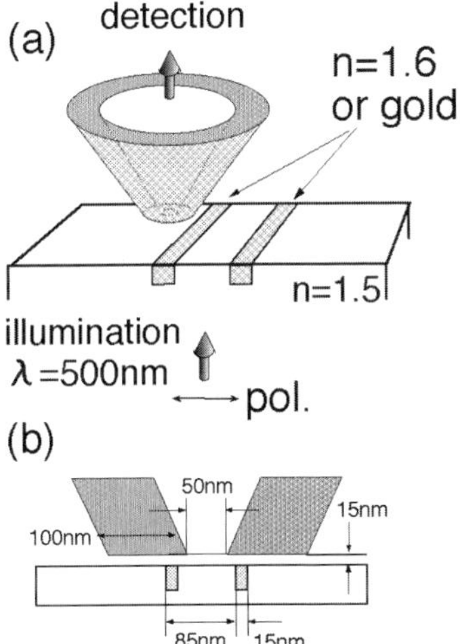

Fig. 2.13. Three-dimensional model for simulationg NSOM imaging

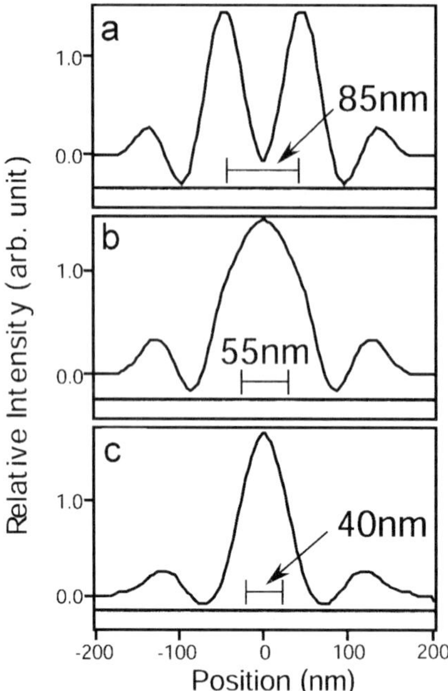

Fig. 2.14. Calculated NSOM images of the samples embedded with two dielectric strips. **a** 85 nm, **b** 55 nm, and **c** 40 nm separation of the two strips

sample is illuminated by a linearly polarized light (wavelength is assumed to be 500 nm) from the bottom, and detected at the top through an aperture probe. The polarization of the illumination is perpendicular to the strips. A fiber probe (refractive index = 1.5) with a metal coating (100 nm thick) with an aperture (diameter = 50 nm) scans on the sample surface at a constant height.

We used a finite-difference time-domain (FD–TD) method [3,16] for obtaining the field distribution for this three-dimensional model, and calculate the intensity of the light transmitted through the probe.

Figure 2.14 shows the calculated results of NSOM images by FD–TD. The strips are dielectrics with a refractive index of 1.6. The separations between two strips are 85, 55 and 40 nm, for Fig. 2.14a, b and c, respectively. The two strips are resolved in bright contrast only for the strip-to-strip separation of 85 nm as shown in Fig. 2.14a, but not for the narrower separations shown in Fig. 2.14b,c.

2.2.3 Deconvolution of Dielectric Strips with Nonnegativity Constraint

To reconstruct the refractive-index distribution in a sample, we deconvolve the observed image with a point spread function (PSF) or a unit response. Figure 2.15 shows an image of a single-strip sample, and we used this image as the PSF, or more precisely the line spread function. Deconvolution was made under a non-negative constraint [21,22] with the Gauss–Seidel method for the calculation of a matrix inversion. Figure 2.16 shows the deconvolution results for the distances 85 nm (a), 55 nm (b) and 40 nm (c). Two strips are perfectly resolved as two peaks in the case of an 85-nm separation between the two strips (in Fig. 2.16a), and are also reconstructed in the case of a 55-nm separation (in Fig. 2.16b) with some small noise. In the case of a 40-nm

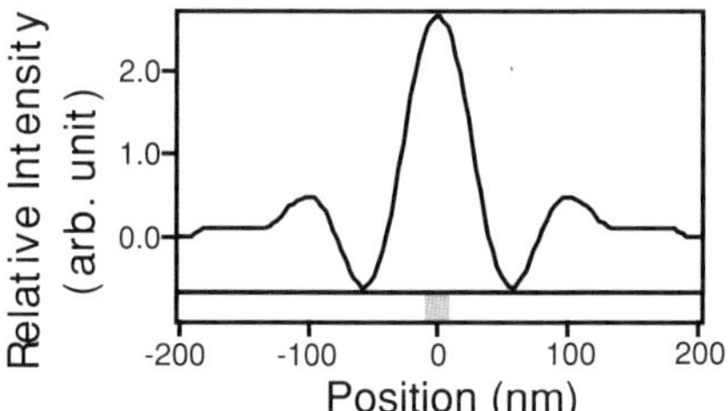

Fig. 2.15. A calculated NSOM image for a single dielectric strip

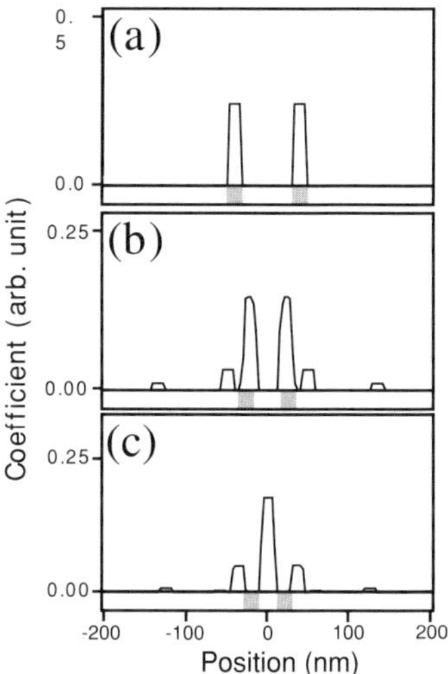

Fig. 2.16. Deconvolution results of dielectric strips for separations of **a** 85 nm, **b** 55 nm, and **c** 40 nm. The two strips are resolved in **a** and **b**, but not in **c**

separation, two strips are not resolved, as shown in Fig. 2.16c. This is because the separation between the strips is smaller than the aperture size, 50 nm.

2.2.4 Reconstruction of Metal Strips

We have also simulated NSOM imaging with metallic strips embedded in a glass substrate. The geometry is the same as that with the dielectric strips, as shown in Fig. 2.13. Figure 2.17 shows the calculated result for NSOM images of gold strips. The separations between the two strips are 85, 55 and 40 nm, for Figs. 2.17a, b and c, respectively.

The deconvolution did not successfully resolve the metal strips from the images shown in Fig. 2.17. To know the reason for this failure, we convolve the NSOM image of a single metal strip (shown in Fig. 2.18) with the original metal distribution. Figure 2.19a–c show the convolution images as solid curves, for different separations between the strips. The broken lines

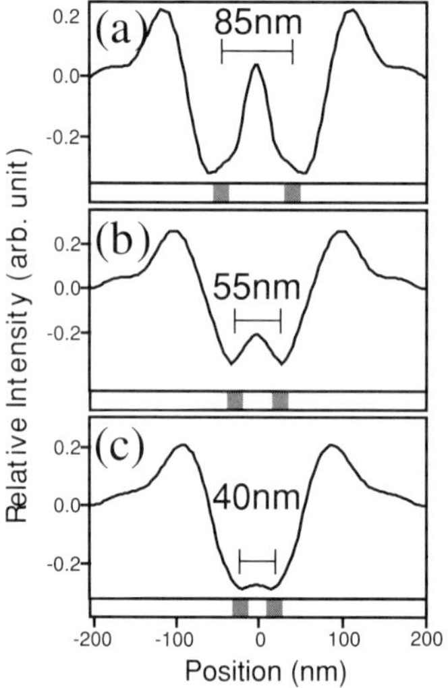

Fig. 2.17. Calculated NSOM images of the samples embedded with two metal strips. **a** 85 nm, **b** 55 nm, and **c** 40 nm separation of the two strips

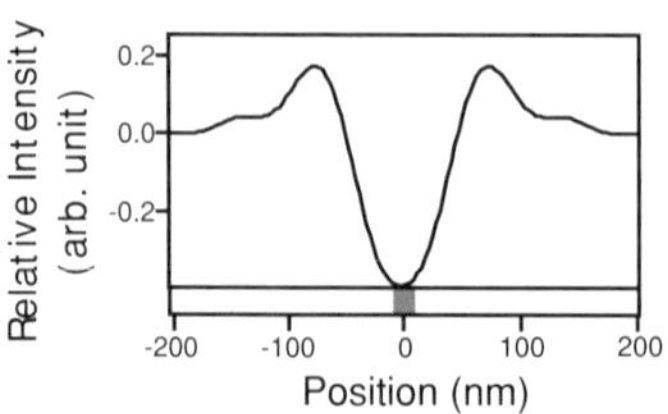

Fig. 2.18. A calculated NSOM image for a single metal strip

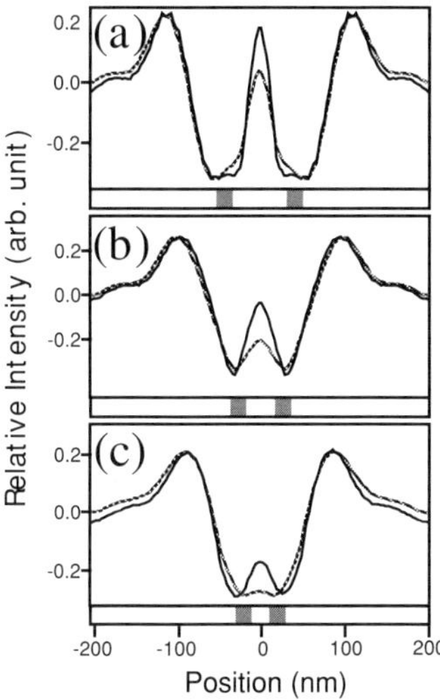

Fig. 2.19. Comparisons of convoluted images (*solid lines*) and observed images (*broken lines*) for a metal-strip sample with a separation of **a** 85 nm, **b** 55 nm, and **c** 40 nm

in Fig. 2.19 represent the observed images. A significant difference can be seen in the center region in each figure. This is caused by the interaction between the two metal strips and the metal part of the probe. The interaction made the linear deconvolution unsuccessful.

In order to reconstruct the sample, a more strictly constrained inversion method is to be used. We used a nonlinear optimization method for this purpose. The constraint is that the allowable sample structure is two metal strips with free parameters of positions and heights. Using the single strip image shown in Fig. 2.18, we find the optimal positions and heights of two metal strips using the Gauss–Newton method. Figure 2.20 shows the result of the reconstruction. The reconstructed strips agree quite well with the original distribution of the strips for all three cases, including the case of a small separation shorter than the size of the probe aperture.

2.3 Radiation Force Exerted near a Nano-Aperture

2.3.1 Radiation Force to Trap a Small Particle

In 1970, Ashkin demonstrated that two counterpropagating beams of laser light can trap a dielectric sphere of a few micrometers diameter [23], and in 1986 Ashkin et al. showed that even a single laser beam which is focused into the sphere can pull up and trap this sphere at the position of the focus point [24]. This technology has been particularly well applied in the biophysical

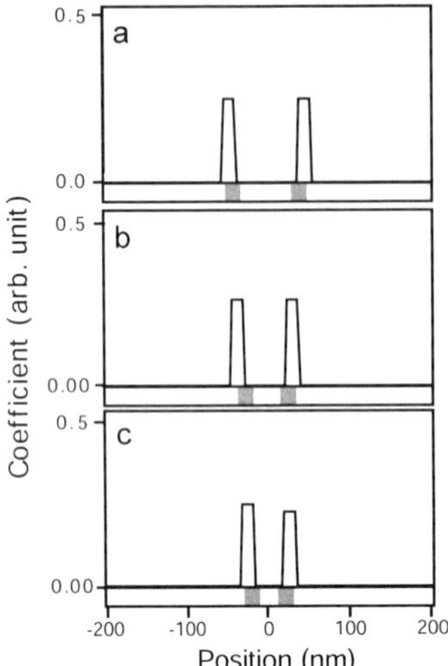

Fig. 2.20. Reconstructed results of metal strips for separations of **a** 85 nm, **b** 55 nm, and **c** 40 nm

sciences to trap living cells [25], and measure the force associated with the transcription of RNA [26], and move single DNA molecules in viscous flows [27], and so on.

The force exerted on a particle is given by the total change in the momentum of the incident photons due to scattering, absorption and spontaneous emission by the sphere. If the particle is small enough, e.g. like an atom [28,29], compared with the wavelength of the incident light, the analysis is not very difficult because the distribution of the field due to the particle is negligible.

For a sphere with a size near the wavelength of the incident light, the electromagnetic field (EMF) can be determined by the Mie scattering theory [30]. We found that the force exerted on a dielectric layer near a prism, by illumination under total internal reflection conditions, is a complicated function of the incident angle and the polarization of the incident light [31]. We also found experimentally that a sphere can be moved on a channel waveguide by an evanescent field [32,33].

In this section, we describe the photon force exerted on a subwavelength dielectric sphere near a small aperture by the interaction between the aperture and the sphere via evanescent photons [34]. This configuration corresponds to a near-field scanning optical microscope (NSOM) with an aperture [35], or a scattering probe [36,37].

Figure 2.21 shows the model which we used to perform our numerical analysis. The surface of a glass substrate ($\varepsilon = 2.28$) is coated with a metallic layer which is assumed to be a perfect conductor in the calculation, and which

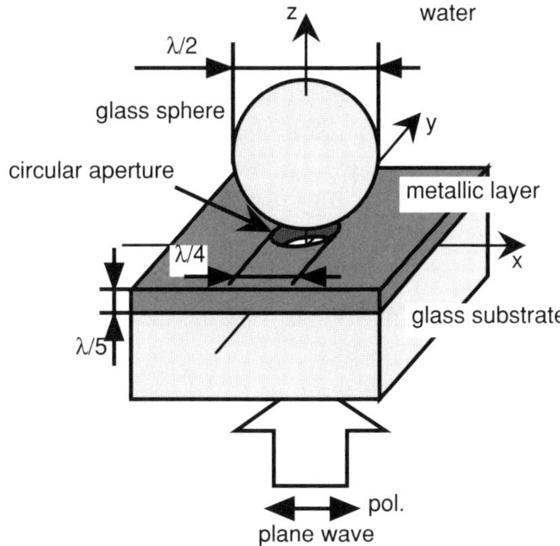

Fig. 2.21. The geometry of the model used for the numerical analysis. Incident light is x-polarized and propagates along the z-axis

is in contact with water ($\varepsilon_1 = 1.77$). A circular aperture is made into this metallic layer, the diameter of which is $\lambda/4$. λ is the wavelength of the incident light in a vacuum. A glass sphere ($\varepsilon_2 = 2.28$), the diameter of which is $\lambda/2$, is located in water near the aperture. The thickness of the metallic layer is $\lambda/5$. The incident light is a plane wave x-linearly polarized and propagating along the z-axis towards the metallic layer through the glass substrate. The x-, y- and z-axes are shown in Fig. 2.21.

We have made a series of numerical calculations to obtain the force dependence on the position of the sphere, in order to know where the sphere is finally trapped. For each calculation, the position of the sphere was changed near the aperture, and the electric field distributions (EFD) were calculated using the finite-difference time-domain (FD–TD) method [7,15,38]. From each calculated EFD, the radiation force was obtained from the Maxwell stress tensor on the surface of the sphere.

The radiation force calculated using the Maxwell stress tensor is an electromagnetic expression of the force which is due to the conservation of momentum of the electric field. This force occurs on the boundary of two materials with different permittivities, and its direction is perpendicular to the surface of the boundary between the two mediums, while its magnitude per unit area can be expressed as

$$F_{\mathrm{R}} = \frac{\varepsilon_2 - \varepsilon_1}{2} {E_{\mathrm{p}}}^2 + \frac{\varepsilon_1}{2} {E_{\mathrm{s1}}}^2 - \frac{\varepsilon_2}{2} {E_{\mathrm{s2}}}^2 \,. \tag{2.7}$$

Equation (2.7) holds for a force directed from medium 2 towards medium 1, where ε_1 and ε_2 are the relative permittivities of mediums 1 and 2, respectively. E_{s1} and E_{s2} are the components of the electric fields perpendicular to the surfaces of mediums 1 and 2, respectively. E_{p} is the component of the electric field parallel to this surface. The boundary condition of the electric field imposes the same E_{p} for both materials. F_{R} is always positive when $\varepsilon_2 > \varepsilon_1$ is fulfilled. One can obtain the radiation force exerted on the sphere by integrating F_{R} over the whole surface of the sphere.

2.3.2 Force Distribution Exerted on the Sphere near a Subwavelength Aperture

Figure 2.22 shows the calculated result of the spatial distribution of the light intensity near the aperture and the sphere. The sphere is absent in Fig. 2.22a,b, and it is placed right above the aperture in Fig. 2.22c,d. Figure 2.22a,c shows the intensity distribution in the x–z plane, which is the

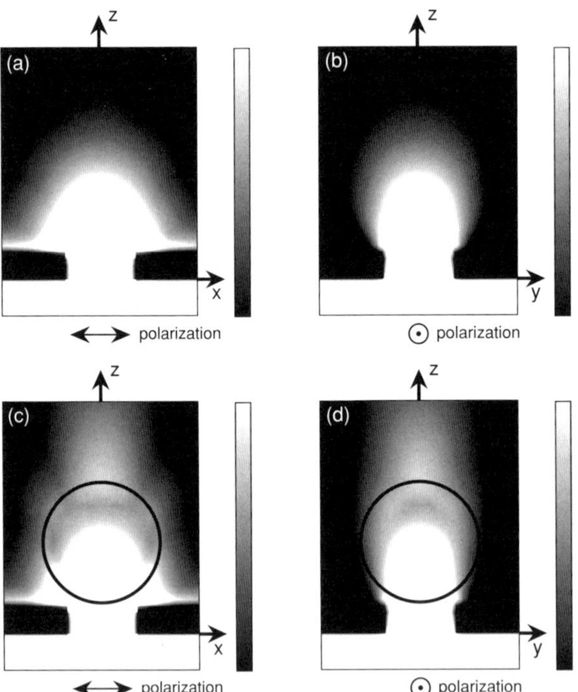

Fig. 2.22. Spatial distribution of the light intensity near the aperture (diameter: $\lambda/4$) and the sphere (diameter: $\lambda/2$). The dielectric sphere is absent in **a** and **b**, and present in **c** and **d** (shown as a *black circle*). The observation planes are the polarization plane of the incident for **a** and **c**, and perpendicular to it for **b** and **d**

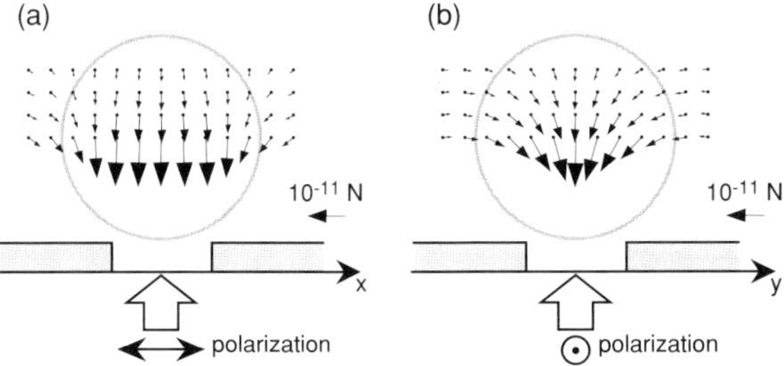

Fig. 2.23. Spatial distribution of the photon force exerted on the sphere near a subwavelength aperture. The origin of each arrow represents the center of the sphere and vectors represent the direction and the magnitude of the forces

plane of polarization of the incident light, and Fig. 2.22b,d shows the intensity distribution in the y–z plane. In Fig. 2.22c,d it is shown that light is converged at the top of the sphere because of its lens effect. In the x–z plane, the evanescent field spreads horizontally on the upper side of the metallic surface. This is because the condition that the electric field must be perpendicular to the metallic surface is satisfied in this plane.

The arrows in Fig. 2.23 show how the photon force exerted on the sphere changes with the position of the sphere in the three-dimensional space. The observation planes, e.g. the x–z and y–z plane, of Fig. 2.23a and b correspond to those of Fig. 2.22c and d, respectively. The origin of each arrow represented by a dot in the figure corresponds to the position of the center of the sphere. Gray circles represent the sphere placed at the lowest and central position, and correspond to the sphere in Fig. 2.22c,d. In Fig. 2.23a the force near the aperture is perpendicular to the plane of the aperture, and in Fig. 2.23b the force is directed towards the center of the aperture. In both the x–z and y–z planes, the sphere is attracted by the strong field near the aperture in both the horizontal, e.g. the x–y plane, and the vertical direction, e.g. z. This situation resembles the conventional optical trapping scheme in which the particle is trapped near the focus spot.

We have confirmed that this trapping effect is due to the optical near field by performing calculations with an aperture diameter equal to 2λ, so that the aperture transmits propagating light; a situation in which the near-field light intensity is negligible compared to that of the far-field light. Every other condition being the same as for the calculations performed with an aperture size of $\lambda/4$, the result indicates that the sphere is repulsed from the aperture in the z-direction, even though it is horizontally attracted towards the aperture. The difference between the calculations performed with a 2λ versus a $\lambda/4$ aperture size is due to the difference of the initial momentum of the photons. The sphere receives an upward momentum by scattering a propagating photon, because the propagating photon initially has an upward momentum. On the contrary, evanescent photons have no upward momentum, thus the

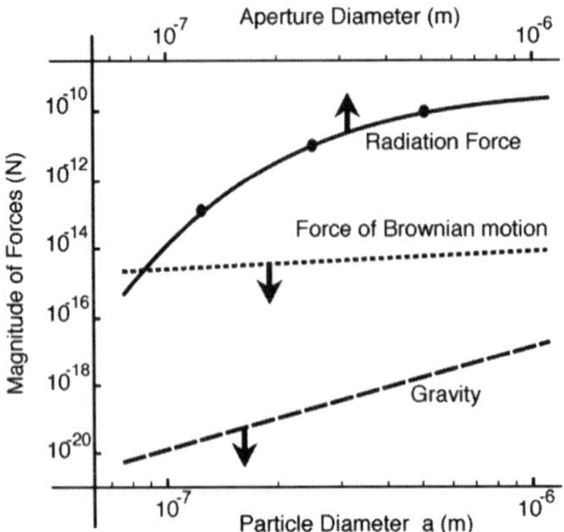

Fig. 2.24. Forces given to the particle in the model of Fig. 2.21. The radiation force is a function of the aperture diameter, while the force of the Brownian motion (*dotted line*) and the gravity (*dashed line*) are functions of the particle diameter. The radiation force with an aperture larger than 100 nm is superior to the thermal fluctuation and the gravity for particles smaller than 1 μm

sphere receives a downward momentum by scattering an evanescent photon as a reaction of giving an upward momentum to the photon. The sphere is trapped only when the evanescent photons are dominant behind the aperture.

Figure 2.24 shows the plots of three forces exerted on the particle, the radiation force, gravity, and the force due to Brownian motion, as functions of the size of the particle or the aperture. The radiation force must be the dominant force among these forces in order to trap the particle. We regarded these three forces as effectively exerted on particles in water, and compared their magnitude.

The radiation force shown as a solid line in Fig. 2.24 is the most dominant among the three forces when the laser intensity of the incident light is assumed to be 0.2 W/μm^2. The magnitude of the radiation force is proportional to the laser intensity. The radiation force depends on the diameter of the aperture much more than it depends on the size of the particle. The gravity including the buoyancy is shown as a dashed line in Fig. 2.24. The specific gravity of a glass particle is assumed to be 2.5 g/cm^3. Both the gravity and the buoyancy processes are proportional to the third power of the diameter of the particle. The Brownian force is due to the thermal fluctuations. The fluctuation–dissipation theorem of Einstein states that the Brownian force is equal to $6\pi a\eta k_B T$, where a is the diameter of the particle, η is the viscosity of water, k_B is the Boltzmann constant, and T is the temperature of the water (300 K in this calculation).

By comparing the magnitude of the radiation, Brownian and gravity forces, we found that the radiation force obtained with a 100-nm aperture is equivalent to the Brownian force of a 1-μm particle. This means that we can trap a particle smaller than 1 μm with an aperture larger than 100 nm. Apertures smaller than 100 nm cannot generate sufficient radiation force to trap particles in water.

2.3.3 Force Exerted on Two Spheres in the Near Field of a Small Aperture

We have also investigated the radiation force with the presence of two spheres in the optical near field near a small aperture. We assumed that the first sphere is trapped near the aperture as shown in Fig. 2.23. Our new interest is where the second sphere is finally trapped. Both spheres are assumed to have the same refractive indices and diameters as those of Fig. 2.21.

Figure 2.25a,b shows photon forces exerted on the second sphere in the x–z and y–z planes, respectively. Black circles represent the first sphere which doesn't move because it is trapped near the aperture. Gray circles indicate the highest and lowest position of the second sphere, whose position is changed for each calculation. In both planes, the force is attractive towards the top of the first sphere. In addition, at the lowest position in Fig. 2.25a the force is strongly attractive towards the aperture. This position dependence of the force corresponds to the spatial distribution of the localized fields shown in Fig. 2.22. The sphere is attracted towards the strong part of the field distribution.

As a necessary check of our methodology, a comparison is made in Fig. 2.26 between the FD–TD results and the numerical solution using the extended Lorenz–Mie theory (ELMT) [39,40] for a radiation force exerted on two spheres in a plane wave. The model includes two spheres ($\varepsilon_2 = 2.28, \phi = 5\lambda$) in water ($\varepsilon_1 = 1.77$) irradiated by a linearly polarized plane wave propagating in the z-direction. In this model, the two spheres are

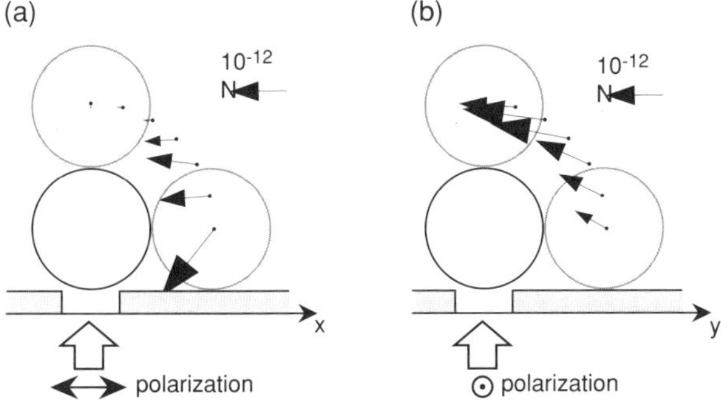

Fig. 2.25. The radiation force exerted on the second sphere (*gray circles*). Forces towards the top of the first circle (*black circles*) are exerted on the second sphere. In addition, at the lowest position in Fig. 2.25a the second sphere is strongly attracted by the evanescent field near the surface

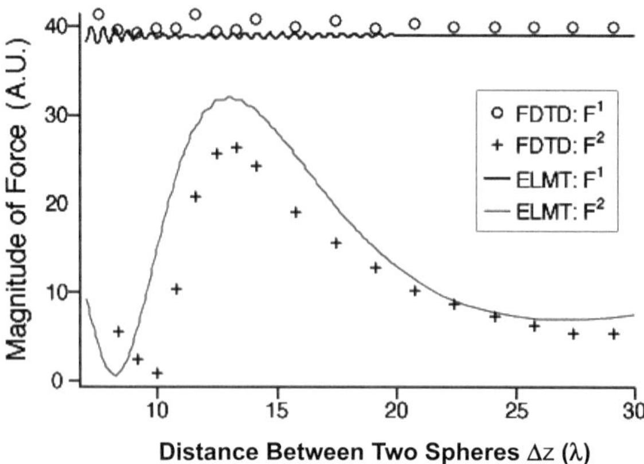

Fig. 2.26. Comparison of FD–TD results with the numerical solutions of the extended Lorenz–Mie theory (ELMT) for the radiation force exerted on two spheres in the plane wave. The horizontal axis is the distance between two spheres Δz normalized by a wavelength λ in water. The first sphere (experiences F_1) scatters the plane wave, and the scattered light exerts force on the second sphere (experiences F_2)

arranged on the z-axis, and the distance between the two spheres is varied. F_1 and F_2 are the z-components of the forces exerted on the first sphere and the second sphere, respectively. F_1 is almost constant, and F_2 varies with Δz. This is because the second sphere lies in the shadow of the first sphere against the plane wave and hence F_2 is strongly influenced by the electric field scattered by the first sphere. The good qualitative agreement between the results for the two methods confirms the validity of the FD–TD method for the two-sphere system.

A subwavelength scatterer can also generate locally confined evanescent photons. In particular, it has been determined that the tip of a metallic needle can enhance the local electric field [7,38]. The magnitude of the enhanced field is sufficient to trap a particle as small as 10 nm at its tip [41]. With such an effect, the radiation force of evanescent photons can be expected to exceed the limit of the Brownian motion.

References

1. A. Sommerfeld: *Optics* (Academic Press, New York, 1964)
2. C. J. Bouwkamp: Philips Res. Rep. **5**, 401 (1950)
3. H. Furukawa and S. Kawata: Opt. Commun. **132**, 170 (1996)
4. H. Hatano and S. Kawata: J. Microsc. **194**, 230 (1999)
5. M. Tanaka and K. Tanaka: J. Opt. Soc. Am. A **15**, 101 (1998)
6. K. Tanaka, M. Tanaka, and T. Omoya: J. Opt. Soc. Am. A **15**, 1918 (1998)
7. Y. Inouye and S. Kawata: J. Microsc. **178**, 14 (1995)
8. D. Courjon, K. Sarayaddine, and M. Spajer: Opt. Comm. **71**, 23 (1989)

9. t. Nakano and S. Kawata: J. Mod. Opt. **39**, 645 (1992)
10. C. Girard and D. Courjon: Phy. Rev. B **42**, 9340 (1990)
11. C. A. Brebbia: *The Boundary Element Method for Engineers* (Pentech Press, 1978)
12. C. Hafner: *The Generalized Multiple Multipole Technique for Computational Electromagnetics* (Artech, Boston, MA, 1990)
13. L. Novotny, D. W. Pohl, and P. Regli: J. Opt. Soc. Am. A **11**, 1768 (1994)
14. O. C. Zienkiewicz and K. Morgan: *Finite Elements and Approximation* (John Wiley, 1983)
15. K. S. Yee: IEEE Trans. Antennas Propagat. AP-14, 302 (1966)
16. D. A. Christensen: Ultramicroscopy **57**, 189 (1995)
17. S. Kawata, H. Takaoka, and H. Furukawa: J. Spectro. Soc. Jpn. **45**, 93 (1996)
18. J. M. Vigoureux and D. Courjon: Appl. Opt. **31**, 3170 (1992)
19. B. Hecht, H. Bielefeldt, Y. Inouye, D. W. Pohl, and L. Novotny: J. Appl. Phys. **81**, 2492 (1997)
20. H. Hatano, Y. Inouye, and S. Kawata: Opt. Lett. **22**, 1532 (1997)
21. O. Nakamura, S. Kawata, and S. Minami: J. Opt. Soc. Am. A **5**, 554 (1988)
22. S. Kawata and Y. Ishioka: J. Opt. Soc. Am. **70**, 762 (1980)
23. A. Ashkin: Phys. Rev. Lett. **24**, 156 (1970)
24. A. Ashkin, J. M. Dziedzic, J. E. Bjorkholm, and S. Chu: Opt. Lett. **11**, 288 (1986)
25. Y. Liu, G. J. Sonek, M. W. Berns, and B. J. Tromberg: Biophys. J. **71**, 2158 (1996)
26. H. Yin, M. D. Wang, K. Svoboda, R. Landick, S. M. Block, and J. Gelles: Science **270**, 1653 (1995)
27. T. T. Perkins, D. E. Smith, R. G. Larson, and S. Chu: Science **268**, 83 (1990)
28. V. I. Balykin, V. S. Letokhov, Y. B. Ovchinnikov, and A. I. Sidorov: Phys. Rev. Lett. **60**, 2137 (1988)
29. M. A. Kasevich, D. S. Weiss, and S. Chu: Opt. Lett. **15**, 607 (1990)
30. M. Born and E. Wolf: *Principles of Optics*, 6th ed., pp. 633 (Pergamon, Oxford, 1993)
31. T. Sugiura and S. Kawata: Bioimaging **1**, 1 (1993)
32. S. Kawata and T. Sugiura: Opt. Lett. **17**, 772 (1992)
33. S. Kawata and T. Tani: Opt. Lett. **21**, 1768 (1996)
34. K. Okamoto and S. Kawata: Phy. Rev. Lett. **83**, 4534 (1999)
35. E. A. Ash and G. Nicholls: Nature **237**, 510 (1972)
36. T. Sugiura, T. Okada, Y. Inouye, O. Nakamura, and S. Kawata: Opt. Lett. **22**, 1663 (1997)
37. K. Sasaki, H. Fujiwara, and M. Masuhara: J. Vac. Sci. Technol. B **15**, 2786 (1997)
38. H. Furukawa and S. Kawata: Opt. Commun. **148**, 221 (1998)
39. J. P. Barton, D. R. Alexander, and S. A. Schaub: J. Appl. Phys. **65**, 2900 (1989)
40. J. P. Barton, D. R. Alexander, and S. A. Schaub: J. Appl. Phys. **66**, 4594 (1989)
41. L. Novotny, R. X. Bian, and X. S. Xie: Phys. Rev. Lett. **79**, 645 (1997)

3 High-Resolution and High-Throughput Probes

M. Ohtsu and K. Sawada

The basis of the near-field optical microscope (NOM) is short-range electromagnetism between two antennas, a probe antenna and a sample antenna, which are much smaller than the wavelength of the driving field [1,2]. It is apparent that fabricating and manipulating small antennas are the most important factors in the successful development of NOM. One of the most realistic and commonly used methods for preparating a small antenna is sharpening an optical fiber to a very small apex. By employing the scanning technique already established in a scanning tunneling microscope (STM) and an atomic force microscope (AFM), the antenna at the apex of a sharpened fiber works as a probe on a sample surface under precise distance control.

A conventional optical fiber is sharpened by chemical etching in buffered hydrofluoric acid. An apex (diameter $d = 2a$), which is the antenna of the fiber probe, can easily be made small, and a minimum apex size of only a few nanometers has already been achieved. In NOM, only the apex works as an antenna, which interacts electromagnetically with the sample. The tapered part has a simple conical shape, and the cone angle θ is clearly defined. The apex radius and the cone angle can be varied by controlling the etching conditions. The exterior surface of the probe is coated with opaque metal, such as aluminum or gold, to avoid the illumination of excitatory light or the detection of the background signal through the sidewall of the tapered part. The very end of the metal coating is removed to allow the sharpened glass part, including the apex, to protrude. The small aperture with a diameter of d_{f} (also called the foot diameter) makes an important contribution, as will be discussed later. We call the protruded glass part and the metal-coated part the tapered core and the metallized tapered core, respectively.

The metallized tapered core is regarded as a metal-clad (metallic) waveguide through which excitation or signal light passes. Since the metallic waveguide has a complex loss mechanism, e.g., the existence of a cutoff diameter and absorption by the metal cladding, optimization of the structure of the waveguide is essential.

The important components for optimizing of the probe are summarized as follows:

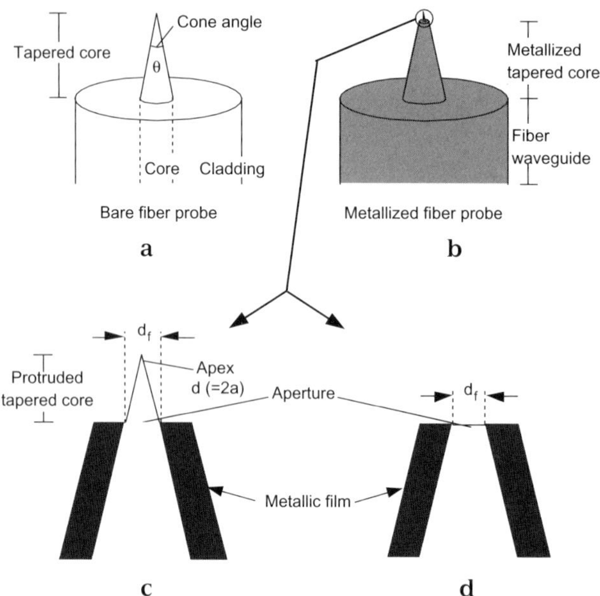

Fig. 3.1. Schematics of typical fiber probes

1. The apex size should be the same as the characteristic size of the sample to be investigated.
2. The smaller the cone angle, the smaller the undesirable background signal due to the dipole coupling between the sample and the tapered part.
3. The coupling can also be eliminated more efficiently by combining a metal aperture with a glass tapered part.
4. By proper metallization of the apex, the signal intensity is much enhanced compared with that from a bare glass apex.

Following these optimization criteria, high-quality fiber probes have been fabricated [1,2]. Using them, a near-field optical image of a single string of deoxyribonucleic acid (DNA) has been obtained by a collection mode operated under a constant-distance mode with the optical near-field intensity as the feedback signal [3]. As shown by Fig. 3.2, the observed width of the narrowest string is around 4 nm, even though the pixel size is as large as 2×2 nm. To our knowledge, this is the first and only successful optical observation of a single string of DNA by purely optical means and with such high resolving capability. This high resolution is attributed to the special care taken during the preparation of the sample and probe. The probe efficiently picks up the high spatial Fourier frequency component of the scattered near-field and simultaneously rejects the lower components effectively. With further improvement in the system, including a scanner and other mechanical component, it will be possible to observe the double-stranded or double helix of DNA by purely op-

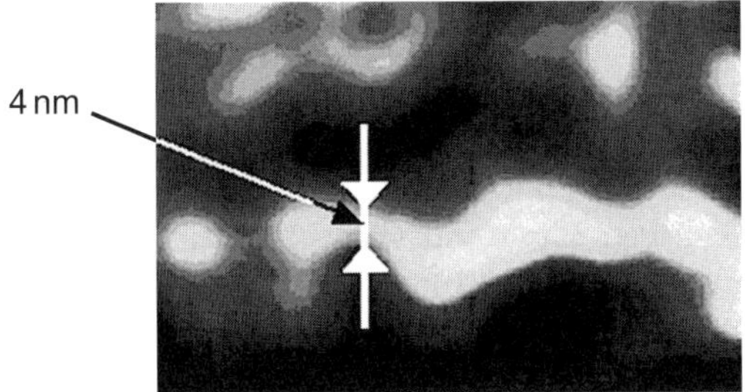

Fig. 3.2. Collection-mode optical near-field image obtained under a pixel size of 2×2 nm

tical means. Because the control is optical, it is easier to extend the operation into liquid or fluid environments which opens up the possibility of directly manipulating the DNA structure or investigating individual segments of the double helix structure itself [4].

The improvement of the sensitivity by a metal probe is frequently employed in a near-field plasmon microscope since it can easily be realized by using metal or metal-coated glass fiber probes for STM. In collection-mode NOM, however, when the protrusion part is completely metallized, the transmission intensity of the signal light into the fiber is generally decreased. Therefore, the thickness of the metal film, which also influences the resonant behavior of the plasmon probe, is an important design factor.

A serious problem of the fiber probe is its low throughput. The essential cause of low throughput is the guiding loss along (or inside) the metallized tapered core. To study this loss mechanism and to realize high throughput probes, we focus our discussion on the characteristics of the tapered core. Since the nanometric protruded part at the top of the fiber probe does not contribute to this loss, this section treats only the fiber probe without a protruded part, i.e., a flat apertured probe. Its cross-sectional profile is shown in Fig. 3.1d, for which the foot diameter d_{f} of the protruded probe in Fig. 3.1c can be called an aperture diameter.

As an example, the throughput of the probe of Fig. 3.1d decreases rapidly with decreasing probe size [1,2]. To increase the throughput, i.e., to generate a strong optical near-field under illumination-mode operation and to realize high collection efficiency in collection-mode operation, this section describes the possible excitation of the plasmon mode in the metallized tapered core. We also demonstrate highly efficient excitation of the optical near-field on a probe. Further, we demonstrate double- and triple-tapered probes fabricated to increase the throughput by shortening the tapered core.

3.1 Excitation of a HE-Plasmon Mode

3.1.1 Mode Analysis

Mode analysis has been carried out by approximating a tapered core as a cylindrical core with a metal cladding. The result shows that this core can guide the light, even if its diameter is smaller than half the wavelength. It also shows that the cutoff core diameter of the HE_{11} mode is as small as 30 nm, whereas that of the EH_{11} mode is 450 nm [5]. It means that only the HE_{11} mode can excite the optical near field efficiently when the apex size of the probe is in the sub-100 nm range. It is also found that the equivalent refractive index of the HE_{11} mode approaches that of a surface plasmon which is between those of glass and gold. This means that the origin of the HE_{11} mode in a metallized core is the surface plasmon. Thus, we call the HE_{11} mode the *HE-plasmon mode* from now on. However, the HE-plasmon mode is not easily excited in a conventional core because its coupling efficiency with the lowest order optical fiber guided mode is very low due to mode mismatching between the HE_{11} and HE-plasmon modes at the foot of the tapered core.

3.1.2 Edged Probes for Exciting a HE-Plasmon Mode

An effective way of exciting the HE-plasmon mode is to use the coupling of the plasmon by scattering at the edge of the metal [5]. If the tapered core has a sharp edge at its foot, part of the guided light inside the single-mode fiber can be scattered at this edge and converted to the HE-plasmon mode [6]. We call the probe with such a core *an edged probe.*

Figure 3.3 shows scanning electron micrographs of a fabricated edged probe with a flat aperture at the top (the aperture diameter d_f is 500 nm). An arrow R in Fig. 3.3a represents the direction parallel to the surface from which a part of the core was removed. The x and y axes are the directions normal and parallel to the arrow R, respectively. Figure 3.3b,c represents the profile of the tapered core buried in a gold metallic film. Part of the foot of the core was removed to form a sharp edge; the height of the removed part was 1.5 μm. This probe was fabricated by selective chemical etching [7] and by using a focused ion beam (FIB).

Probes with an aperture diameter d_f as small as 30 nm have been realized by these steps. Figure 3.4 shows the throughput of these probes. This figure confirms the realization of high throughput by the edged probe. We compared the output light power of the edged and the conventional fiber probes. Figure 3.5a,b shows the measured spatial distributions of the EH_{11} and HE-plasmon mode powers for an edged probe with $d_f = 500$ nm, respectively. They agree with the calculated results. As a reference, the spatial distribution for an axially symmetrical fiber probe with $d_f = 500$ nm was also measured. It had a double-peaked profile corresponding to the EH_{11} mode,

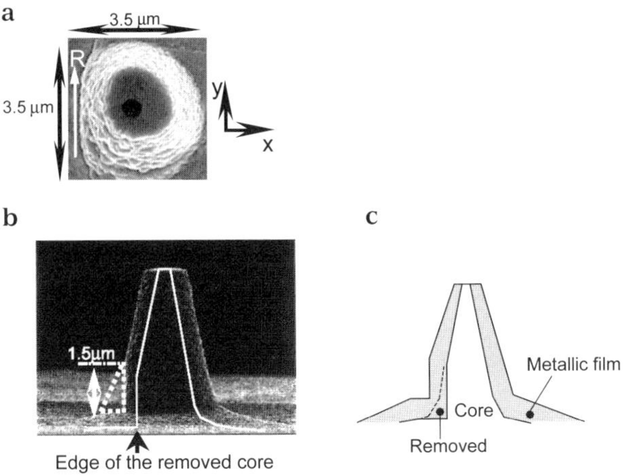

Fig. 3.3. Scanning electron micrographs of an edged probe with a flat aperture at the top. **a** Top view. **b** Side view. An arrow R in **a** represents the direction parallel to the surface from which a part of the core was removed. The x and y axes are the directions normal and parallel to the arrow R, respectively

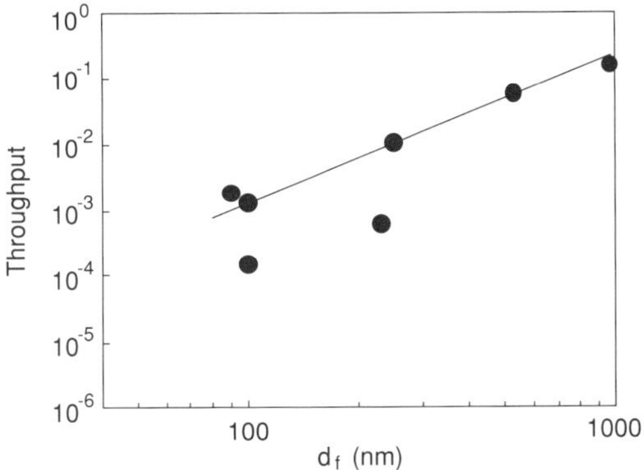

Fig. 3.4. Measured relation between the aperture diameter d_f and the throughput of the edged probe

and the positions of the peaks moved when the direction of the incident light polarization was varied. These results indicate that the edge at the foot of the tapered core successfully excites the HE-plasmon mode and the excitatory efficiency depends on the direction of the incident light polarization.

To check whether the output light power on the aperture was enhanced due to the edged structure, its spatial distributions on the probes with and without the edge were compared for $d_\mathrm{f} = 100$ nm. Note that the EH_{11} mode is not guided inside the probe with $d_\mathrm{f} = 100$ nm because it is smaller than its cutoff diameter (450 nm). The cross-sectional profiles of the measured

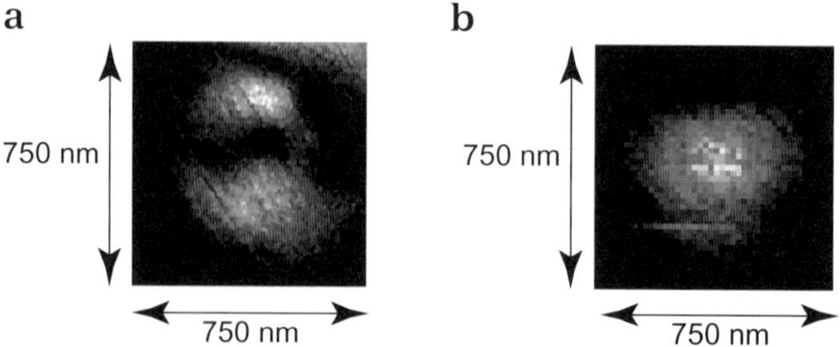

Fig. 3.5. Spatial distribution of the output light power on top of an edged probe with $d_\mathrm{f} = 500$ nm. **a** The EH_{11} mode. **b** The HE-plasmon mode

light power distributions for edged probes are ten times larger than those of probes without an edge. The full width at half maximum is 150 nm, which is comparable to that of the HE-plasmon mode (= 120 nm) for $d_\mathrm{f} = 100$ nm estimated by mode analysis. These results indicate effective excitation of the HE-plasmon mode by the edge at the foot of the tapered core.

3.2 Multiple-Tapered Probes

Note that an increase of throughput is possible by tailoring the probe structure. Since the guiding loss in the core is due to loss in the metal cladding, the easiest way to decrease the loss is to shorten the length of the tapered core.

3.2.1 Double-Tapered Probe

Chemical etching is a powerful technique for shortening the enormous absorption region because it enables reliable design and fabrication of an optimum structure for the tapered core. First, we demonstrate the strong dependence of throughput on the cone angle θ and hence the resultant length l of a tapered core. For probes with different cone angles, throughput is evaluated for a wide range of aperture diameters ($80 \leq d_f \leq 900$ nm). On the basis of these results, we optimize the shape of the probe taking into account the experimental utilities [8]; for example, by increasing the length of the tapered core, we can avoid contact between the cladding and the bumpy surface structure of the sample.

We briefly describe the fabrication technique of the probe by a selective etching process. The cone angle θ can be controlled by buffering the etching solution, which is adjusted by the volume ratio X of NH_4F maintaining that of HF to H_2O at 1 : 1. Here, the composition of the solution is expressed as $X : 1 : 1$. To evaluate the dependence of throughput on the cone angle, two types of probes were prepared using a one-step etching technique with a

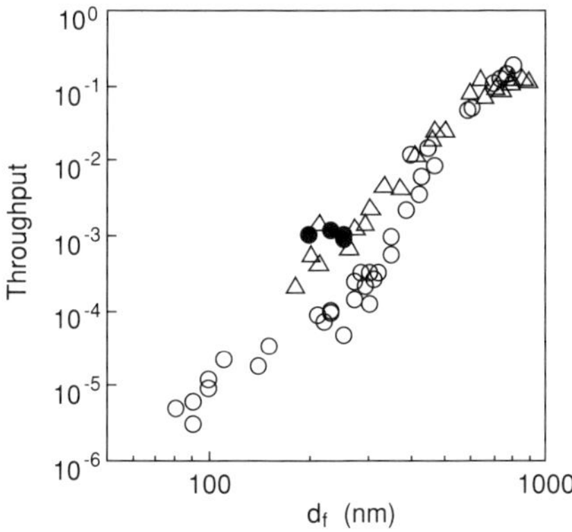

Fig. 3.6. Measured relations between the aperture diameter d_f and the throughput for probe A with a single tapered core whose cone angle is 20° (*open circles*), probe B with a single tapered core whose cone angle is 50° (*open triangles*), and a double-tapered probe (*closed circles*)

solution of $X = 10$ (probe A) and $X = 2.7$ (probe B). The cone angle θ and the length l of the tapered core of probe A were 20° and 6 µm, respectively, and those of probe B were 50° and 2.5 µm.

Using a sputter coating method, the exterior surface of the etched core was coated with a gold film 200 nm thick, which is seven times the skin depth at a wavelength of 633 nm. The selective resin coating (SRC) method is employed to fabricate a small aperture. Acrylic resin dissolved in an organic solvent is coated as a guard layer over the sides of the tapered part, leaving the top of the tapered core free of the acrylic resin. Here, the surface tension of the resin is used. On removing the gold film from the top of the core with the commonly used KI–I_2 etching solution, a small protruding type aperture (with the foot or aperture diameter d_f, c.f. Fig. 3.1) is fabricated. The aperture diameter d_f can be varied by controlling the etching time in the KI–I_2 etching solution and the coating condition which can be adjusted by changing the cladding diameter and the viscosity of the resin solution.

To estimate the throughput, the light from a He–Ne laser (633-nm wavelength) of 130 µW power is coupled into the fiber probe with a coupling efficiency of 60%. The far-field light ejected from the aperture is collected with a 0.4 NA objective. The output light power is measured with an optical power meter. The geometric size of the aperture is estimated with a scanning electron microscope after the throughput measurement.

Figure 3.6 shows the measured value of the throughput as a function of the aperture diameter d_f for probes A (open circles) and B (open triangles) of different cone angles. For the diameter d_f that is larger than the wavelength of He–Ne laser light in glass ($\lambda_c \sim 400$ nm), the throughput of probes A and B are almost the same when the aperture diameters are the same. This

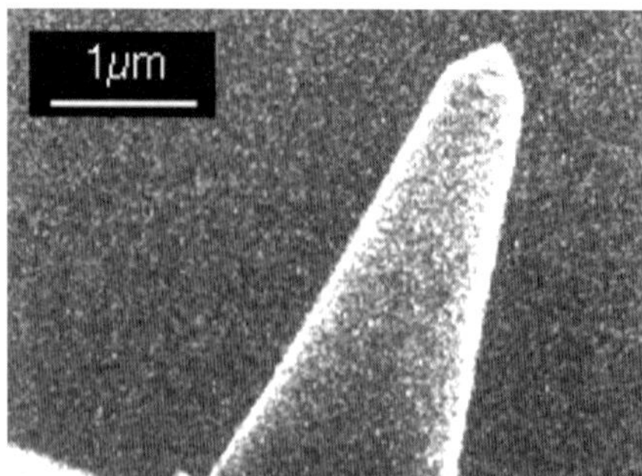

Fig. 3.7. Scanning electron micrograph of a double-tapered core prior to metal coating

means that the efficiency of delivering light into the region $d_f > \lambda_c$ is not so strongly dependent on the cone angle θ and the length l of the tapered core, which is 5 μm long at most. On the other hand, in the region $d_f < \lambda_c$ the difference between the two types of probes is remarkable due to the influence of the metal coating. For $d_f = 200$ nm, the throughput of probe A is ten times lower than that of probe B. Although the throughput depends on the spatial mode characteristics of the tapered core, the difference in the guiding lengths between probes A and B can also contribute considerably to the throughput. The guiding lengths of the core diameters of 400 nm ($= \lambda_c$) and 200 nm ($= d_f$) are 570 nm for probe A and 210 nm for probe B, respectively. It is theoretically estimated that, in the EH_{11} mode, the absorption of light by the coated metallic film is drastically influenced in the region where the core diameter is smaller than λ_c. In such a strong loss region, a difference in the guiding length of as much as 360 nm results in a decline of throughput by an order. We can conclude that, to increase throughput, it is reasonable to shorten the strong loss region in the tapered core by increasing the cone angle.

However, a short tapered core with a large cone angle leads to contact between its cladding and the bumpy surface of the sample during the actual scanning operation. To avoid this inconvenience, it is necessary to lengthen the tapered core while maintaining high throughput. For this purpose, we successfully developed a two-step etching process for fabricating a double-tapered probe with high reproducibility. As shown in the scanning electron micrograph of Fig. 3.7, the resultant cone angle is about 90°. The fabrication technique for the aperture is the same as that described above. Again, we measured the throughput of some of the fabricated double-tapered probes with $d_f = 200$ nm. The results are also plotted in Fig. 3.6 (closed circles). Throughput ten times higher than that of probe A (open circles) is achieved. This comparison implies that one of the most important factors in determining throughput is the length of the loss region in the narrow tapered core.

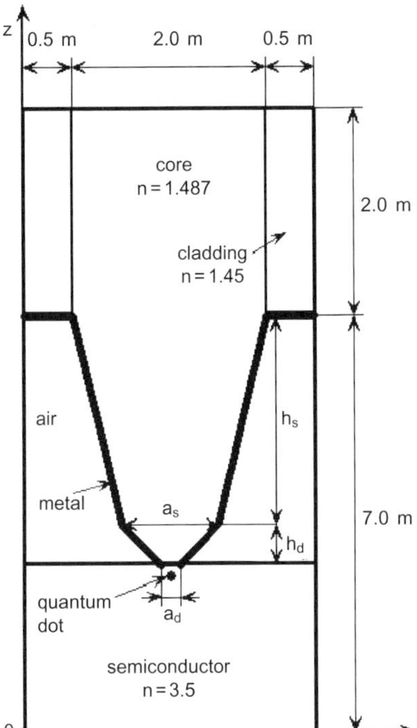

Fig. 3.8. Cross-sectional as diagram of the simulation model. Sample 1: h_s = 3.6 µm, a_s = 0.2 µm. Sample 2: h_s = 0.9 µm, a_s = 0.2 µm. Sample 3: h_s = 2.0 µm, a_s = 1.0 µm, h_d = 0.4 µm, a_d = 0.2 µm

In double-tapered probes, realization of apertures smaller than 100 nm is not feasible due to its large cone angle. However, due to the limitations in measurement sensitivity, high throughput is more crucial than realizing a very small aperture for weak signal detection in the advanced spectroscopy of semiconductors.

Recently, with the progress of computational work, numerical analysis has come to be recognized as a powerful tool for deep understanding of the electromagnetic field in a tapered waveguide and in the vicinity of an aperture and for the systematic design of highly efficient probes. The finite-difference time-domain (FDTD) method [9–11] is one of the most promising methods for these purposes because it can be easily applied to actual three-dimensional problems. We have reported an FDTD simulation for the double-tapered structure in [12], in which spatial resolution beyond that of a conventional NOM is successfully reproduced by simulation.

Here, we show an FDTD simulation, which corresponds to the experimental configuration in [8]. Figure 3.8 illustrates the FDTD geometry of the probe model. A fiber probe with a double- or single-tapered structure collects luminescence ($\lambda = 1$ µm) from a quantum dot (x-directed dipole radiation) buried $\lambda/40$ beneath the semiconductor surface. The radiation caught by the aperture with a diameter of $\lambda/5$ propagates in the tapered part clad with a perfectly conducting metal. The signal intensity is evaluated by calculating the light power finally coupled to the ordinary waveguide (optical fiber) region. The simulation box consists of a $120 \times 120 \times 360$ grid in the x, y, and z directions. The space step is $\lambda/40$. The calculation is done for three types of probes whose dimensions are summarized in the Fig. 3.8 caption. Sample 1 is a single-tapered probe. Sample 2 is a single-tapered probe which has a shorter tip length than that of sample 1. Sample 3 is a double-tapered probe whose cone angle at the tip is the same as that of sample 2. The calculation shows that the ratio of signal intensity for samples 1–3 is 1 : 32 : 100. It is clear that a large cone angle contributes to high collection ability since this structure minimizes the length of the cutoff region. In addition, the double-tapered probe makes the coupling of light into the normal waveguide more efficient than that of the single-tapered probe with the same cone angle. The collection efficiency and spatial resolution of the double-tapered probe in [13] are also reproduced by this FDTD simulation.

3.2.2 Triple-Tapered Probe

We have demonstrated above that by using a double-tapered structure in the probe, throughput can be increased more than ten times. However, a remaining problem of the double-tapered probe is the deterioration of resolution by the larger apex diameter of the double-tapered core due to its large cone angle. To solve this problem, the triple-tapered probe shown in Fig. 3.9 was proposed and fabricated [14]. The lowest right of this figure shows the

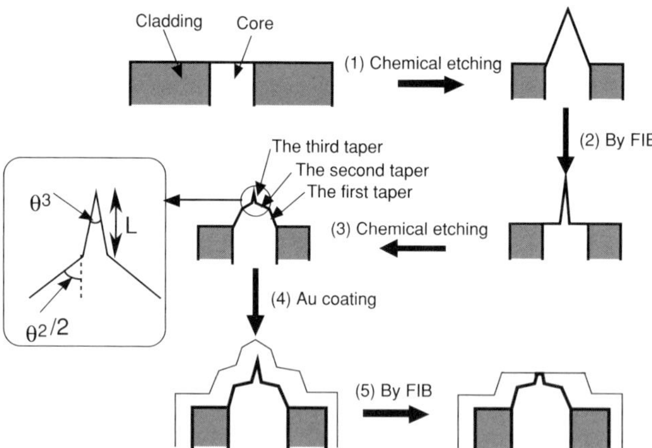

Fig. 3.9. Schematic of fabricating a triple-tapered probe with an aperature at a flat top

result of the fabrication which has a very sharp third taper at the top of the double-tapered core. Due to this sharp taper, generation and localization of a very high spatial Fourier frequency component of the optical near field are expected. Its fabrication consists of five steps.

1. chemical etching to sharpen the core and to reduce the cladding diameter;
2. irradiating the FIB to form a very sharp tapered core on a flat floor, which is used as the third taper;
3. chemical etching to form the first and second tapers at the foot of the third taper;
4. coating a gold film; and
5. removing a gold film from the top of the third taper to form an aperture.

Chemical etchings for steps 1 and 3 are the established techniques. Step 5 was used to form a flat aperture to compare the throughput with other apertured probes described in previous sections. However, it is also possible to maintain the top of the third taper after removing the gold film by using a $KI–I_2$ etching solution to produce a protruded triple-tapered probe, which can realize very high resolution while maintaining high throughput.

Figure 3.10 shows a scanning electron micrograph of the results of step 2 obtained by using a Ga+ ion beam of 30-nm diameter. Figure 3.10b shows a scanning electron micrograph of the triple-tapered core formed by step 3. A schematic of the triple-tapered core is given by the inset of Fig. 3.9, where θ_2 and θ_3 represent the cone angles of the second and third tapers, respectively. The length of the third taper is represented by L. Although Fig. 3.10b shows $\theta_2 = 150°$, the value of θ_3 realized by this step can be controlled within the range of 90–150° with an error of ±5°. Figure 3.10c is a magnified picture of the third taper of $\theta_2 = 30°$ and $L = 350$ nm with a fabrication error of 3° and 30 nm, respectively. It also shows the nanometric apex diameter of the third taper. Figure 3.10d,e shows the side and magnified top views of the probe, respectively, fabricated by steps 4 and 5. The value of d_f is 60 nm, and the gold film is 300 nm thick.

Figure 3.11 shows the measured results of the dependence of throughput on d_f. Closed circles, open circles, and open triangles represent the values for the conventional probe, the edged probe, and the triple-tapered probe, respectively. It is confirmed by this figure that the triple-tapered metallized fiber probe has a throughput 1000 times higher than that of the conventional probe for 60 nm $\leq d_f \leq$ 100 nm. In other words, a throughput as high as 1×10^{-4}–1×10^{-2} has been realized across the above range of values for d_f.

Realization of such an extremely high throughput can be attributed to several causes.

1. The overall profile of the first and the second tapers is similar to a convex lens, which enables focusing the guided light from the fiber at the foot of the third taper.
2. The third taper is much shorter than the total length of the conventional fiber probe, which reduces the guiding loss.

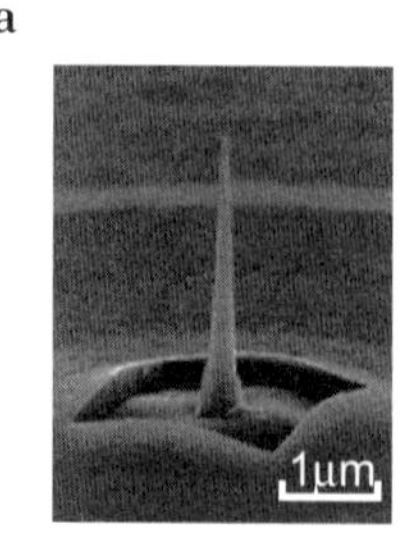

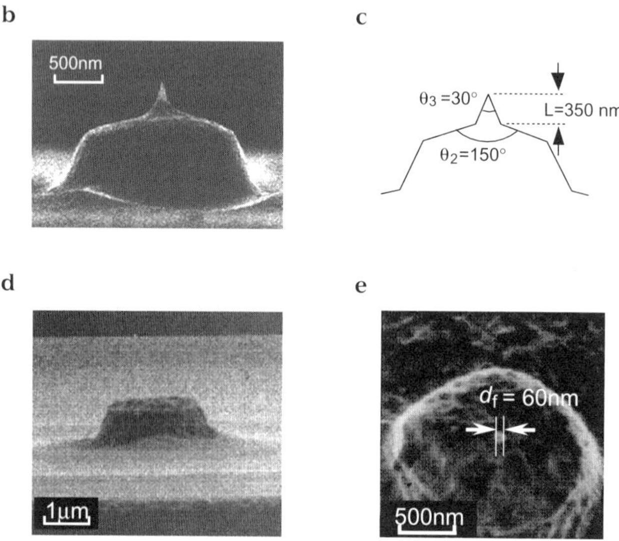

Fig. 3.10. Scanning electron micrographs of a triple-tapered probe. **a** Results of step 2 obtained by using a 30-nm diameter Ga+ ion beam. **b, c** Profile of the triple-tapered core formed by step 3 and a magnified image of the third-taper. **d, e** Side view and magnified top view of the probe produced by steps 4 and 5

3. Figure 3.10 also shows that the relation between d_f and the throughput of the triple-tapered probe is similar to that of the edged probe for 80 nm $\leq d_f \leq$ 300 nm. This similarity reveals the possible excitation of the HE-plasmon mode, even in the triple-tapered probe. This is attributed to the edge formed between the second and third tapers due to the drastic change of the cone angle.

Figure 3.12 shows the measured cross-sectional profile of the spatial distribution of the output light power on the planar aperture ($d_f = 60$ nm) at the top of the triple-tapered probe. Its full width at half maximum is 160 nm which is the convolution between the aperture diameter (60 nm) and the size of the probe used for the measurement. A remarkably sharp peak at the center

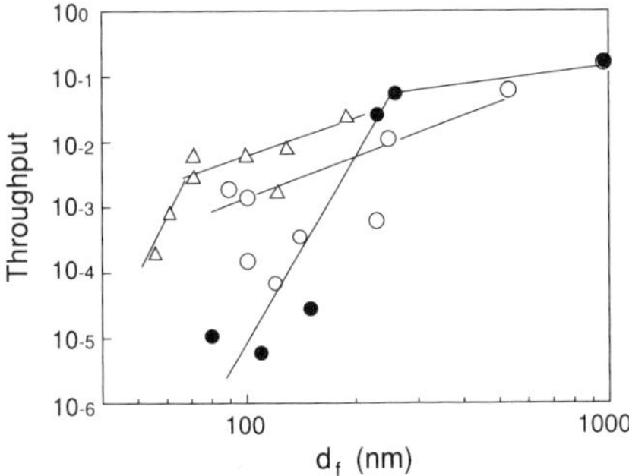

Fig. 3.11. Measured relations between the aperture diameter d_{f} and throughput. Closed circles, open circles, and open triangles represent the values for the conventional probe, the edged probe, and the triple-tapered probe, respectively

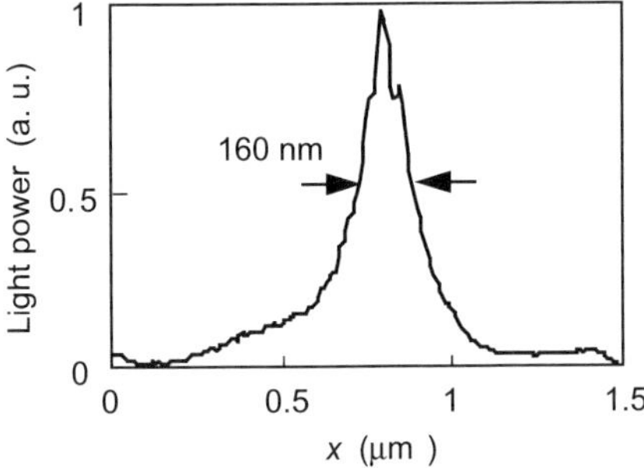

Fig. 3.12. Measured cross-sectional profile of the spatial distribution of the output light power on the planar aperture of the triple-tapered probe with $d_{\mathrm{f}} = 60$ nm

of this curve is attributed to localization of the optical near-field due to the third taper on the top of the core.

References

1. M. Ohtsu: *Near-Field Nano/Atom Optics and Technology* (Springer-Verlag, Berlin, Tokyo, New York 1998)
2. M. Ohtsu and H. Hori: *Near-Field Nano-Optics* (Kluwer Academic/Plenum, New York 1999)
3. Uma Maheswari Rajagopalan, S. Mononobe, K. Yoshida, M. Yoshimoto, and M. Ohtsu: Jpn. J. Appl. Phys. **38**, 6713 (1999)
4. M. Naya, R. Micheletto, S. Mononobe, R. Uma Maheswari, and M. Ohtsu: Appl. Opt. **36**, 1681 (1997)
5. D. Marcuse: *Light Transmission Optics*, Chap. 10 (Van Nostrand Reinhold, New York 1972)
6. T. Yatsui, M. Kourogi, and M. Ohtsu: Appl. Phys. Lett. **71**, 1756 (1997)

7. T. Pangaribun, K. Yamada, S. Jiang, H. Ohsawa, and M. Ohtsu: Jpn. J. Appl. Phys. **31**, L1302 (1992)
8. T. Saiki, S. Mononobe, M. Ohtsu, N. Saito, and J. Kusano: Appl. Phys. Lett. **68**, 2612 (1996)
9. K. S. Yee: IEEE Trans. Antennas Propagat. **AP-14**, 302 (1966)
10. A. Taflove: in *The Finite-Difference Time-Domain Method* (Artech House, Norwood, MA 1995)
11. H. Furukawa and S. Kawata: Opt. Commun. **132**, 170 (1996)
12. H. Nakamura, K. Sawada, H. Kambe, T. Saiki, and T. Sato: Suppl. Prog. Theor. Phys. (in press)
13. T. Saiki and K. Matsuda: Appl. Phys. Lett. **74**, 2773 (1999)
14. T. Yatsui, M. Kourogi, and M. Ohtsu: Appl. Phys. Lett. **73**, 2090 (1998)

4 Apertureless Near-Field Probes

S. Kawata, Y. Inouye, T. Kataoka, and T. Okamoto

Apertureless probes made of metal, dielectrics, and semiconductors were investigated for use in near-field scanning optical microscopy (NSOM) [1–5]. In the configuration, the electric field localized near the tip of the sample is scattered by the fine structure of the sample or the tip and is detected with external collection optics. From the viewpoint of electromagnetism, the intensity of the scattered field detected in the far-field changes on a nanometric scale as the boundary conditions of the electromagnetic field vary on a nanometric-scale while scanning the tip or sample of a fine structure with nanometric resolution. Apertureless probes have the following advantages compared to probes with small apertures:

1. Since a probe does not have an aperture or opaque coating surrounding the aperture but just the scattering point at the apex, the resolution of an apertureless-probe NSOM can be much higher and can reach a few nanometers.
2. Since the metal-coated dielectric waveguide is not used for sending (or receiving) photons to (or from) the probe apex, large optical throughput can be gained without any loss in waveguide propagation near the apex where the diameter of the dielectric is much shorter than the wavelength.
3. By using metal as a probe material, field enhancement is anticipated with the local mode of the surface plasmon-polariton at the apex of the probe [6].
4. The spectral response of near-field detection with a metallic tip ranges from ultraviolet to infrared because of the use of external optics (e.g., a Cassegrain objective mirror or a lens using an appropriate material), whereas the spectral response of an apertured probe is limited by the component material of the waveguide. The scattering efficiency of a metal is higher in the infrared region than in the visible, then the use of a metallic probe tip can be beneficial in infrared microspectroscopy [5,7].

In this chapter, we present the current progress in NSOM using an apertureless probe, especially a metallic probe. In Sect. 4.1, a local plasmon induced on a metallic nanoparticle is described for demonstrating how much the field is enhanced by a small metallic particle. Section 4.2 demonstrates a combination of an apertureless probe with laser trapping techniques which is suitable for observing biological samples. Section 4.3 shows the field enhancement factor of a metallic tip using numerical simulations and the application of the tip to near-field Raman spectroscopy. In Sect. 4.4, a scattering near-field optical microscope with a microcavity is discussed for realizing highly sensitive detection of an evanescent field.

4.1 Local Plasmon in a Metallic Nanoparticle

We describe the optical properties of metallic nanoparticles and their applications as sensors. We also describe gold nanoparticle probes, where a gold nanoparticle is fixed onto the tip of a silicon nitride cantilever.

4.1.1 Local Plasmon Resonance in a Metallic Nanoparticle

It is well known that an aqueous suspension of a gold colloid has a red color. The absorption spectrum of this suspension has a peak around the 520-nm wavelength. This absorption is caused by the resonant oscillation of free electrons in the gold nanoparticles. This phenomenon is called surface plasmon resonance (SPR) or sometimes local plasmon resonance to distinguish from the SPR at plane metal–dielectric interfaces. This resonant absorption and the additional resonant scattering can be explained by solving a scattering problem for a spherical particle.

The scattering problem for an arbitrary sphere has been solved exactly by Mie [8]. For clear physical insight, however, the quasi-static approximation is easier to understand. This approximation can be employed when the diameter of the sphere is much smaller than the wavelength. Quasi-static implies that the field is static spatially, i.e., there are no retardation effects, but temporally it oscillates by $e^{-i\omega t}$. The range of the validity of the electrostatic solution is discussed by Kerker [9]. In this approximation, the potential satisfies Laplace's equation ($\nabla^2\Phi = 0$). The equation in spherical polar coordinates is

$$\frac{1}{r^2}\frac{\partial}{\partial r}\left(r^2\frac{\partial\Phi}{\partial r}\right) + \frac{1}{r^2\sin\theta}\frac{\partial}{\partial\theta}\left(\sin\theta\frac{\partial\Phi}{\partial\theta}\right) + \frac{1}{r^2\sin^2\theta}\frac{\partial^2\Phi}{\partial\phi^2} = 0\,. \tag{4.1}$$

The general solution of this equation is given by [10]

$$\Phi = \sum_{n=0}^{\infty}\sum_{m=0}^{n}\left(a_{nm}r^n + \frac{b_{nm}}{r^{n+1}}\right)P_n^m(\cos\theta)e^{im\phi}\,. \tag{4.2}$$

The potentials in each region are solved to satisfy the boundary condition at the sphere surface. Then, the electric field is given by $\boldsymbol{E} = -\boldsymbol{\nabla}\Phi$. When the external field is uniform, that is, plane wave incidence, the scattering field $\boldsymbol{E}_\mathrm{s}$ and the field inside the sphere $\boldsymbol{E}_\mathrm{i}$ are given by

$$\boldsymbol{E}_\mathrm{s} = \frac{\varepsilon_1-\varepsilon_2}{\varepsilon_1+2\varepsilon_2}\frac{r_1^3}{r^3}E_0(2\cos\theta\boldsymbol{e}_r + \sin\theta\boldsymbol{e}_\theta)\,, \tag{4.3}$$

$$\boldsymbol{E}_i = \frac{3\varepsilon_2}{\varepsilon_1 + 2\varepsilon_2} E_0 \boldsymbol{e}_z \,, \tag{4.4}$$

where r_1 is the sphere radius; ε_1 and ε_2 are the dielectric constants of the sphere and the ambient, respectively, and $\boldsymbol{e}_z$, $\boldsymbol{e}_r$, and $\boldsymbol{e}_\theta$ are the unit vectors. The induced scattering field is identical to that of the dipole moment $\boldsymbol{p}$ placed at the center of the sphere given by

$$\boldsymbol{p} = \varepsilon_2 \alpha E_0 \,, \tag{4.5}$$

$$\alpha = 4\pi r_1^3 \frac{\varepsilon_1 - \varepsilon_2}{\varepsilon_1 + 2\varepsilon_2} \,. \tag{4.6}$$

Thus, the polarizability of the sphere is expressed by the above α.

From the polarizability, the scattering cross section C_{sca} and the absorptive cross section C_{abs} are given by [11]

$$C_{\mathrm{sca}} = \frac{k^4}{6\pi} |\alpha|^2 \,, \tag{4.7}$$

$$C_{\mathrm{abs}} = k \mathrm{Im}(\alpha) \,, \tag{4.8}$$

where k is the wave vector in the ambient. The scattering cross section is defined in the far-field. Near the sphere, however, an additional electric field component exists that has no energy flow outward but is very large. Here, we define the near-field scattering cross section C_{nf} by the integration of the intensity of the electric field on the surface of the sphere, and it is given by

$$C_{\mathrm{nf}} = \frac{\alpha^2}{6\pi} \left(\frac{3}{r_1^4} + \frac{k^2}{r_1^2} + k^4 \right) \,. \tag{4.9}$$

Figure 4.1 shows the spectra of the scattering cross section of gold spheres

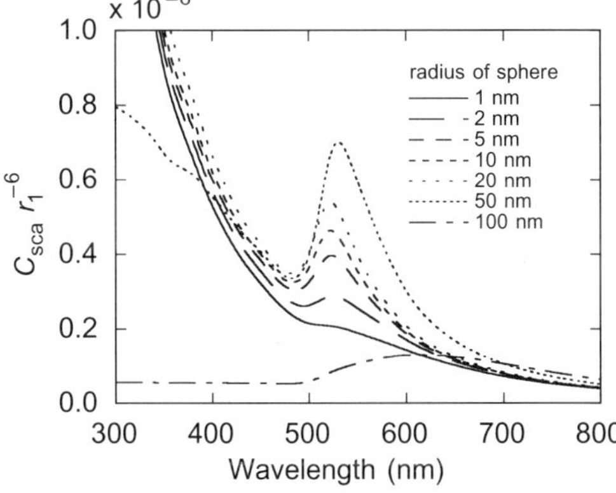

Fig. 4.1. Scattering cross section of a gold sphere in air normalized by r_1^6

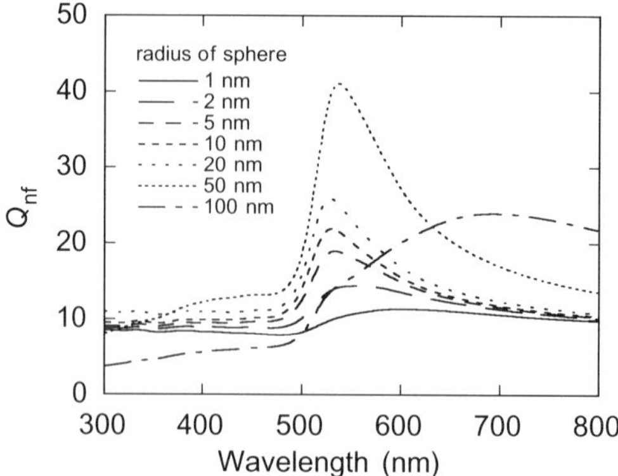

Fig. 4.2. Near-field scattering efficiency of a gold sphere in air

of various radii in air. These spectra were calculated by using the Mie scattering formula, rather than the quasi-static approximation. To clarify the size effect and the retardation effect, the scattering cross section is normalized by r_1^6 because the scattering cross section C_{sca} is proportional to r_1^6, as described by (4.7). The size effect implies the dependence of the dielectric function of the metallic particle on its radius and is notable when the free path length of the electron is longer than the particle radius.

Peaks between 500 and 600 nm are caused by a local plasmon resonance. At the peak wavelength, which is sometimes called the Fröhlich wavelength [12], the real part of the denominator of the polarizability, i.e., $\mathrm{Re}(\varepsilon_1 + 2\varepsilon_2)$, is zero. The low peak height for small spheres is caused by the size effect, where the imaginary part of the dielectric constant increases as the particle diameter decreases. The low and broadened peak for a 100-nm radius sphere is caused by the retardation effect.

Figure 4.2 shows the near-field scattering efficiencies of gold spheres in air, calculated according to Messinger [13]. The near-field scattering efficiency is much larger than the far-field scattering efficiency. For example, the peak of the near-field scattering efficiency for the sphere of 50-nm radius is $Q_{\mathrm{nf}}(\lambda = 537\ \mathrm{nm}) = 41.1$, which is 30 times larger than the far-field scattering efficiency, $Q_{\mathrm{sca}}(\lambda = 531\ \mathrm{nm}) = 1.39$.

The scattering and the absorptive cross sections also depend on the dielectric constant of the ambient [see (4.7)]. Figure 4.3 shows the dependence of the absorption efficiency of a gold sphere of 20-nm diameter on the refractive index of the ambient. The absorptive efficiency increases and the resonance or the Fröhlich wavelength becomes longer as the refractive index of the ambient increases. The scattering and the near-field cross sections show similar dependence.

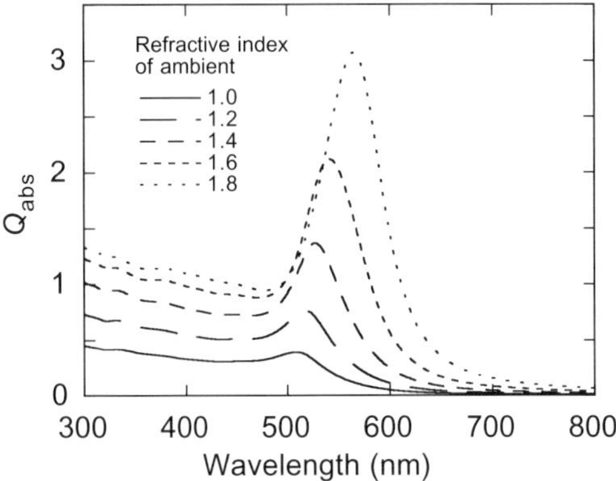

Fig. 4.3. Absorption efficiency of a 20-nm gold sphere in ambients of various refractive indexes

4.1.2 Local Plasmon Resonance in a Metallic Nanoparticle above a Substrate

In this section, we discuss the scattering problem of a metallic nanoparticle above a flat substrate. There are no analytic solutions corresponding to the Mie scattering formula for this configuration. The simplest method for solving this problem is using the dipole approximation, where a dipole is generated at the center of the sphere and the image dipole is also induced in the substrate. Under the quasi-static approximation, this problem has been solved by Aravind and Metiu [14] and Wind et al. [15].

The characteristic property in this system is that a much enhanced field is generated at the gap between the sphere and the substrate. We calculated the field around a gold sphere according to Aravind and Metiu [14]. Figure 4.4 shows the contour plot of the cross-sectional distribution of the calculated scattered intensity around a gold sphere on a BK7 glass ($n_{\mathrm{d}} = 1.516$) on the incident plane, i.e., the x,y plane [16]. It is assumed that a p-polarized plane wave (wavelength: $\lambda = 548$ nm, intensity: $I_0 = 1.0$) illuminates the sphere at a 45° incident angle, the ratio of the gap to the sphere radius is 0.01, and the size effect in the dielectric constant is ignored. The dark regions correspond to high-intensity areas. The intensity at the gap is much enhanced. In contrast, the maximum intensity for s-polarized incidence is only 1.9.

Figures 4.5–4.7 show the intensity profile of each polarization component on the surface of the substrate. The polarization components $I_x = |E_x|^2$ (Fig. 4.5) and $I_y = |E_y|^2$ (Fig. 4.6) have two peaks each, whereas the component $I_z = |E_z|^2$ has a single peak, as shown in Fig. 4.7. The respective peak values of each component are $I_{xmax} = 3.8$, $I_{ymax} = 1.9$, and $I_{zmax} = 79$, so that the field on the surface of the substrate is almost perpendicularly polarized to the surface. The full width at half maximum of the intensity peak in I_z is $0.39r_1$ and $0.38r_1$ along the x and y axes, respectively. Thus, the

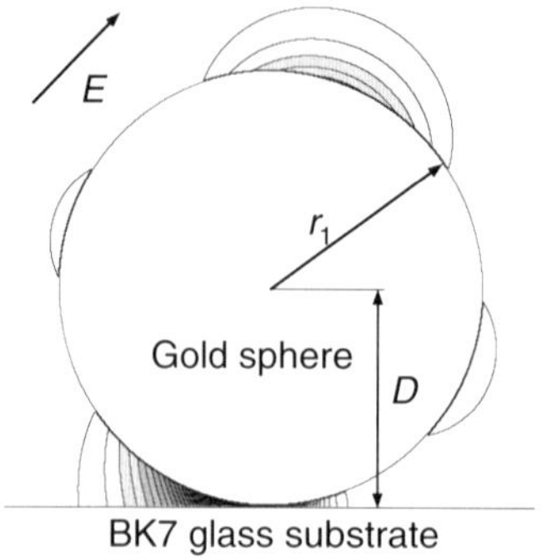

Fig. 4.4. Cross-sectional intensity distribution of the scattered local field around a gold sphere on a BK7 glass substrate. The ratio of the gap and the radius is assumed to be 0.01. A p-polarized field with a 548-nm wavelength is assumed to be incident at 45°. The maximum intensity is 79 times higher than the intensity of the incident field [16]

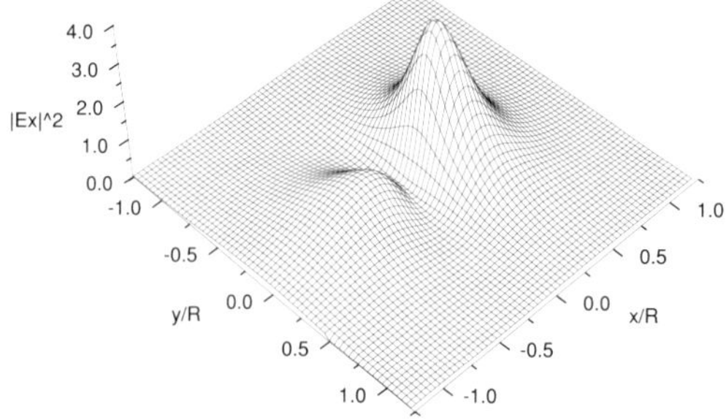

Fig. 4.5. The intensity distribution for an x-polarized component on a substrate [16]

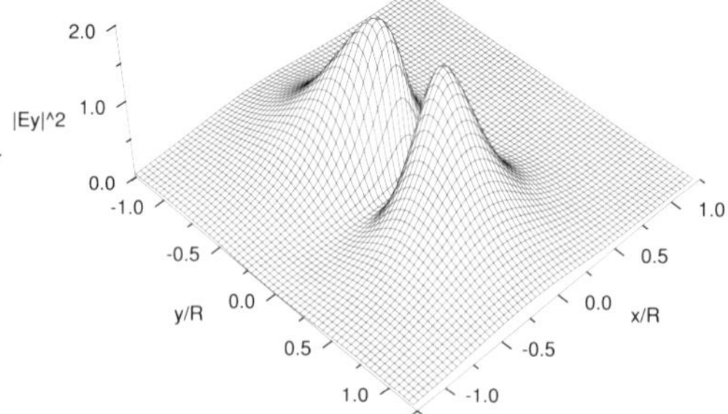

Fig. 4.6. The intensity distribution for a y-polarized component on a substrate [16]

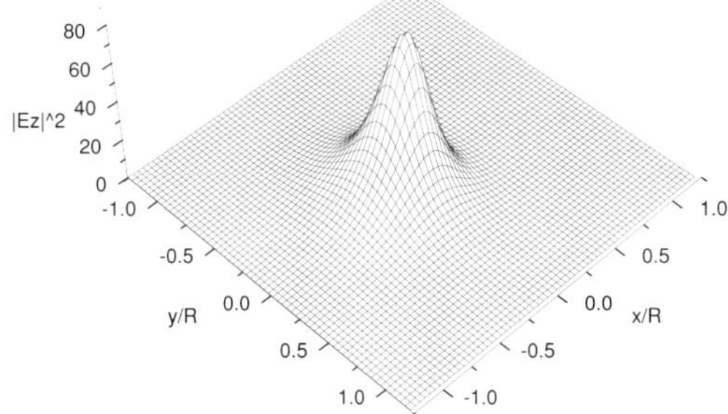

Fig. 4.7. The intensity distribution for a z-polarized component on a substrate [16]

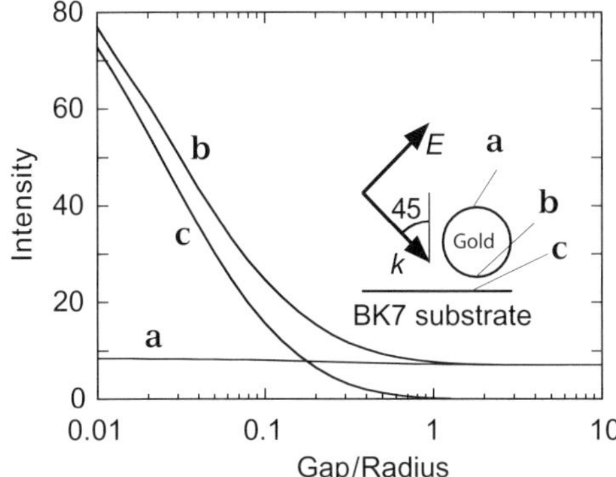

Fig. 4.8. Intensity of the scattered local field as a function of the gap distance calculated **a** at the top of the sphere, **b** at the bottom, and **c** on the surface of the substrate just below the center of the sphere [16]

enhanced field area is very localized, and its diameter is less than one-fifth of the sphere diameter.

Figure 4.8 shows the dependence of the intensity on the gap distance calculated at the top and the bottom of the sphere and at the point on the substrate surface just below the center of the sphere. At the top of the sphere, the intensity is almost independent of the gap distance and is low. The intensity at the gap increases as the distance decreases.

Figure 4.9 shows the dependence of the scattered intensity on the incident wavelength calculated on the substrate just below the center of the sphere [16]. Three substrates are assumed: an LaSFn17 glass (Sumita Optical Glass, n_d=1.883), a BK7 glass, and air, i.e., no substrate. The resonant wavelengths are 552, 548, and 542 nm for LaSFn17, BK7, and air, respectively. Though the dependence of the resonant wavelength on the refractive index of the substrate is not notable, the intensity at resonance is highly dependent on it. The higher the refractive index of the substrate, the larger the intensity of

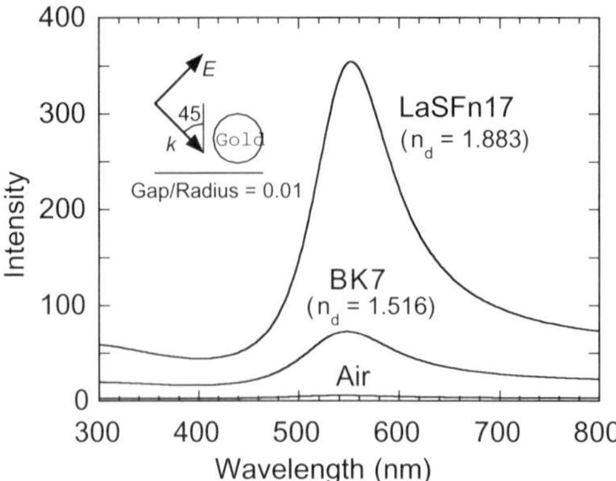

Fig. 4.9. Intensity of the scattered local field as a function of the incident wavelength calculated on the surface of the substrate just below the center of the sphere for a LaSFn17 glass substrate ($n_\mathrm{d} = 1.883$), a BK7 glass, and air, i.e., no substrate. The other parameters are the same as in Fig. 4.4 [16]

the peak. An LaSFn17 substrate $0.01r_1$ apart from the sphere enhances the intensity by a factor of 60 compared with that of an isolated sphere.

4.1.3 Optical Sensor Using Colloidal Gold Monolayers

As an application of metallic nanoparticles, a local plasmon sensor [17] was proposed using colloidal gold monolayers where colloidal gold particles a few ten of nanometers in diameter are immobilized on a glass slide using a functional organic coupling agent [18–22]. This sensor uses the property that the peak in the absorption spectrum of gold nanoparticles sensitively depends on the dielectric constant of the ambient near the particles.

Figure 4.10 shows the scanning electron micrograph of a colloidal gold monolayer deposited on a glass surface using the coupling agent, 3-amino-

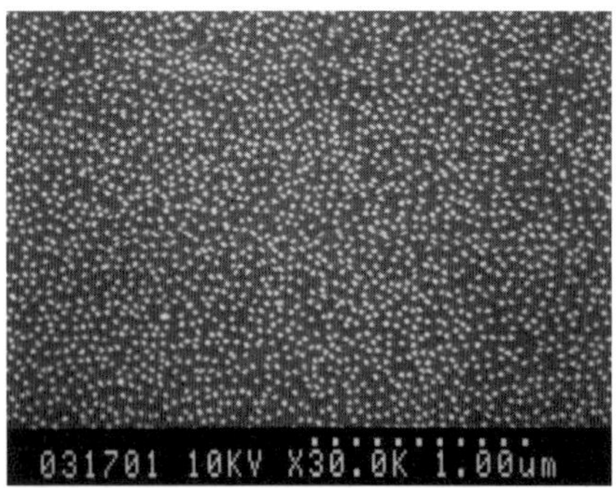

Fig. 4.10. Scanning electron micrograph of a 20.2-nm colloidal gold monolayer [17]

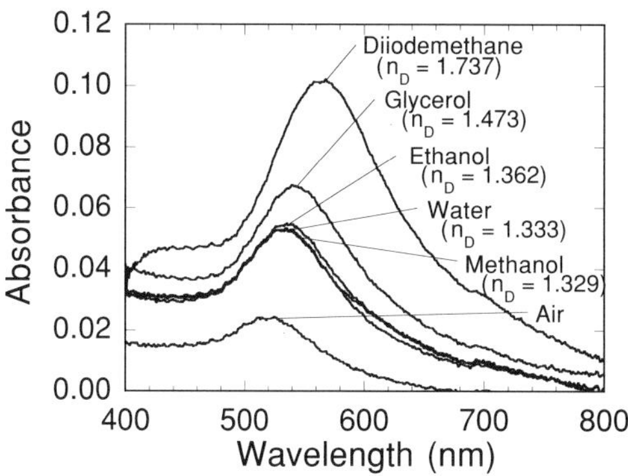

Fig. 4.11. Measured absorption spectra of a 20.2-nm colloidal gold monolayer immersed in liquid samples of various refractive indexes [17]

propyl-trimethoxysilane [17]. Gold particles are immobilized uniformly and are isolated from each other. The mean diameter of the deposited particles is 20.2 nm. The immobilized gold particles were not washed off by ultrasonic cleaning even in an organic solvent (methanol and ethanol). The density of the particles was roughly 370 particles in 1 μm^2.

This sensor intrinsically detects the refractive index of samples surrounding the gold particles, as shown in Fig. 4.3. It can also detect the thickness of the film coated on the particles. This sensor has several advantages in comparison with conventional surface-plasmon-resonance (SPR) sensors [23]. It requires no glass prism for attenuated total-reflection geometry. The colloidal monolayer can be deposited on an inner surface such as a capillary wall because vacuum deposition that requires open surfaces is not used. Furthermore, the sensor can also consist of only one nanoparticle, in principle. This could essentially reduce the dimension of the sensor and the required sample size. For these reasons, this sensor may be applied to a wider field where conventional SPR sensors cannot be used.

Figure 4.11 shows the typical absorption spectra of a 20.2-nm colloidal gold monolayer immersed in several kinds of liquid samples: methanol, ethanol, water, glycerol, and diiodemethane [17]. The indexes of refraction of the liquids at a wavelength of 589.3 nm are also shown in Fig. 4.11. The peak absorbance increases, and the resonant wavelength becomes longer as the refractive index of the sample liquid increases.

For biosensors, especially for affinity sensors, the change in the thickness or the refractive index of the film sample at the sensor surface must be sensitively detected. To investigate this sensitivity poly(methyl methacrylate) (PMMA) films with several thicknesses were deposited on the colloidal gold monolayers. The refractive index of PMMA is 1.496 at a wavelength of 520 nm. Figure 4.12 shows the measured absorption spectra for several thick-

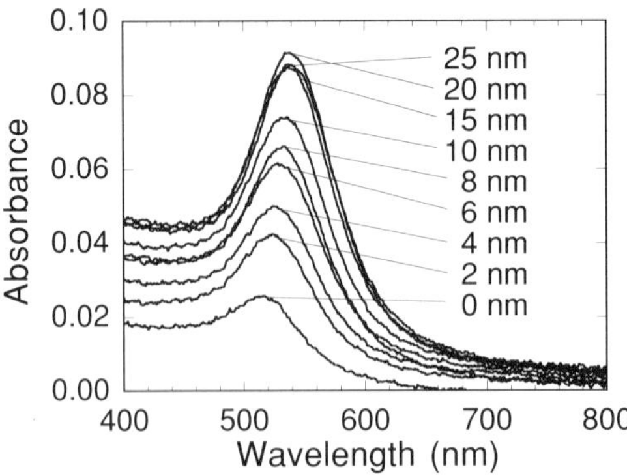

Fig. 4.12. Measured absorption spectra of a 13.9-nm colloidal gold monolayer coated by PMMA films of various thicknesses [17]

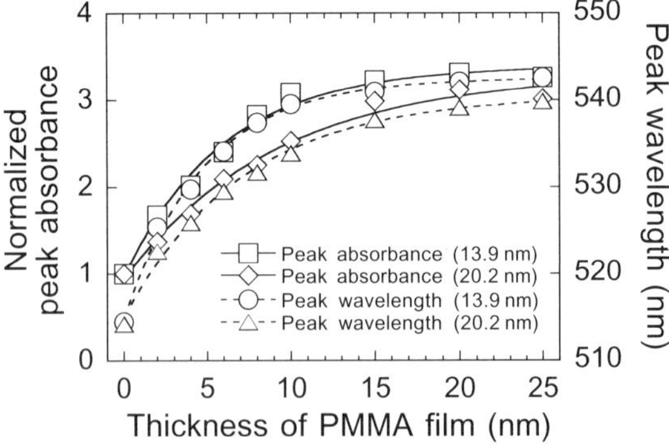

Fig. 4.13. Peak absorbance and resonant wavelength as a function of the thickness of coated a PMMA film for 13.9- and 20.2-nm colloidal gold monolayers. The solid or the dashed curves show the fitted single exponential functions [17]

nesses of coated PMMA [17]. Both resonant wavelength and peak absorbance increase as the thickness increases.

Figure 4.13 summarizes the dependences of peak absorbance and resonant wavelength on the thickness of PMMA [17]. The peak absorbance is normalized with that without a coating. Single exponential functions fitted to the experimental values are also shown in the figure. The peak absorbance and the resonant wavelength sensitively depend on the thickness of the PMMA film in the thickness region less than ~10 nm, and they are saturated in the thicker film region. The maximum sensitivities of the resonant wavelength to the film thickness, i.e., the slopes of the fitted curves in Fig. 4.13 at zero thickness, are 5.7 and 3.6 for 13.9- and 20.2-nm particles, respectively. The decay thicknesses for the resonant wavelength, at which the slopes become 1/e, are 4.8 nm and 6.3 nm for 13.9- and 20.2-nm particles, respectively.

The decay thicknesses for the peak absorbance are 6.0 nm and 6.6 nm for 13.9- and 20.2-nm particles, respectively. These values are comparable to the dimensions of biological samples such as antibodies or antigen proteins.

4.1.4 Gold Nanoparticle Probe

In the experiments described in the previous section, we observed the optical properties of an ensemble of gold nanoparticles rather than those of a single nanoparticle. Although each nanoparticle has a shape and environment different from others, we cannot observe its individual properties. But it is more desirable to observe the optical properties of a single nanoparticle rather than those of an ensemble. This may be accomplished by diluting the concentration of nanoparticles to a single particle in the field of view of a microscope. By this method, however, we cannot manipulate the particle and change the environment around the particle so easily. To realize this, we tried to fix a gold nanoparticle onto the tip of a SiN cantilever [24].

Figure 4.14 shows a scanning electron micrograph of a nanoparticle probe that we fabricated. The diameter of the attached gold particle is 1.0 μm. A technique for attaching a colloidal particle to a cantilever was proposed by Ducker et al. [25,26] They used thin wires to manipulate glue and a colloidal particle. We developed an alternative technique. We attached a gold particle to an uncoated SiN cantilever (Olympus) using an epoxy resin, Araldite Rapid (Chiba-Geigy). The procedure is as follows. We dispersed a diluted suspension of colloidal gold particles on a slide glass using a spin coater. Under a microscope, it was observed that the particles were spread uniformly. We poured the epoxy resin onto another slide glass and then scraped it off with a razor blade until the thickness of the resin layer was $\sim$ 1 μm. We let the tip of the cantilever touch the resin layer and then detached it from the layer. Next, we replaced the resin slide glass by the slide glass on which the gold particles were spread, and we let the tip contact the slide glass and scanned it across a gold particle. If the particle became attached to the tip, it

Fig. 4.14. Scanning electron micrograph of a gold particle attached to the tip of a SiN cantilever [24]

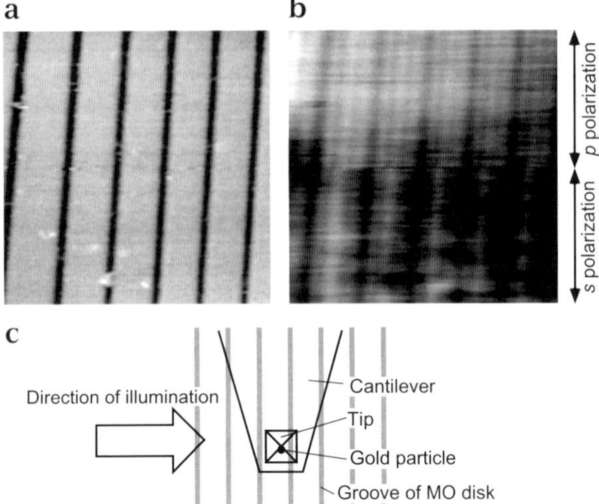

Fig. 4.15. **a** AFM and **b** NSOM images of a magneto-optical disk. The polarization state of the illumination was changed from p to s in the middle of scanning. The scanning area is 8.8 × 8.8 μm^2 for both images. The geometry of the probe, the direction of illumination, and the MO grooves are also shown **c** [24]

moved with the tip. Gold particles were routinely positioned on the shoulders of the tips, as shown in Fig. 4.14. These processes were done under an optical microscope.

We used this nanoparticle probe for an NSOM [24]. Recently, metallic tips were introduced into NSOMs, i.e., apertureless NSOMs, where a metallic tip converts the evanescent waves generated on the sample surface into propagating waves [3,27]. The apertureless NSOM realizes high optical throughput because it possesses the advantage of higher scattering efficiency of metals than dielectrics and is free from coupling loss to the optical fiber used in the conventional NSOMs. However, this NSOM has a drawback: a part of the probe other than the very tip also converts evanescent waves to propagating waves. To reduce these undesired scattered components, some probe modulation techniques were proposed [3,28].

The nanoparticle probe may also solve this problem. Since the scattering efficiency of a gold particle is much higher than that of dielectric particles with the same diameter [3], evanescent waves are more efficiently converted to scattered waves by the particle than by the cantilever.

We observed NSOM and AFM images with a home-made NSOM combined with a contact mode AFM [24]. Figure 4.15 shows simultaneously obtained (a) an AFM image and (b) an NSOM image of a transparent polycarbonate plate for a magneto-optical (MO) disk that is grooved with 1.6-μm spacing. The sample placed on a glass prism was illuminated by a focused

Ar^+ laser beam (wavelength: 514.5 nm, power: 4 mW) from the prism side. First the sample was illuminated by a p-polarized beam. In the middle of the scanning, we changed the polarization state into an s polarization. Figure 4.15c shows the geometry of the probe, the direction of illumination, and the MO grooves. Some differences are observed between NSOM images obtained with p-polarized and s-polarized beams. This difference might be caused by the dependence of the amplitude and the phase of the diffracted evanescent waves on the polarization. The thick dark lines corresponding to the grooves for s polarization slightly shift to those for p polarization. Two dim lines are observed between thick dark lines for p polarization, whereas they are not so clear for s polarization. The lateral resolution obtained is not very high ($\sim$ 400 nm) because the diameter of the particle is larger than the wavelength of the illumination beam. The use of smaller particles may improve the lateral resolution.

The diameter of 1 µm in the above particle probe is still too large to observe local plasmon resonance. If the diameter can be reduced to less than 100 nm, we can obtain the local plasmon resonance and observe resultant enhanced optical signals such as Raman, two-photon, and second-harmonics [6]. However, it is very difficult to attach such particles smaller than 500 nm in diameter. Here, we proposed an alternative method to fabricate gold nanoparticle probes. In this method a nanoparticle is grown directly onto the tip using the photocatalytic effect of TiO_2 [29].

First, a SiN cantilever (Olympus: OMCL-RC800PN-1) was coated by a Ti thin film 8–10 nm thick by vacuum evaporation. Then the cantilever was heated at 400°C for 30 minutes in air to oxidize it, resulting in an anatase TiO_2 coating layer. The TiO_2-coated cantilever was placed onto a glass prism as shown in Fig. 4.16 and was immersed in an aqueous solution of $HAuCl_4$. An evanescent wave, generated by a He–Cd laser beam (wavelength: 325 nm, power: 50 mW) incident from the prism side with an incident angle larger than the critical angle, illuminated the top of the cantilever tip. The photocatalytic effect of the TiO_2 film reduced gold from the $AuCl_4$ ion onto the illuminated region. Figure 4.17 shows a scanning electron micrograph of the cantilever tip, on which a 300-nm gold particle is grown. So far, we obtained gold particles with diameters of 100–300 nm by controlling the illuminating time. This nanoparticle probe may be used for surface-enhanced microscopy with an NSOM resolution.

4.2 Laser-Trapping of a Metallic Particle for a Near-Field Microscope Probe

To use a metallic nanoparticle as a near-field probe, the particle has to be suspended in the near field of the sample surface. Radiative force can be used to trap a particle in solution, and such a method is called laser trapping

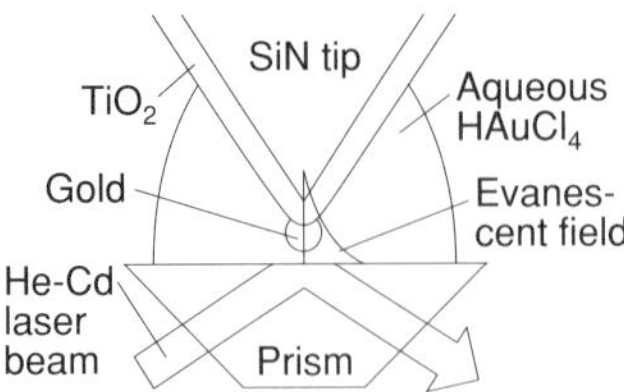

Fig. 4.16. Schematic diagram of the optical system to reduce gold onto the top of a SiN cantilever tip

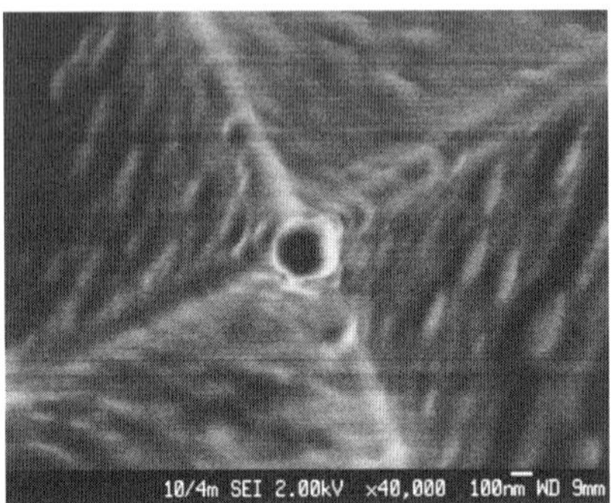

Fig. 4.17. Scanning electron micrograph of a gold nanoparticle probe

[30,31]. In this section, we show the results of laser-trapped-probe, near-field scanning optical microscopy [33–35].

4.2.1 Mechanism of Laser Trapping

A metallic particle smaller than the wavelength of light can be trapped in three dimensions by a focused laser beam [32,33]. When the converging laser beam is irradiated on a small particle, scattering force $\boldsymbol{F}_{\mathrm{scat}}$ and gradient force $\boldsymbol{F}_{\mathrm{grad}}$ [30,32,33] are exerted on the particle. Scattering force $\boldsymbol{F}_{\mathrm{scat}}$ works always to push the particle in the propagative direction of light, whereas gradient force $\boldsymbol{F}_{\mathrm{grad}}$ works to pull up the particle according to the gradient of the electromagnetic field around the particle. The gradient force $\boldsymbol{F}_{\mathrm{grad}}$ on a particle of polarizability α in a medium of refractive index n_1 is given by

$$\boldsymbol{F}_{\mathrm{grad}} = \frac{1}{4} n_1 \, \alpha \, \mathrm{grad} \left(|\boldsymbol{E}|^2 \right) , \tag{4.10}$$

where $\boldsymbol{E}$ is the electric field. If the particle is spherical, the polarizability α is given in the following form

$$\alpha = 4\pi\varepsilon\epsilon_0 \, \frac{m^2 - 1}{m^2 + 2} \, r^3 , \tag{4.11}$$

where m is relative refractive index of the particle compared to the surroundings and r is the radius of the particle. Polarizability α has a complex value in general. When the real part of polarizability is positive, the particle is attracted into the stronger light field according to the gradient of the force distribution. The gold particle is pulled into the laser beam spot by the gradient force because the real part of α is positive for gold. The gradient force on a gold particle can be much larger than that of the scattering force.

4.2.2 Laser Trapping of a Probe for NSOM

Figure 4.18 shows an optical configuration of a NSOM with a metallic particle as a probe [34,35]. An intense near-infrared laser beam is focused with a microscope objective to trap a metallic particle. A single metallic particle is suspended near the focal spot, and approaches a sample surface from beneath it. A gold colloidal particle, which has a diameter of several tens of nanometers, is used as a near-field probe.

A visible laser is also focused onto the particle collinearly with the near-infrared laser beam. By focusing the visible laser on this particle, the surface plasmon is excited at the particle to extremely enhance the field locally near the probe particle. This laser beam is scattered by the probe particle, and the scattered light works as a point light source as small as the diameter of the particle to excite the fluorescent sample molecules. When the probe particle is sufficiently close to the sample surface, a nanometric laser spot enhanced by the local mode of surface plasmon interacts with the sample. The fluorescence is collected by the objective and then is detected by a far-field detector. This light intensity contains local information about the sample in its fluorescent spectrum, and the spatial resolution exceeds the diffraction limit.

Since the spring constant to trap a probe is weak, the sample is not physically damaged. Regulating the distance between the probe particle and the sample is not necessary because the probe particle touches the sample surface during scanning.

An NSOM with a laser-trapped particle was first demonstrated by Kawata et al. [36]. In their experiment, a polystyrene particle 1 μm in diameter was

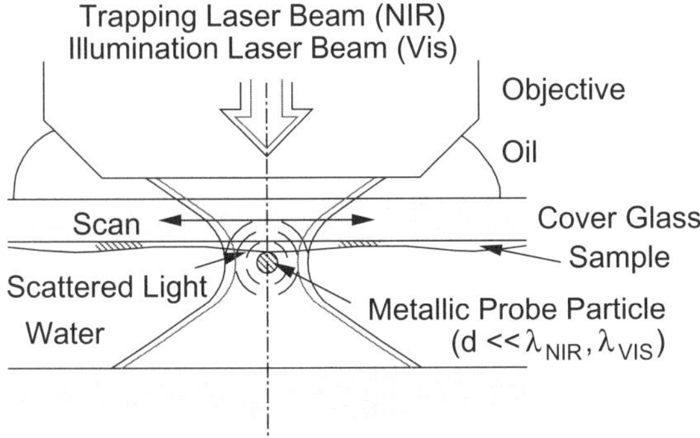

Fig. 4.18. Schematics of a laser-trapping NSOM with a metallic probe particle

used as a near-field probe. Sasaki et al. [37] applied this probing method for a tiny chemical sensor which could detect the pH in the nanometer region. In 1997, Sugiura et al. developed a laser-trapping NSOM with a three dimensionally trapped metallic particle [33,34]. This probe scans in water, so that biological samples in water can be imaged.

4.2.3 Experimental Setup

Figure 4.19 shows an experimental setup for the NSOM with a laser-trapped probe. The laser for trapping the probe is a Nd:YLF laser (2.5 W, $\lambda = 1047$-nm wavelength). The laser beam is focused with an oil immersion objective (100×, NA 1.35). Dichroic mirror DM1 reflects the light of the Nd:YLF laser and transmits visible light.

An Ar ion laser (20 mW, $\lambda = 488$ nm) is used for sample illumination. This laser is also focused onto the probe particle. The pinhole in front of the photomultiplier is placed for confocal detection of the fluorescence from the sample excited by the metal plasmon polariton.

4.2.4 Feedback Stabilization of a Particle

According to the thermal energy, a probe particle continuously moves in a laser-trapping potential by collisions with water molecules (Brownian motion). Due to the Brownian motion, the intensity of the light scattered from the particle changes in time. This will cause noise in measured signals when we use the trapped particle as a near-field probe. This noise affects the maximum resolving power of a laser-trapping NSOM. Consequently, it is necessary to reduce the fluctuation of the probe position to enhance the resolution of

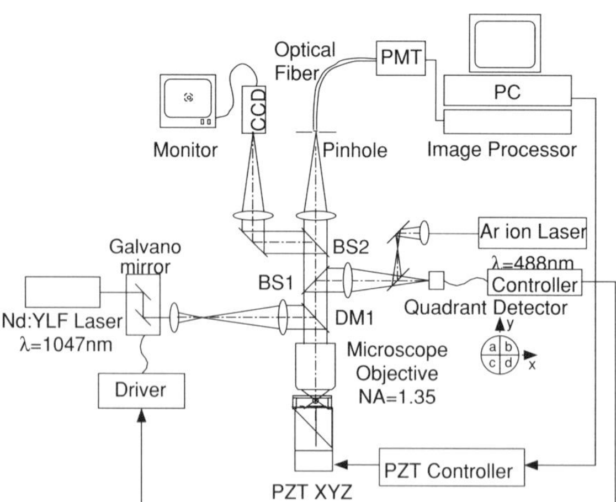

Fig. 4.19. Experimental setup of a laser-trapping NSOM

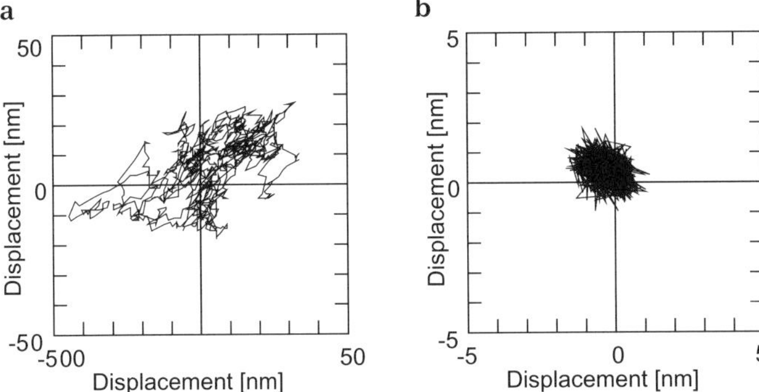

Fig. 4.20. Particle displacement in an optical trap (experimental data). **a** Without feedback stabilization of a probe particle; **b** with feedback stabilization

the NSOM. A feedback stabilization technique is effective in reducing the fluctuation of the particle [34].

In the feedback stabilization technique, information about the particle position is given back to the spot position of the trapping laser beam. The position of the probe particle is monitored by the detection of the spot position. The spot position can be sensitively detected with a quadrant detector and two sets of differential amplifiers to calculate the displacement signals of the spot in x and y directions. Two galvanomirrors are used to shift the spot position of the trapping laser beam. The angles of the galvanomirrors are changed to minimize the displacement signals in both the x and y directions. The displacement signals in the x and y directions are captured by a personal computer.

Figure 4.20 shows displacement diagrams that depict movements of a probe particle. (a) is the result taken without feedback of the displacement signal to the galvanomirror and (b) is the result with feedback. The time interval of the displacement measurement was 10 ms. The probe particle was a polystyrene latex particle whose diameter was 1 µm. The standard deviation of the particle position was about 13 nm without the feedback (a), although with the feedback the standard deviation of the particle position was reduced to 0.45 nm. This result shows that the feedback method is effective in stabilizing the position of the probe particle.

4.2.5 Experimental Results

We observed fluorescent beads on a glass substrate to demonstrate fluorescence imaging capability. Figure 4.21 shows an observed result with fluorescent beads which had a diameter of 50 nm. The excitation wavelength of the fluorescence was 488 nm of an argon ion laser, and fluorescent emission was filtered with a dichroic mirror and an absorption filter (cutoff wavelength

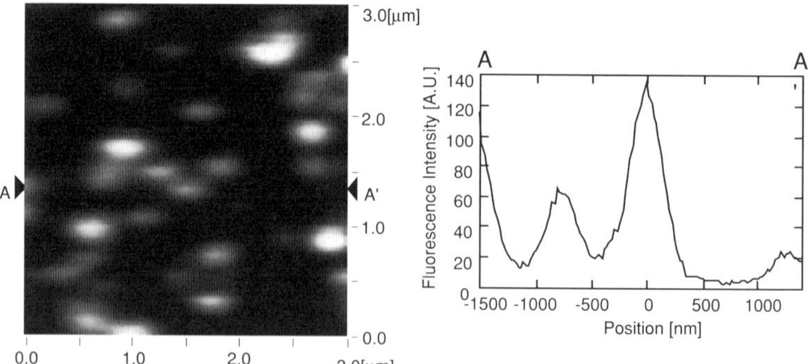

Fig. 4.21. Single fluorescent beads ($d = 50$ nm) and a line plot of the data

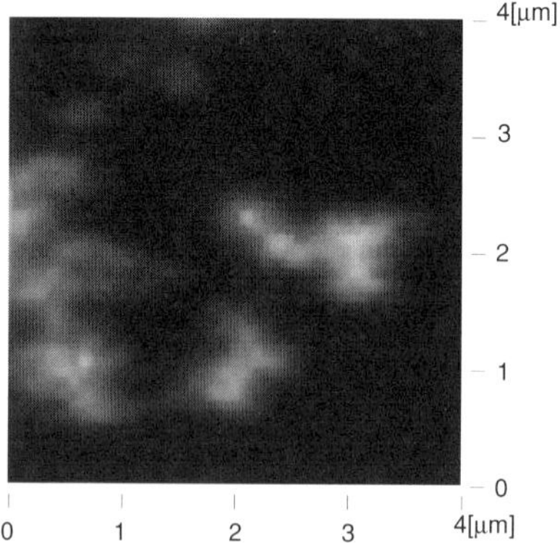

Fig. 4.22. DNA molecules observed with laser-trapping NSOM

$\lambda_{\text{cut}} = 515$ nm) to cut the light scattered by a probe particle. The probe particle was a colloidal gold particle 40 nm in diameter.

There are many bright spots in the image. Each spot corresponds to a single fluorescent particle. The line plot shown at the right of the image is a scanned result across one of these bright spots. The center of the plot corresponds to a bright spot. From this line plot, we see that the spot has a sharp peak in the center. We guess that the sharpness of the spot is from the effect of a near-field component around a probe particle.

DNA molecules were also observed [34]. DNA molecules were stained with YOYO-1 iodide dye for visualization by fluorescent emission. The dye has an absorption peak at a wavelength of 491-nm, an excitation peak at the 509-nm wavelength, and has high affinity to bind DNA molecules. The efficiency of fluorescent emission from the dye molecule significantly increases after binding to DNA.

DNA molecules from calf thymus were used as a test sample. The sample was prepared as follows: first the DNA was dissolved in water and stained with YOYO-1 iodide. Then the DNA solution was dripped on a silane-coated cover glass. After 5 minutes, the glass was rinsed with deionized water 10 times to wash away unadsorbed DNA molecules.

Figure 4.22 shows the result of DNA molecules stained with YOYO-1 iodide. An entangled DNA molecule can be seen in the middle of the picture. And many aggregated DNA molecules are also observed in the upper part. These show that the laser-trapping NSOM can image biomolecules fixed on a substrate.

4.3 Near-Field Enhancement at a Metallic Probe

In the previous section, we described how photons couple with collective oscillation of electrons or surface plasmons in a nanoparticle. Such coupling is induced at a metallic apex. In this section, we will demonstrate by numerical analysis that a localized surface plasmon is generated and the electric field is confined locally around the tip of a metallic needle that has a nanometric radius. We also show the experimental detection of near-field Raman scattering using localized field enhancement at the tip for vibrational spectroscopy.

4.3.1 Field Enhancement at the Tip

We analyzed the scattered field at a metallic tip that had a nanometric-scale radius to present the merits for a NSOM using the tip [38]. Figure 4.23 shows the geometry of the three-dimensional model for our computer simulation. A conical probe made of platinum–iridium with the complex refractive index of $n = 2.7 + i\,5.1$ at a wavelength of $\lambda = 800$ nm is located near a flat surface of a dielectric substrate whose refractive index is $n = 1.6$. The solid angle of the cone was 3.9×10^{-2}, the diameter of the tip was 20 nm, and the distance between the probe and the substrate surface was 15 nm. Illumination light

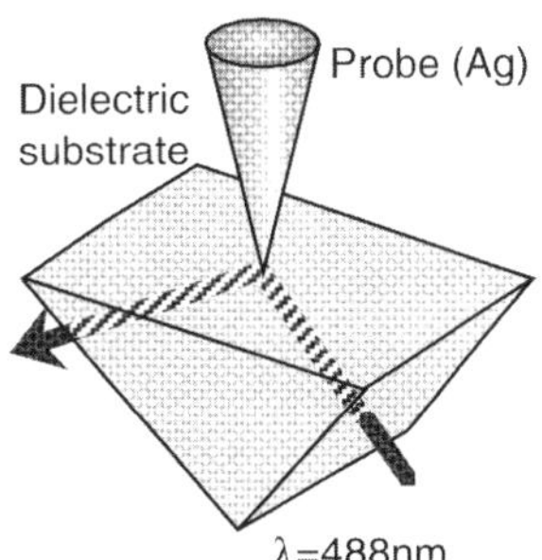

Fig. 4.23. Three-dimensional configuration of a model system for the calculation

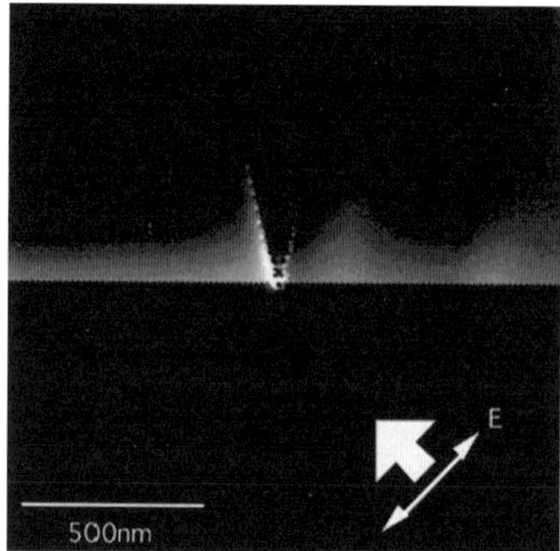

Fig. 4.24. Sectional field distribution, including a probe (platinum–iridium) and a sample (flat glass substrate) for p-polarized illumination

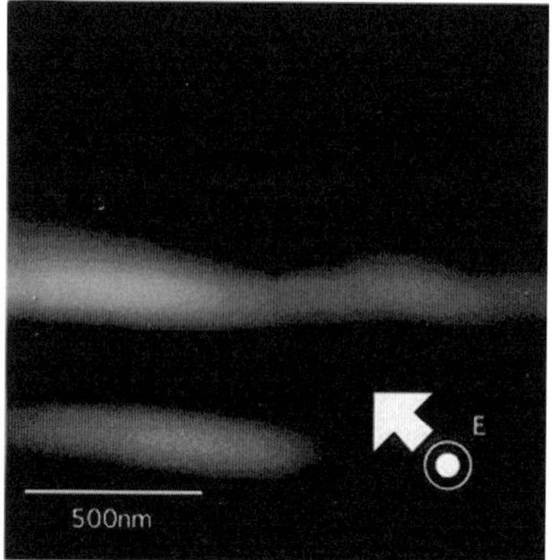

Fig. 4.25. Sectional field distribution, including a probe (platinum–iridium) and a sample (flat glass substrate) for s-polarized illumination

was incident on the probe through the dielectric substrate at 45° more than the critical angle. The wavelength of light was assumed to be $\lambda = 800$ nm.

Figures 4.24–4.26 show the calculated sections of the three-dimensional electric field. Figures 4.24 and 4.25 show vertical cuts of it including the plane of incidence parallel to the electric field oscillation (p polarization) and perpendicular to the field (s polarization), respectively. The size of each image is $1.5 \times 1.5\ \mu m^2$ with a voxel of $5 \times 5 \times 5\ nm^3$. Figure 4.26 shows a lateral cut parallel to the substrate surface at a height of 7 nm from the substrate when p-polarized illumination is employed. The area displayed is $250 \times 250\ nm^2$ below the probe. The extent of the light spot under the probe in Fig. 4.26

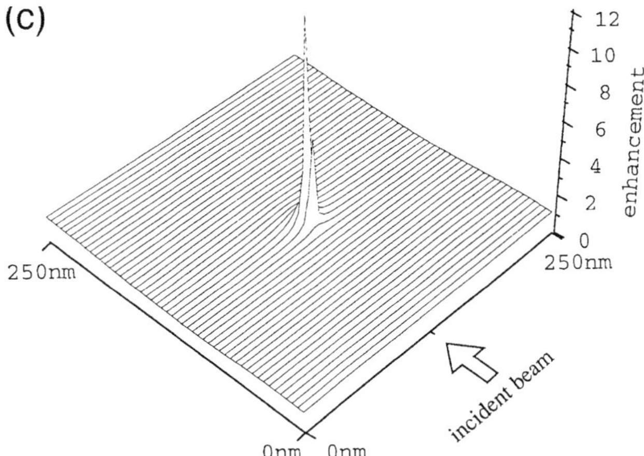

Fig. 4.26. Section parallel to the substrate at a 7-nm height

is almost approximately the same as the probe size (about 30 nm), or 27 times shorter than the wavelength (800 nm) of light. Calculations were made with the finite-difference time-domain (FDTD) method, which is a discrete straightforward calculation of Maxwell's electromagnetic equations [39–42].

This result implies that the resolution of an apertureless probe NSOM is given by the size of the probe apex, which is its great advantage. In a NSOM with an aperture in a metal coating [42], the penetration of photons through the metal coating blurs the spot, so that the practical minimum size of the spot is at least practically 50 nm in diameter even if an extremely small aperture is used. In addition, the surface plasmon on the coating thin-film metal is excited at the probe tip by evanescent photons and propagates nearly parallel to the substrate surface, resulting in enlargement of the spot size. Novotny et al. showed by numerical simulation that the extent of the light field through a 50-nm wide slit in the vicinity of the apex of a probe is 100 nm [43].

It is found in Fig. 4.24 that the field is amplified at the probe apex and is about 44 times higher than the incident field without a probe. The field enhancement is due to the surface plasmon polariton, as described above. Considering the boundary condition, the incident electric field should have a component vertical to the metallic surface at the apex of the tip to excite the localized-mode surface plasmon [44]. Because of this, enhancement is not observed for s-polarized illumination, as shown in Fig. 4.25. The intensity of the images is normalized individually in a gray scale so that the intensity distribution can be seen well in each image. The gray scale in Fig. 4.25 is 10 times larger than that of Fig. 4.24; an interference fringe as a standing wave between the incident wave and the totally internal reflected wave can be recognized in the lower half of this figure.

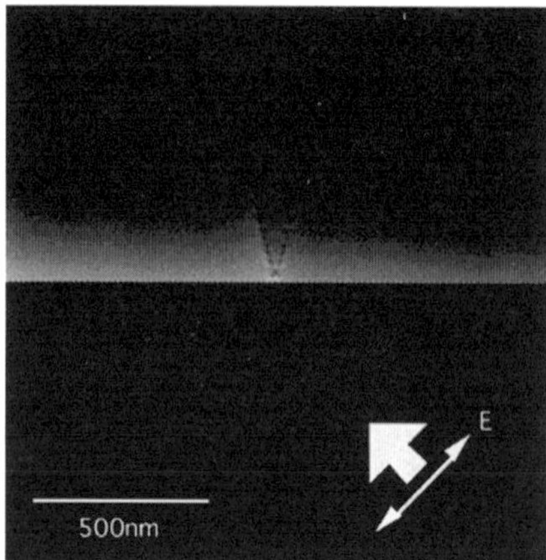

Fig. 4.27. Same as Fig. 4.24, but with a dielectric tip

Figure 4.27 shows the sectional image of the p-polarized field with a dielectric probe without metal and the substrate. The geometry is the same as that in Fig. 4.24, except that the probe material is glass with refractive index $n = 1.6$. Although a plasmon polariton is not involved in a dielectric probe without metal, the field under the dielectric probe is also enhanced. The enhancement is not as high as that under the metallic probe but is still seven times higher than that of the incident field. This is due to multiple scattering between probe and sample.

The field enhancement of a metallic probe is caused mainly by the localized surface plasmon mode excited at the apex of the probe by an evanescent field. To enhance the localized electric field effectively, a sharper cone angle [45] and a smaller diameter at the apex are necessary for increased electric flux density. The conductivity of the tip material should also be high (mostly noble metal) for exciting the plasmon.

4.3.2 Near-Field Raman Spectroscopy

Combination of near-field optics with vibrational Raman spectroscopy makes it possible to assign molecules, to analyze the chemical behavior of a molecule, and to observe molecular dynamics on a nanometric or molecular scale. Lasers and detectors of the visible region are available in Raman spectroscopy, and the quenching phenomenon and photobleaching can be avoided; hence near-field Raman spectroscopy is suitable for molecular sensing with nanometric spatial resolution [46]. Although the cross section of Raman scattering is much smaller than a fluorescent or infrared absorptive cross section, the field enhancement may compensate for the low-scattering cross section to attain an adequate signal-to-noise ratio by using a metallic tip for detection [47].

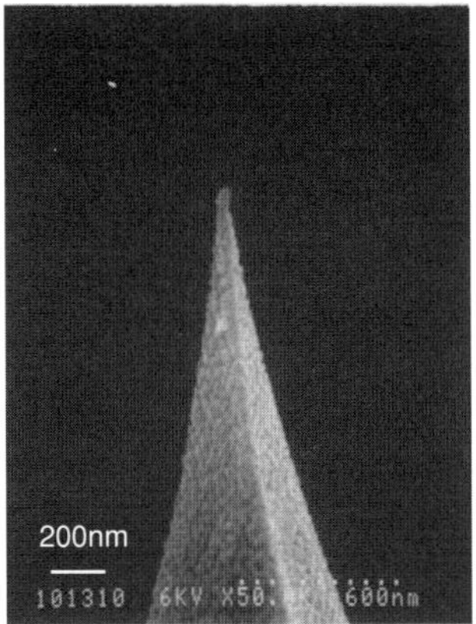

Fig. 4.28. A SEM image of a metallized cantilever

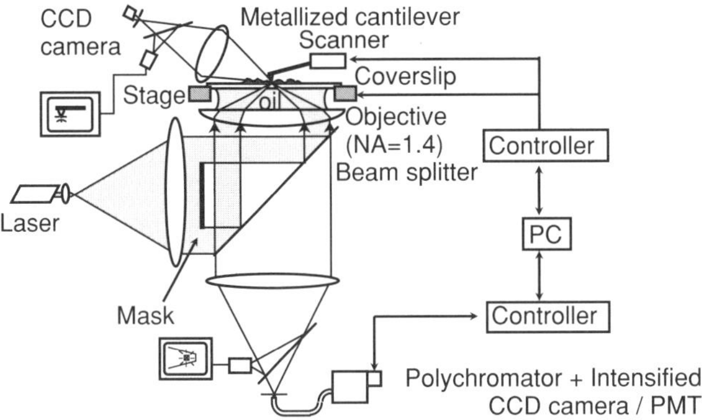

Fig. 4.29. Experimental setup for near-field Raman spectroscopy

A silver-film-coated cantilever is employed for near-field Raman detection. Figure 4.28 shows a scanning electron microscope image of a metallized cantilever. The evaporation rate of silver is 0.3 Å/s; this value is optimized to avoid undesirable bending or deforming of the lever. The silver is 40 nm thick, and the resultant diameter of the tip is 40 nm.

The experimental setup is shown in Fig. 4.29 [47,48]. The metallized cantilever approaches the sample surface by an AFM operation. An argon-ion laser beam is focused onto the sample surface with an oil immersion objective and goes through the mask which stops the components of far-field illumination. The sample is hence illuminated by the evanescent spot, in which the cantilever is inserted to produce field enhancement. Raman scattering is induced by the molecules at the apex of the metallized cantilever probe. Raman scattering is collected by the objective lens and is sent to a polychro-

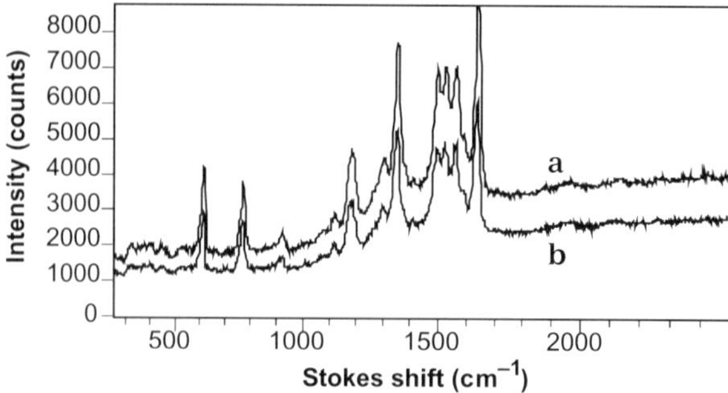

Fig. 4.30. Raman spectra obtained **a** with and **b** without a silver-coated cantilever

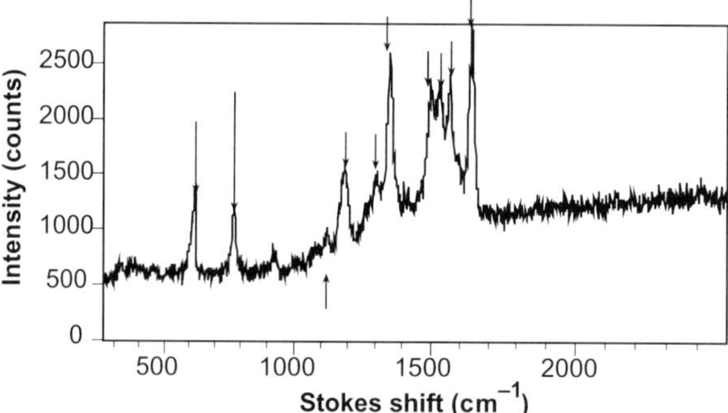

Fig. 4.31. A near-field enhanced Raman spectrum with a silver-coated cantilever

mator equipped with an intensified CCD camera. Rayleigh scattering of the illumination is eliminated by a notch filter.

The Raman spectra of Rhodamine 6G obtained with the system developed are shown in Fig. 4.30. Spectrum (a) is obtained with a silver-coated cantilever, and spectrum (b) is obtained without it. The intensity of spectrum (a) is higher than that of (b) due to the local field enhancement at the metallized tip. Several Stokes-shifted Raman lines are observed in the spectrum. The background component of the spectrum is due to fluorescence. The acquisition time of the spectrum was 5 seconds, and no accumulation was done. The sample was made by coating a solution of Rhodamine 6G (6×10^{-4} wt%) onto the 8-nm silver-island coated coverslip and depositing Rhodamine 6G by evaporating the solution. Raman scattering is hence doubly enhanced due to surface enhanced Raman scattering (SERS) with silver islands film coated on both the substrate and metallized tip [49].

The Raman spectrum Fig. 4.30a includes near-field and also far-field components. The near-field spectrum is obtained by subtracting the spectrum obtained when the tip position is far from the sample surface, from the spectrum shown in Fig. 4.30a. Figure 4.31 shows the results of subtraction.

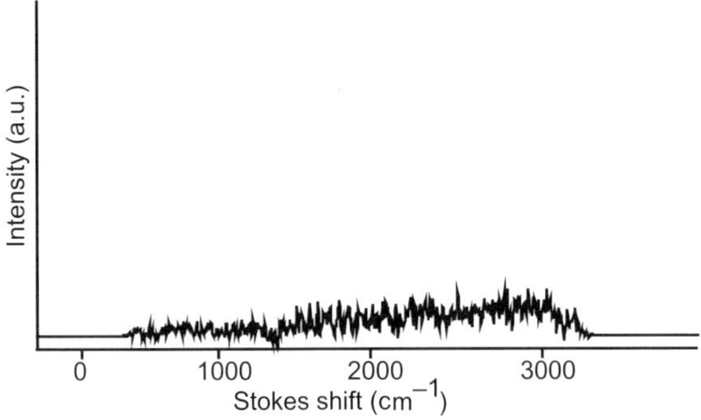

Fig. 4.32. A near-field enhanced Raman spectrum with a silicon cantilever

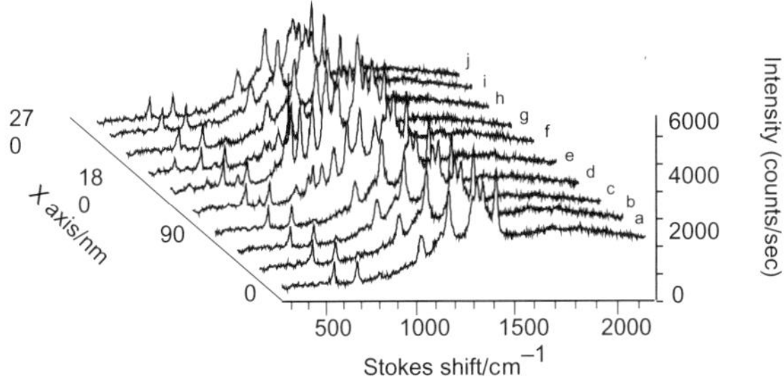

Fig. 4.33. Mapping of near-field enhanced Raman spectra

The enhancement factor F due to the silver tip is defined as the ratio of the intensity of the near-field scattering to that of far-field intensity per unit area:

$$F = \frac{[(P_\mathrm{n} - B_\mathrm{n}) - (P_\mathrm{f} - B_\mathrm{f})]}{P_\mathrm{f} - B_\mathrm{f}} \frac{d_\mathrm{f}}{d_\mathrm{n}}, \tag{4.12}$$

where P_n and P_f are the intensity at the Raman peak at 1653 cm^{-1} in the near-field spectrum and that in the far-field spectrum, respectively, and B_n and B_f are the background intensities at the same wave number for near-field and far-field spectra, respectively; d_f and d_n are the spot size, i.e., d_f is the focused spot size (400 nmϕ), and d_n is the tip-enhanced field size (50 nmϕ). In the experimental result shown in Figs. 4.30 and 4.31, F is ~ 40.

Figure 4.32 shows the spectrum of the same sample as that in Fig. 4.31 but with a silicon tip (the silicon AFM cantilever without a silver coating). No enhancement is observed with a silicon tip.

Figure 4.33 shows a one-dimensional near-field image of the Rhodamine 6G distribution with spatial steps every 30 nm [50]. The near-field Raman spectrum is detected at each position. The spectra in Fig. 4.33 are all enhanced by the tip apex, and some of them, e.g., spectra e, f, and g, exhibit remarkable features that are quite different from others or a far-field spectrum. The distinguishable near-field features in those spectra replotted in

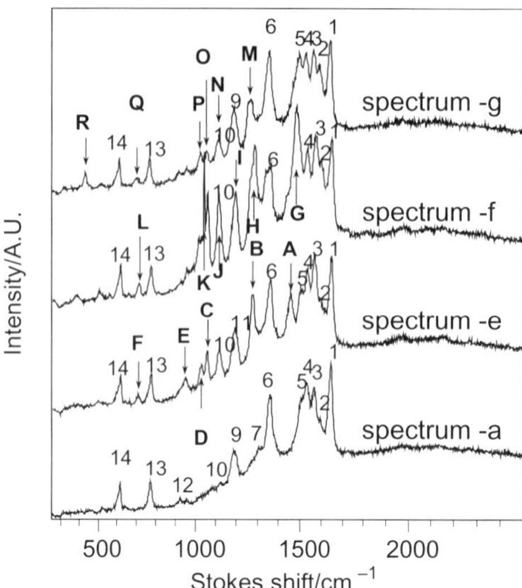

Fig. 4.34. Near-field characteristic Raman spectra

Fig. 4.34 are marked with alphabetic capital letters, and the numbers in the different spectra correspond to the same assignments. Some peaks are slightly shifted, and their relative intensities are quite different from those of the far-field spectrum. Furthermore, some additional peaks appear that are not observed in the far field.

For the aromatic C-C stretching vibrational mode, the peaks from 1350 to 1650 cm^{-1} are heavily modified. New shifted peaks, e.g., A: 1457 cm^{-1} in spectrum e, G: 1483 cm^{-1} in spectrum f, appeared. For the C–H out-of-plane bending-vibration mode (number 13), peak intensities are constant throughout the scanning, whereas in-plane bending (number 11) weak peaks are strongly enhanced and slightly shifted (see, e.g., J: 1112 cm^{-1}). For the C–O–C bending vibrational mode, the new Stokes-shifted peaks and those are largely enhanced (see peaks, e.g., B: 1278 cm^{-1}, H: 1286 cm^{-1}, and M: 1275 cm^{-1}). Furthermore, other new peaks which have not yet been assigned also appear in the spectra (see, e.g., C: 1054 cm^{-1}, D: 1027 cm^{-1}, E: 946 cm^{-1}, F: 902 cm^{-1}, K: 1040 cm^{-1}, L: 705 cm^{-1}, O: 1057 cm^{-1}, P: 1027 cm^{-1}, Q, 700 cm^{-1}, and R: 425 cm^{-1}). Some of the peaks can be identified as the same vibrational modes (e.g., C, and K, and O; D and P; L and Q). Although some of the Stokes-shifted peaks are strongly enhanced, the fluorescence intensity is constant throughout the positions. This can be explained by the chemical mechanism of surface-enhanced Raman scattering (SERS) [51,52], due to charge transfer excitation between the molecules and the metal.

4.4 Scattering Near-Field Optical Microscope with a Microcavity

We describe the method of measuring the topography of a sample surface by detecting the variation of the intensity of the scattered light from a scanning probe. The lateral spatial resolution of NSOM is not affected by the diffractive limit of light and is determined by the size and the shape of the probe. The evanescent field generated around a small aperture is one of the major probing techniques for NSOM. The spatial resolution of this technique increases with a decrease in the diameter of the aperture [53–56]. In some cases, the aperture was formed at the tip of a metallized glass pipette to measure the recessed regions of the rough surface of a sample [57]. Later, the local plasmon around a small spherical protrusion, which is vacuum coated with metal, was used as a probe [58]. In recent years, it has been reported that when a transparent sample on the substrate is irradiated by a laser beam under total internal reflection, evanescent waves are generated according to the topography of the sample. The evanescent waves are picked up by a sharpened optical fiber [59–63].

4.4.1 Resonant Microcavity Probe

We also tried to develop several types of probes with good characteristics for our NSOM system. A single dielectric sphere 500 nm in diameter is located on top of a quartz substrate with a pyramidal shape [64]. It was not a perfect sphere, and there were many protrusions on the surface. One of the protrusions might act as a "real" probe. The typical scanned data are shown in Fig. 4.35. The vertical and lateral resolutions that are 1 and 10 nm,

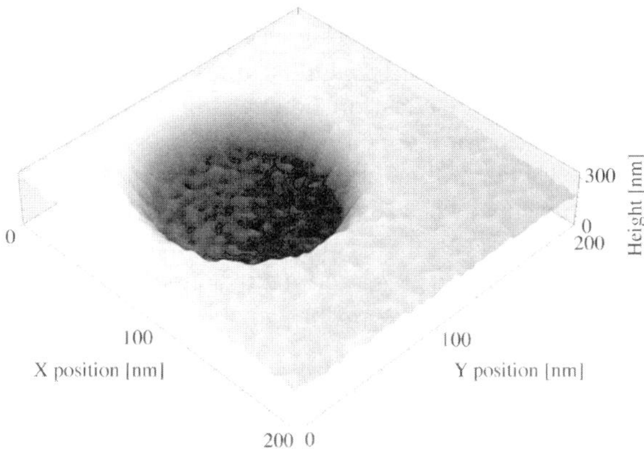

Fig. 4.35. NSOM images of the standard sample. The diameter and the depth of the pit are 100 and 20 nm, respectively. The scanning range is 200 × 200 nm

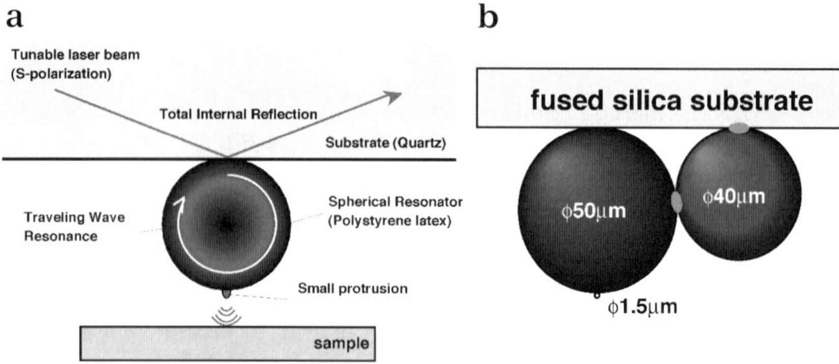

Fig. 4.36. Concept of a "resonant microcavity probe." A 50-µm resonant sphere is supported by a 40-µm sphere. An evanescent wave generated with WGMs inside a 50-µm sphere could illuminate a several-micrometer protrusion

respectively, are obtained for a standard sample which is prepared by vacuum coating of silver. The probe sphere was not coated with metal in this experiment, hence the plasmon [58] is not used.

The 500-nm-probe was unstable and the reproducibility of nanometer resolution was below 5%. Thus, lately we proposed a new type of NSOM probe which might have high spatial resolution [65,66]. The probe is named a "resonant microcavity probe." It consists of a several-micrometer protrusion on a sphere measuring a few tens of micrometers, as shown in Fig. 4.36. In Fig. 4.36a, a larger sphere acts as an optical resonator of whispering gallery modes (WGMs) or morphology-dependent resonances (MDRs) [67], and a smaller protrusion interacts with the sample surface. When the probe comes close to a sample surface, the smaller protrusion interacts with the surface and may scatter an evanescent field. The scattered lights are used as a reference to scan across the sample surface in the constant-height mode. Figure 4.36b shows the actual configuration of this probe. A 40-µm sphere fixes a spherical resonator on a quartz substrate. This probe may have several advantages. (1) MDRs in the larger sphere transfer much laser power to the protrusion, and (2) an intense and localized evanescent field could be formed around the protrusion. (3) The signal-to-noise ratio of the scattered light signal from the protrusion could be better, and (4) the probe can be handled with a micromanipulator.

4.4.2 FDTD Simulation of a Resonant Microcavity Probe

The finite-difference time-domain method (FDTD) is a good tool for electromagnetic field analysis [39]. We used a specialized "PLANC-FDTD-2D"

(IMS Lab Inc.) in a PC. This software can treat the total internal reflection of a plane wave at a semi-infinite boundary [38,68,42] to simulate a probe configuration like Fig. 4.36. The absorptive boundary condition of the analysis area is a first-order Mur algorithm [69]. The mesh size for the calculation is 40 nm. The calculation is performed with a TM wave (s-polarized). The amplitude of the incident wave is 1 V/m.

The size of the resonant sphere is reduced because of the limitation of the computer memory. The actual condition of adhesion of the protrusion beneath the sphere is not clear. Therefore, the calculation model makes 50-nm gaps at the two adhesion points shown in Fig. 4.36a. In the glass plate region, the analytical solution of total internal reflection at a semi-infinite boundary is used.

The resonant condition of the larger sphere is found by changing the wavelength, polarization, and incident angle of the incident wave (i.e., laser light). In a series of simulations, the maximum intensity of the resonant wave inside the sphere is achieved at 70°. Therefore, the following experiments and simulations use the incident angle of 70°. TM waves drive more intensive resonant waves than TE waves inside the cavity in the simulation. This result is consistent with the experimental result, which recorded the polarization-dependent emission of the large sphere. Thus, the polarization of the incident wave is fixed in the TM mode.

Figure 4.37 shows a summarized result of the simulation. In Fig. 4.37a, electric field distribution is drawn along axis "A," as shown in Fig. 4.37b. The microcavity with a 500-nm protrusion, of refractive index 1.5, forms a localized evanescent field beneath itself. However, the lateral width of the localized field is the same as the diameter of the protrusion. This means that

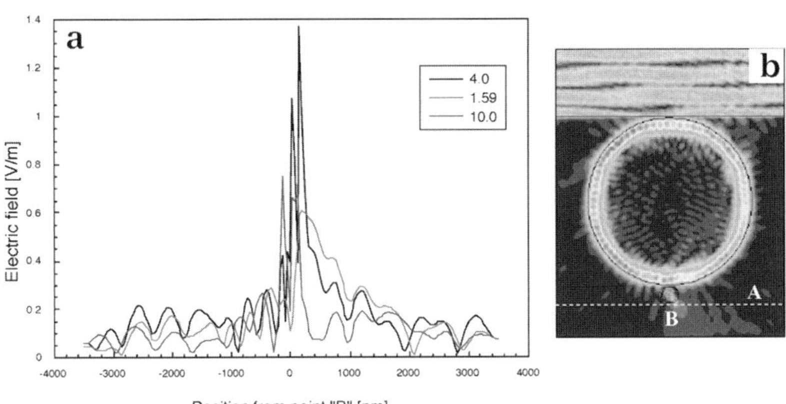

Fig. 4.37. FDTD simulation shows that the refractive index of the microprotrusion is effective in the distribution of an electric field for probing. A protrusion with an appropriate refractive index could generate a narrow peak in the lateral distribution of the electric field

the lateral resolution may be comparable to 500 nm. But a protrusion with a high refractive index of 5.5 shows a more favorable result. The evanescent field of the resonant microcavity is strongly combined with the protrusion. The lateral distribution of the electric field beneath the protrusion narrows and its peak intensity is enhanced in Fig. 4.37a. The peak electric field is 1.4 times that of the incident plane wave. Figure 4.37b shows a cross section of the probe. The resonant wave propagates on the inner surface of the resonant sphere, and the wave generates an evanescent field on the outer surface of the sphere. The evanescent field couples with small protrusion "B." Therefore, the simulation suggests that a protrusion with a high refractive index is needed on the surface of the resonant microcavity to attain high spatial resolution. We have also observed that the roughly estimated far-field intensity increases by decreasing the gap between the probe and the sample surface. This is a basic characteristic of this type of NSOM.

4.4.3 Fabrication of a "Resonant Microcavity Probe"

A high-magnification inverted microscope (Nikon: TE300-NT) coupled with a micromanipulator (Narishige: NT88-2A4) is used to fabricate the probe. A single small protrusion (Soken Chemical & Engineering Co., Ltd.: PMMA sphere, refractive index: 1.49), 1.5 μm in diameter, is put on the cover glass. Then a single latex sphere (Sekisui Chemical Co,. Ltd.: polystyrene latex sphere, refractive index: 1.59), 50 μm in diameter, fixed on the mirror-polished surface of a plate glass (fused silica) is brought near the protrusion by the micromanipulator. Then, the protrusion adheres to the latex sphere. The adhesion of the protrusion is identified by the disappearance of the scattered light from the protrusion on the cover glass.

The location of the protrusion is identified by light scattering from the laser beam. The tunable laser system consists of a high-power diode-pumped laser (Spectra-Physics: Millennia VU) and a tunable solid-state source (Spectra-Physics: Model 3900S).

The experimental setup is shown in Fig. 4.38. The output beam of the laser system is introduced into a polarization-maintaining (PM) optical fiber, which has a pigtail style fiber collimator (OZ Optics). The fiber guides the laser light to the sample stage of the micromanipulator. Part of the laser light is introduced to a wavelength meter (ADVANTEST: TQ8325) to monitor the wavelength. The divergent output light of the fiber is collimated with a receptacle style collimator (OZ Optics). The collimated beam diameter is about 600 μm, and the power is about 200–800 nm. This beam illuminates the protrusion on the cover glass. The scattered light from the protrusion is observed with a dual-mode cooled CCD camera (Hamamatsu Photonics: C4880-10-12A).

A basic experiment with a 50-μm polystyrene latex sphere is performed. The latex sphere is put on the surface of the glass plate and irradiated by

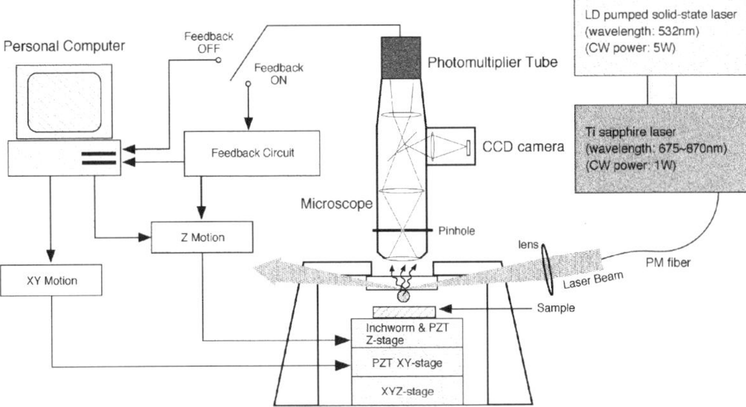

Fig. 4.38. Schematic view of the NSOM system

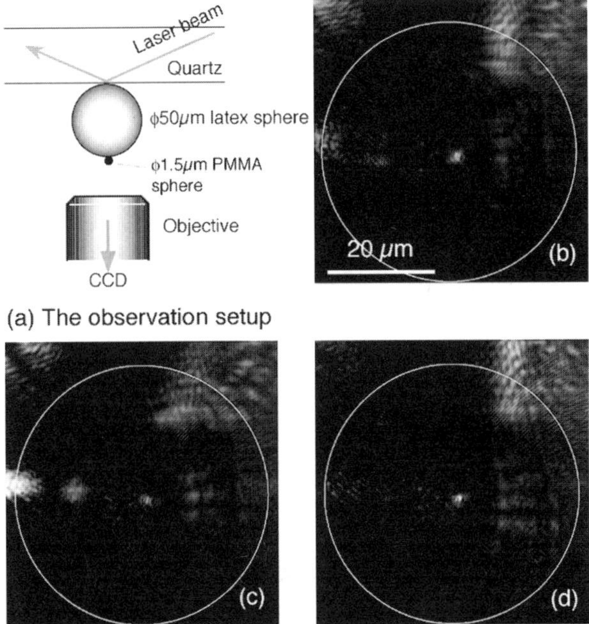

Fig. 4.39. A microprotrusion on the resonant sphere scatters the evanescent field, which is generated by WGMs inside the sphere. **a** Experimental setup for the observation. From **b** to **d**, the wavelength of the laser varies in 0.4-nm steps. The clear image of the small protrusion on the resonant sphere is observed. The intensive spot on the left hand side in **c** may indicate the MDRs inside the resonant sphere

the s-polarized laser beam under total internal reflection. Then the evanescent field illuminates the sphere. As the wavelength of the laser is scanned continually, fluctuation of the emitted light from the sphere is recorded with a photomultiplier (Hamamatsu Photonics: R374) through the objective. The wavelength-dependent intensity fluctuation of the latex sphere has a period of intense peaks which is equal to a free spectral range (FSR) calculated from the diameter ($D = 50$ μm) and the refractive index ($n = 1.5$) of the sphere.

Then our "resonant microcavity probe" is tested. As mentioned, the experimental schema is shown in Fig. 4.39. The collimated laser beam comes from the right side with s polarization. From Fig. 4.39b–d, the laser wavelength varies in 0.4-nm steps. The clear image of the small protrusion on the resonant sphere is observed. The intensive spot on the left-hand side in Fig. 4.39c may indicate the MDRs inside the resonant sphere. The image of the protrusion is hardly observed through the glass plate. Then, an image field in the microscope is covered with a pinhole aperture to filter out unwanted scattered light and to decrease the depth of field. As a result, a clear image of the protrusion is also observed through the glass plate.

4.4.4 Observation of a Vacuum-Evaporated Gold Film

We used a section of an optical fiber as a test sample. Gold is evaporated in a vacuum on the cross section of the fiber to a thickness of 50 nm. Figure 4.40 shows the scanned data with the microcavity probe. The scanning range of (a) is 10×10 μm and of (b) is 5×5 μm, respectively. Figure 4.40b is a rescanned image which corresponds to the area of (0 μm, 0 μm)–(5 μm, 5 μm) in Fig. 4.40a to check the reproducibility of the NSOM system with the same

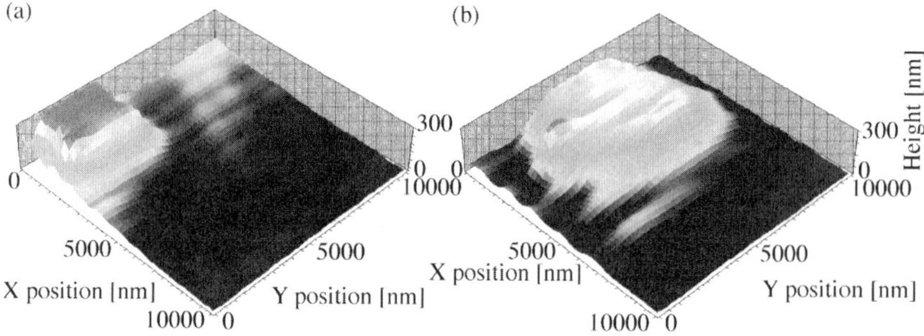

Fig. 4.40. NSOM image of a vacuum evaporated gold film on a cross section of an optical fiber with a prototypical probe. Scanned areas are **a** 10×10 μm, **b** 5×5 μm, respectively. **b** is the rescanned area of (0 μm, 0 μm)–(5 μm, 5 μm) in **a**. Data points are 51×51. A few micrometer humps caused by dust are observed repeatedly. In the x direction, there are some fringes that may be caused by the light scattering noise from a 40-mm sphere

probe. In both (a) and (b), the number of samplings is 51×51, and a "hump" structure is clearly observed. In the next step, we will try to observe the fine slips of plastic deformation formed on the silicon wafer surfaces because their location could be easily observed by the light-scattering method and they contain nanometer-scale fine structures.

References

1. J. M. Vigoureux, C. Girard, and D. Courjon: General principles of scanning tunneling optical microscopy. Opt. Lett., **14**, 1039 (1989)
2. N. F. van Hulst, M. H. P. Moers, O. F. J. Noordman, R. G. Tack, F. B. Segerink, B. and Bölger: Appl. Phys. Lett. **62**, 461 (1993)
3. Y. Inouye and S. Kawata: Optics Lett. **19**, 159 (1994)
4. R. Zenhausern, M. P. O'Boyle, and H. K. Wickramasinghe: Appl. Phys. Lett. **65**, 1623 (1994)
5. A. Lahrech, P. Bachelot, P. Gleyzes, and A. C. Boccara: Opt. Lett. **21**, 1315 (1996)
6. J. Wessel: J. Opt. Soc. Am. B **2**, 1538 (1985)
7. B. Knoll and F. Keilmann: Nature **399**, 134 (1999)
8. G. Mie: Beiträge zur Optik trüber Medien, speziell kolloidaler Metallösungen. Ann. d. Phys. **25**, 377 (1908)
9. M. Kerker: *The Scattering of Light and other Electromagnetic Radiation* (Academic Press, New York 1969) pp. 84–88
10. J. A. Stratton: *Electromagnetic Theory* (McGraw-Hill, New York 1941)
11. C. F. Bohren and D. R. Huffman: *Absorption and Scattering of Light by Small Particles* (John Wiley and Sons, New York 1983)
12. H. Fröhlich: *Theory of Dielectrics* (Oxford University Press, London 1949)
13. B. J. Messinger, K. U. Ravon, R. K. Chang, and P. W. Barber: Phys. Rev. B **24**, 649 (1981)
14. P. K. Aravind and H. Metiu: Surf. Sci. **124**, 506 (1983)
15. M. M. Wind, J. Vlieger, and D. Bedeaux: Physica **141A**, 33 (1987)
16. T. Okamoto and I. Yamaguchi: Opt. Rev. **6**, 211 (1999)
17. T. Okamoto, I. Yamaguchi, and T. Kobayashi: Opt. Lett. **25**, 372 (2000)
18. A. Doron, E. Katz, and I. Willner: **11**, 1313 (1995)
19. R. G. Freeman, K. C. Grabar, K. J. Allison, R. M. Bright, J. A. Davis, A. P. Guthrie, M. B. Hommer, M. A. Jackson, P. C. Smith, D. G. Walter, and M. J. Natan: Science **267**, 1629 (1995)
20. K. C. Grabar, P. C. Smith, M. D. Musick, J. A. Davis, D. G. Walter, M. A. Jackson, A. P. Guthrie, and M. J. Natan: J. Am. Chem. Soc. **118**, 1148 (1996)
21. G. Schmid, S. Peschel, and T. Sawitowski: Z. anorg. allg. Chem. **623**, 719 (1997)
22. T. Sato, D. G. Hasko, and H. Ahmed: J. Vac. Sci. Technol. B **15**, 45 (1997)
23. S. S. Yee (ed.): special issue on Surface Plasmon Resonance (SPR) Optical Sensors, Current Technology and Applications, Sensors and Actuators B **54**(1–2) (1999)
24. T. Okamoto and I. Yamaguchi: Jpn. J. Appl. Phys. **36**, L166 (1997)
25. W. A. Ducker, T. J. Senden, and R. M. Pashley: Nature **353**, 239 (1991)
26. W. A. Ducker, T. J. Senden, and R. M. Pashley: Langmuir **8**, 1831 (1992)

27. S. Kawata and Y. Inouye: Ultramicroscopy **57**, 313 (1995)
28. F. Zenhausen, Y. Martin, and H. K. Wickramasinghe: Science **269**, 1083 (1995)
29. F. Möllers, H. J. Tolle, and R. Memming: J. Electrochem. Soc. **121**, 1160 (1974)
30. A. Ashkin, J. M. Dziedzic, J. E. Bjorkholm, and S. Chu: Opt. Lett. **11**, 288 (1986)
31. A. Ashkin, J. Dziedzic, and T. Yamane: Nature, **330**, 769 (1987)
32. K. Svoboda, S. M. Block: Opt. Lett. **19**, 930 (1994)
33. T. Sugiura, T. Okada, Y. Inouye, O. Nakamura, and S. Kawata: Opt. Lett. **22**, 1663 (1997)
34. T. Sugiura and T. Okada: Proc. SPIE, **3260**, 4 (1998)
35. T. Sugiura, S. Kawata, and T. Okada: J. Microsc. **194**, 291 (1999)
36. S. Kawata, Y. Inouye, and T. Sugiura: Jpn. J. Appl. Phys. **33**, L1725 (1994)
37. K. Sasaki, Z. Shi, R. Kopelman, and H. Masuhara: Chem. Lett. **1996**, 141 (1996)
38. H. Furukawa and S. Kawata: Opt. Comm. **148**, 221 (1998)
39. K. S. Yee: IEEE Trans. Antennas Propagat. **AP-14**, 302 (1966)
40. J. B. Judkins and R. W. Ziolkowski: J. Opt. Soc. Am. A **12**, 1974 (1995)
41. D. A. Christensen: Ultramicroscopy **57**, 189 (1995)
42. H. Furukawa and S. Kawata: Opt. Comm. **132**, 170 (1996)
43. L. Novotny, D. W. Pohl, and P. Regli: J. Opt. Soc. Am. A **11**, 1768 (1994)
44. H. Raether: *Surface Plasmons on Smooth and Rough Surfaces and on Gratings* (Springer, Berlin Heidelberg New York 1988)
45. J. Meixner: IEEE Trans. Antennas Propagation **AP-20**, 442 (1972)
46. V. Deckert, D. Zeisel, R. Zenobi, and T. Vo-Dinh: Anal. Chem. **70**, 2646 (1998)
47. N. Hayazawa, Y. Inouye, Z. Sekkat, and S. Kawata: Opt. Commun. **183**, 333 (2000)
48. N. Hayazawa, Y. Inouye, and S. Kawata: J. Microsc. **197**, 472 (1999)
49. R. K. Chang and T. E. Furtak, eds.: *Surface Enhanced Raman Scattering* (Plenum, New York 1982)
50. N. Hayazawa, Y. Inouye, Z. Sekkat, and S. Kawata: Chem. Phys. Lett. **335**, 369 (2001)
51. A. Otto, I. Mrozek, H. Grabhorn, and W. Akemann: J. Phys. Condens. Matter. **4**, 1143 (1992)
52. P. Kambhampati and A. Campion: Surf. Sci. **427–428**, 115 (1999)
53. U. C. Fischer: J. Vac. Sci. Technol. **B3**, 386 (1985)
54. U. C. Fischer and D. W. Pohl: Phys. Rev. Lett. **62**, 458 (1989)
55. U. Durig, D. W. Pohl, and F. Rohner: J. Appl. Phys. **59**, 3318 (1986)
56. S. Jiang, K. Nakagawa, and M. Ohtsu: Jpn. J. Appl. Phys. **33**, L55 (1994)
57. A. Harootunian et al.: Appl. Phys. Lett. **49**, 674 (1986)
58. D. W. Pohl, U. C. Fischer, and U. T. Durig: Proc. SPIE **897**, 84 (1988)
59. R. C. Reddick et al.: Phys. Rev. **B39**, 767 (1989)
60. D. Courjon, K. Sarayeddine, and M. Spajer: Opt. Commun. **71**, 23 (1989)
61. S. Jiang et al.: Jpn. J. Appl. Phys. **30**, 2107 (1991)
62. F. de Fornel et al.: Ultramicroscopy **42–44**, 422 (1992)
63. G. Chabrier et al.: Opt. Commun. **107**, 347 (1994)
64. T. Kataoka et al.: Ultramicroscopy **63**, 219–225 (1996)
65. Y. Oshikane et al.: Techn. Dig. 5th Int. Conf. on Near Field Optics, Shirahama, 1998, pp. 12–13

66. H. Nakagawa et al.: *ibid.*, 179–180
67. R. K. Chang et al.: *Optical Processes in Microcavities*, Chap. 8 (World Scientific, New Jersey 1996)
68. P. B. Wong et al.: IEEE Trans. Antennas Propagation **44**(4), 504 (1966)
69. G. Mur: IEEE Trans. Electromagn. Compat. **EMC-23**(11), 377 (1981)

5 Integrated and Functional Probes

T. Ono, M. Esashi, H. Yamada, Y. Sugawara, J. Takahara, and K. Hane

After the invention of atomic force microscopy (AFM), micromachining technology was soon introduced into the fabrication of AFM probes [1]. Well-defined probes with sharp tips, soft spring, and high resonant frequency were produced in a batch process for commercialization. In addition, using Si micromachining which is compatible with Si integrated circuits, electronic functions for sensing and regulating the force acting on the tip have been installed in the micromachined AFM probe [2].

As for the probe microscopes developed previously, a crucial point of NSOM is also probe fabrication. Although hand-made probes for NSOM are used in several research laboratories, it is advisable to use batch-fabricated probes for routine measurements. The troublesome point in NSOM probe fabrication by Si micromachining is the installation of optical functions in an AFM probe. AFM function is often required for regulating the gap between the probe and the surface and for detecting surface topography. Some efforts to include the optical function in AFM probes have been attempted. A photodiode was installed on the cantilever of an AFM probe for detecting light scattered from an evanescent field [3]. To transfer light to the tip, a waveguide was fabricated on an AFM cantilever [4]. A few methods have been attempted for generating a microaperture on the tip by Si micromachining [5]. In this chapter, some new techniques for integrating the NSOM function with an AFM probe are described. In addition to the microfabricaton of conventional NSOM probes, novel properties of NSOM are revealed for developing multifunctional probes. In the latter part of this chapter, a new technique for detecting the evanescent light from a force acting on an AFM probe is presented. Moreover, the basic considerations of light transmission through a nano-waveguide are given since it is expected that the transmittance of NSOM aperture will be improved by adopting a coaxial probe tip, whose diameter is much smaller than an optical wavelength [6].

5.1 Micromachined Probes

Together with scanning tunneling microscopy (STM) and atomic force microscopy (AFM), which can observe and modify surfaces on atomic and nanometer scales, near-field scanning optical microscopy (NSOM) has been extensively studied owing to its capabilities for optical imaging with spatial resolution beyond the diffractive limit of light [7], for subwavelength photolithography [8] and for the next generation of optical data storage [9] using near-field light. The most crucial part of an NSOM probe is a subwavelength size aperture at the apex of the tip. Currently, tapered optical fiber probes are most widely used for NSOM. Although improvements have been achieved in fabricating such optical fiber NSOM probes, problems remain. The fiber tip is very fragile. The shape of the tip and the size of the metallic aperture at the fiber tip are not reproducible. The opening angle of the fiber tip is small. Therefore, most of the light is absorbed on the metal wall, which leads to low optical transmission efficiency. Moreover, it is very difficult to fabricate a fiber NSOM probe in mass production.

Contrarily, a Si-based NSOM probe can be easily fabricated in a batch process. It is very easy to combine the microfabricated NSOM probe with a well-developed AFM as an AFM/NSOM probe light [10–14]. Several technological approaches have been developed to create the aperture at the apex of the tip such as coating and selective etching metal at the apex of a Si_3N_4 tip [11], metal molding with an aperture at the pyramidal etched pit on an Si cantilever [12], and creating the metallic aperture at the apex of a metal coated SiO_2 tip on a Si cantilever by field evaporation [13]. However, until recently, a simple method for fabricating the aperture for an NSOM probe with high reproducibility in mass production has not yet been achieved. In this section, a simple method using low temperature oxidation and a selective etching (LOSE) technique to fabricate the aperture for an NSOM probe is presented [15].

5.1.1 Fabrication of a Miniature Aperture

It is known that the thickness of silicon oxide grown at a low temperature at convex and concave corners is thinner than that at a flat surface of Si due to the compressive stress at the corner structures (see Fig. 5.1) [16] The nonuniform Si oxidation effect at a convex corner has been used for fabricating Si conical sharp tips for the AFM or field emitter [17]. The anomaly of thermal oxidation at a concave corner has been used to produce a mold for fabricating a Si_3N_4 sharp pyramidal tip for the AFM [18].

In the other approach, the nonuniform Si oxidation effect at the pyramidal etched pit is useful for fabricating the aperture for an NSOM probe using the LOSE technique. The process flow of the LOSE technique is schematically shown in Fig. 5.2. First, a Si(100) wafer is thermally oxidized. Next, photolithography is done to define the mask patterns for etching SiO_2. To make a perfect pyramidal etched pit [i.e., four (111) surfaces should intersect at a point], electron beam lithography is used to define square patterns on a resist as a mask for a SiO_2 opening. Consecutively, pyramidal etched pits and the structure of cantilevers are formed by anisotropic etching of Si in

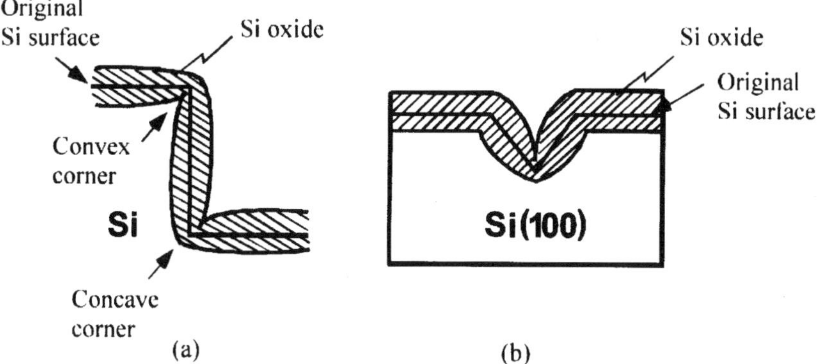

Fig. 5.1. Schematic of Si thermal oxidation **a** at convex and concave corners and **b** at the pyramidal etched pit formed on Si(100)

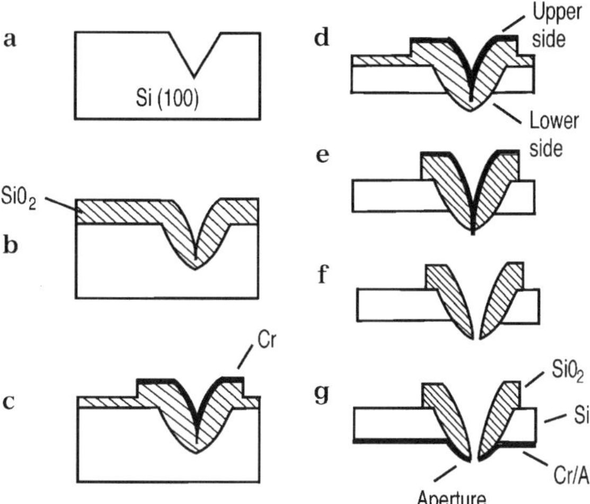

Fig. 5.2. Process flow for making an aperture at the apex of a SiO_2 tip on a Si cantilever for a NSOM probe by the LOSE technique

tetramethyl ammoniumn hydroxide (TMAH) (step a). Next, a SiO_2 film is thermally grown by wet oxidation at 950°C with a thickness of about 1 μm (step b). Subsequently, a chromium (Cr) film about 100 nm thick is sputtered and lifted off to form a protective pattern on the upper side of the oxidized pit (step c). Using this Cr pattern as a mask, SiO_2 about 0.6 μm thick is partly etched, as shown in step c. The remaining oxide about 0.4 μm thick is needed for the following steps. Next, the Si wafer is anisotropically etched from the back side in TMAH solution until it forms a Si cantilever with the SiO_2 tip at the end of the cantilever (step d). The wafer is then dipped into buffered

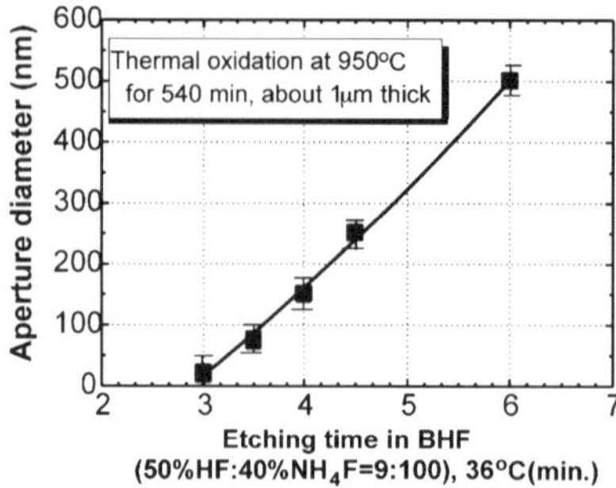

Fig. 5.3. The dependence of aperture size on etching time in a BHF solution at 36°C. The SiO_2 tip was formed by thermal oxidation at 950°C for 9 hours

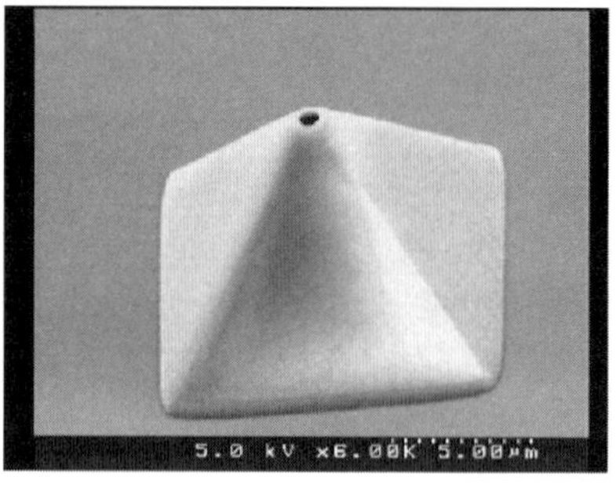

Fig. 5.4. SEM image of a typical NSOM tip with an aperture of about 500 nm

HF (50% HF:40% NH_4F, 9cc:100cc) (BHF) solution for selective etching of SiO_2 until the Cr protrusion comes in sight (step e). At this step, only the exterior wall of the SiO_2 tip is etched in the BHF, whereas the interior wall is protected by the Cr pattern. Finally, a metal film (Cr or Cr–Al) is deposited onto the back side of the cantilever to form an opaque layer (step g).

The dependence of the aperture size formed at the apex of the SiO_2 tip on the etching time in the BHF solution at 36°C is shown in Fig. 5.3. The deviation of aperture size in measuring 300 apertures on a 2×2 cm^2 sample is about 50 nm. This deviation may due to inhomogeneity of the SiO_2 film grown in step b in Fig. 5.2. It should be noted that a dimensional error in the patterning for the tip area has not affected the size of the aperture. This should be an advantage of the LOSE technique for fabricating a highly reproducible aperture and an uniform aperture array.

A SEM image of a typical Cr coated SiO_2 tip with an aperture of about 500 nm diameter fabricated by the LOSE technique is shown in Fig. 5.4. It can be seen that the aperture is situated exactly at the apex of the tip. Pinholes were not found on the sidewalls of the tip. Aperture sizes from 10 to 500 nm are experimentally achieved corresponding to etching times ranging from 3 to 6 min in BHF at 36°C.

If the patterning window for the etched pit in step a, Fig. 5.2, is not very square or round, a double aperture can be formed as shown in Fig. 5.5. The distance between the centers of two apertures is the same as the discrepancy in the opening window of the etch pit. Normally, the sizes of the two apertures in the double one are identical. To fabricate only one small aperture, the patterning window for the etch pit should be done by electron-beam lithography.

A fabricated silicon probe is bonded with a glass substrate on which an electrode is formed. A three-dimensional view of the probe is shown in Fig. 5.6. The probe consists of a Si cantilever with a tiny aperture at the

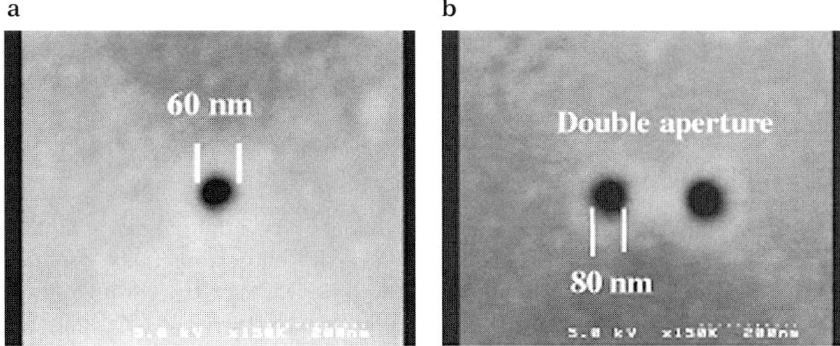

Fig. 5.5. **a** SEM image of a fabricated 60 nm single aperture at the apex of a tip after deposition of approximately 120 nm Cr/Al. **b** SEM image of a fabricated double aperture when the patterning window for the etch pit is rectangular

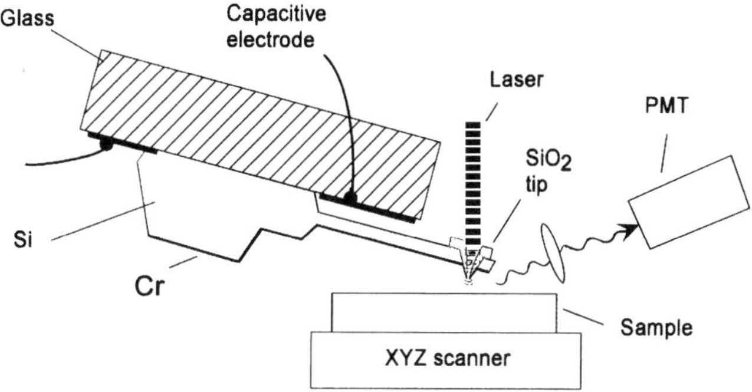

Fig. 5.6. Schematic view of a NSOM/AFM probe

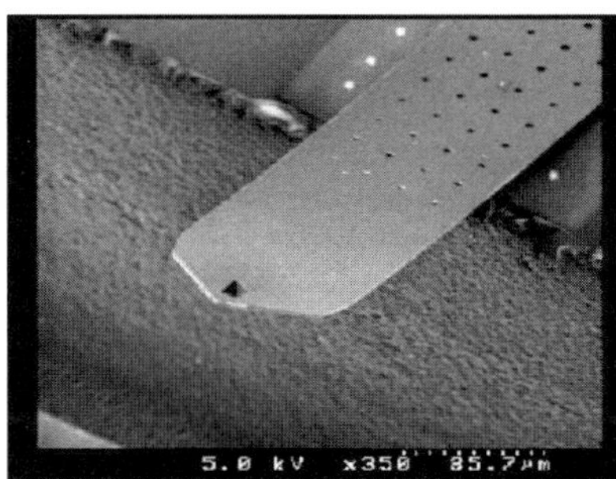

Fig. 5.7. Typical SEM image of a NSOM/AFM probe fabricated by the LOSE technique

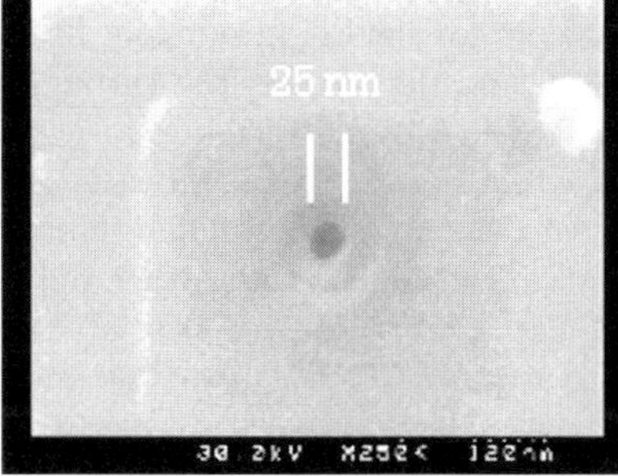

Fig. 5.8. SEM image of an approximately 25-nm diameter aperture at the apex of the tip on a fabricated NSOM probe, as shown in Fig. 5.6

apex of the tip and the opposed parallel electrode with a narrow gap for capacitance detection or electrostatic actuation. SEM images of a fabricated probe and a close-up view of an aperture about 25 nm in diameter are shown in Figs. 5.7 and 5.8.

5.1.2 Throughput Measurement

One of the challenges of the near-field technique, especially in near-field data storage and spectroscopy is how to increase the photon intensity confined at the aperture, or the throughput. It is well known that the throughput of an aperture NSOM probe is drastically decreased with decreasing aperture size. The throughput of a conventional optical fiber tip is quite low (e.g., a 100-nm diameter optical fiber aperture has a throughput of approximately 10^{-5}–10^{-7}) [22]. To evaluate the throughput of a fabricated aperture NSOM probe, we proposed a simple measurement setup, as shown in Fig. 5.9a [20]. A conventional photodiode is used as a detector to collect the near-field light passing through the aperture or the far-field light and is placed on the X,Y,Z scanner of the AFM. A laser diode of 780-nm wavelength and 2-mW power is focused on the tip area with an objective lens. The near-field intensity is defined as the difference between the output signals at the position where the tip is almost in contact with the photodiode surface (the contact position $P1$) and the release point ($P2$), as shown in Fig. 5.9b. The distance between the contact and release positions is about several millimeters and is driven by the piezo scanner of the AFM. After evaluating the near-field intensity ($P1 - P2$), the probe is removed from the laser position without any change in the optical system, so that the far-field light beam directly illuminates the photodiode ($P3$). The throughput of the corresponding aperture is defined as the ratio $(P1 - P2)/P3$. Since the spot size of the laser beam is about 1.6 µm, and the size of the tip is about 15×15 µm^2, therefore, there is no incident far-field light outside the tip area. Using the described measurement setup and definitions, the throughput of several apertures is evaluated and plotted in Fig. 5.10.

From Fig. 5.10, it can be seen that the throughput of our probe is several orders of magnitude higher than that of conventional optical fibers [19,22] and is also higher than other published throughput on micromachined tips [23,24]. The reason for this is that an opaque layer is formed on the outer curvature of the SiO_2 tip. The outer curvature angle of the thermal SiO_2 tip grown at low temperature in the apex area is very large (almost flat in the area of about 0.5×0.5 µm^2 at the apex or nearly equal to the curvature of a sphere about 1 µm in diameter). Therefore, the cutoff effect (if incident light passes through a narrow waveguide whose diameter is smaller than the wavelength, absorption of light by the coated metal results in a decrease in the throughput) is minimized. The laser light beam comes to the aperture without any strong losses inside the tip. Furthermore, the aperture is empty, i.e., there is no glass core as in an optical fiber tip, therefore many modes of

a

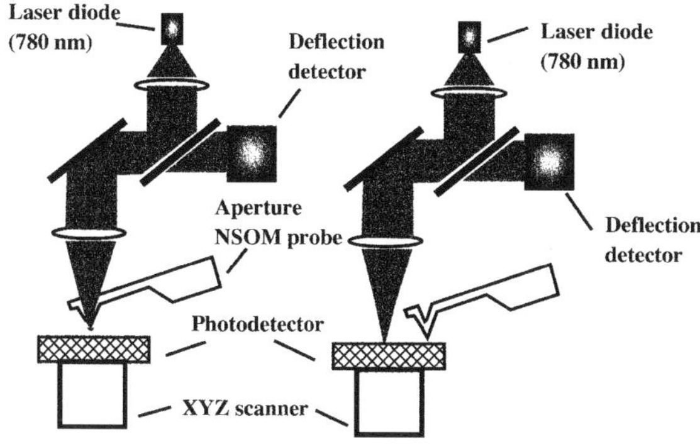

b

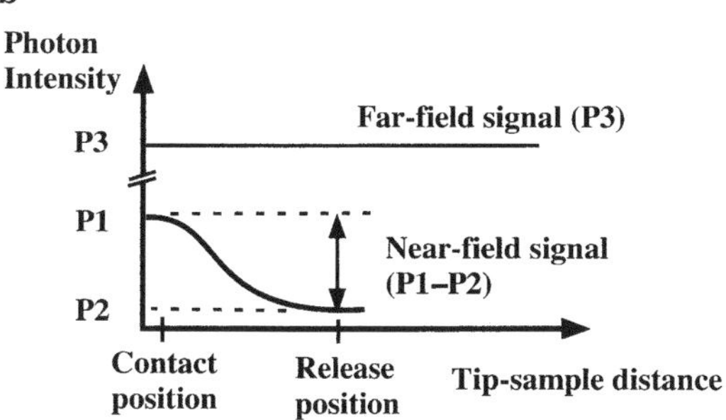

Fig. 5.9. a Schematic diagram of the throughput measurement setup. **b** The near-field intensity is defined as the difference of the output signals at the contact and release positions ($P1 - P2$). The throughput is defined as $(P1 - P2)/P3$

light can reach the aperture and transmit through the aperture in the same way as transmitting through a small aperture in a perfectly conducting plane [25]. However, for an aperture smaller than 100 nm, the throughput is still drastically decreased, as shown in Fig. 5.10.

5.1.3 Fabrication of an Aperture Having a Metal Nanowire at the Center

In many experiments, it was observed that when a protective Cr film on the upper side of the tip (step c, Fig. 5.2) is thick enough (normally thicker than 150 nm), the etching rate of the SiO_2 in the apex area of the SiO_2 tip in BHF (step e, Fig. 5.2) is very much faster than that in other parts. Using this effect, a novel tip for a multipurpose NSOM probe can be fabricated. A SEM image of the SiO_2 tip after etching in BHF at 36°C for 5 min is shown

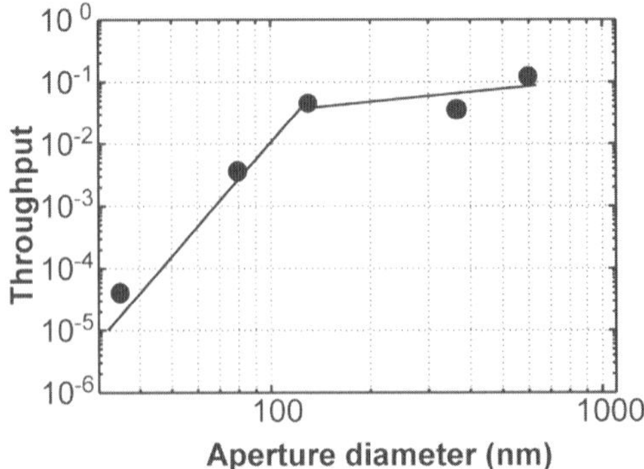

Fig. 5.10. Throughput of fabricated aperture NSOM probes as a function of aperture diameter

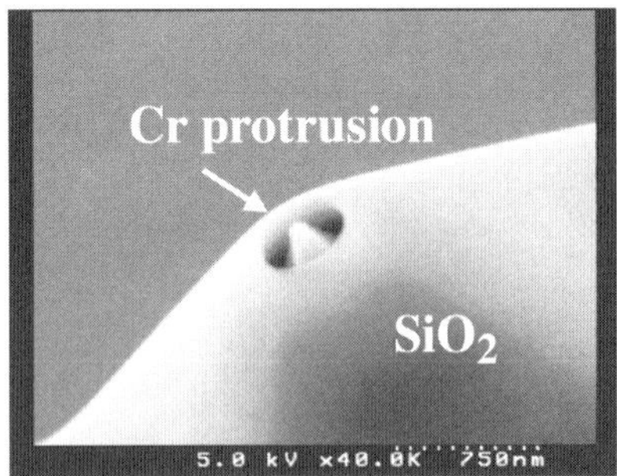

Fig. 5.11. SEM image of a SiO_2 tip after etching in BHF for 5 min. The aperture is created with a Cr protrusion at the center of the aperture

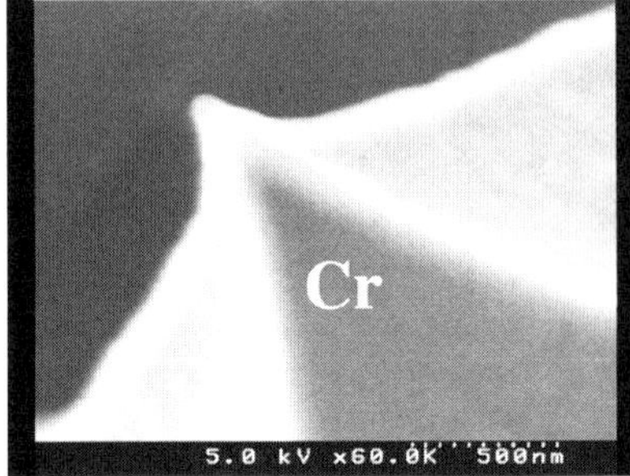

Fig. 5.12. SEM image of the Cr protrusion after completely removing the SiO_2 by etching in BHF. The shape of the Cr protrusion shows a modification in the etch pit after oxidation at low temperature

in Fig. 5.11. By completely etching the SiO_2, the shape of the protective Cr film can be observed, as shown in Fig. 5.11. The shape of the Cr protrusion in Fig. 5.11 clearly shows the modification of the interior contour of the etched pit after low temperature oxidation. The mechanism of this effect is still unclear; however, it seems that when the Cr tip starts to appear in the BHF etchant (see Fig. 5.12), the SiO_2 around the Cr tip is etched at higher speed, leading to an aperture with a Cr protrusion at its center, as shown in Fig. 5.11. The SiO_2 etching speed is highest at the Cr/SiO_2 interface.

The structure shown in Fig. 5.11 suggests an extension of the LOSE technique for fabricating a novel NSOM probe with a tiny aperture and a metal nanowire at the center of the aperture. A near-field photon in the apex area is expected to be enhanced by exciting and coupling with a surface plasmon at the apex of the metal wire [25]. (The metal protrusion can be used as a tip for topographic imaging and it protects the aperture against damage caused by contact with a sample surface. Although it is certainly not easy to do optical near-field imaging or spectroscopy with atomic or molecular resolution, however, it is expected that developing this novel structure will achieve this goal.) This structure should be a good candidate for a multipurpose NSOM probe for high-throughput NSOM, a tunneling-electron luminescence microscope [26], a nanoscillo-scope [27], and a NSOM/STM probe and thermal profiler [28] as well.

5.1.4 Imaging with a Fabricated Aperture Probe

A simple characterization setup for the fabricated NSOM/AFM probe in contact and tapping modes is shown in Fig. 5.13. The fabricated NSOM/AFM probe is mounted on an AFM stage. The tip–sample distance is kept constant by using the well-developed AFM technique. The light beam from the laser diode of 780-nm wavelength, 2-mW power is focused directly on the

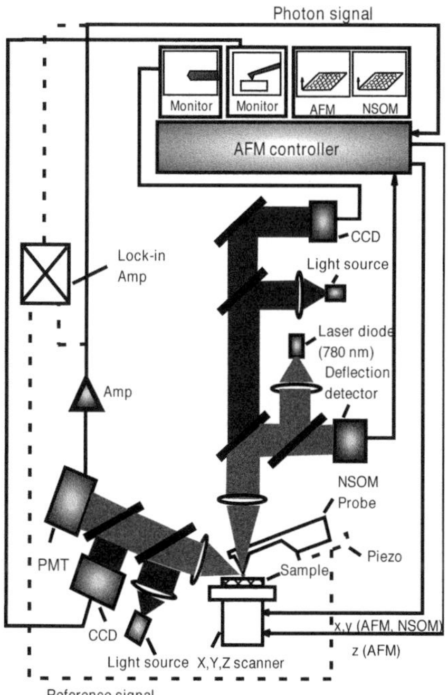

Fig. 5.13. Measurement setup of the NSOM/AFM system for a fabricated probe in contact and tapping modes. The setup is based on a commercial AFM system. The dotted lines in the figure are for the dynamic mode only

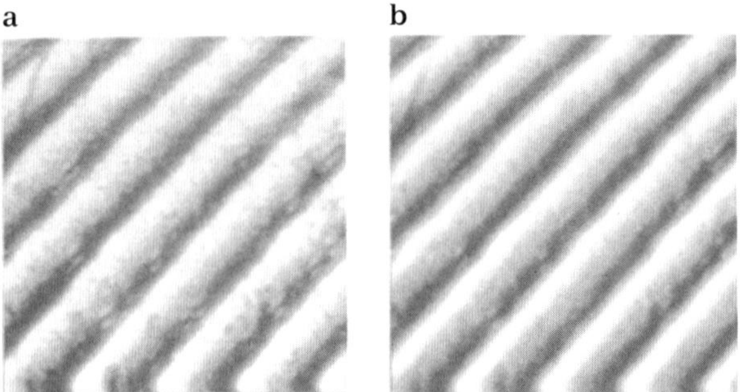

Fig. 5.14. **a** An AFM and **b** corresponding illumination-mode NSOM images in contact mode with a 200-nm Cr grating on a glass substrate using the fabricated NSOM probe whose aperture is about 120 nm in diameter

tip area by using an objective lens. Part of the incident light is reflected by the tip's walls and is monitored by a deflection detector. The near-field light passing through the aperture is reflected by the sample surface, is collected by a long distance working lens and is detected by a photomultiplier (PMT). The signal from the output of the PMT is amplified and acquired by the AFM controller as an optical signal. Using the measurement setup described and the fabricated probe whose aperture is about 120 nm, NSOM and corresponding AFM images in contact mode with a 200-nm Cr grating on a glass substrate are simultaneously observed as shown in Fig. 5.14a,b. From Fig. 5.14, it should be noted, however, that even with a 120-nm aperture size, very small Cr particles (about 20 nm diameter) can be observed. The system presented here is specialized for the illumination mode; however, it is possible to use the fabricated aperture tip as a receiver for collection-mode NSOM. We also expect to develop this setup for optical near-field and Raman near-field spectroscopy as well.

The setup shown in Fig. 5.13 can also be used for NSOM/AFM in noncontact or tapping modes (dotted lines in Fig. 5.13). The probe is mounted on an AFM noncontact stage on which the cantilever can be vibrated by a piezo element. The near-field signal modulated by the vibratory signal is lock-in amplified using the vibratory signal applied to the piezo element as a reference and acquired by the AFM controller as shown by the dotted lines in Fig. 5.13.

5.2 Light Detection from Force

Measurement of an optical near-field localized on a sample surface is one of the important methods for investigating the electromagnetic interaction in the subwavelength region. In conventional methods for detecting the optical near-field, a subwavelength-scale aperture tip or a small scattering tip has been used. As an alternate detection method, the force acting between a semiconductor tip and a glass surface has been proposed [29]. This method derives from a phenomenon called surface photovoltage (SPV) [30,31]. The performance of this method by near-field scanning optical microscopy (NSOM) has been preliminarily demonstrated by imaging a standing evanescent wave in air [29]. To remove the capillary forces caused by a water film on the sample surface and to measure the weak force induced by an optical near field with high sensitivity and high accuracy, we introduced noncontact atomic force microscopy (AFM) [32,33] with the frequency modulation (FM) detection technique [34] operating in a high vacuum. In the FM detection technique, the sensitivity of force detection is increased drastically because of the high Q value of the cantilever in a vacuum. Its usefulness in force detection is demonstrated by achieving true atomic resolution imaging on various surfaces such as InP (110) [35] and Si (111) 7×7 [36] surfaces. Using the force detection method, a 100-nm-diameter polystyrene latex sphere was observed with spatial resolution better than 50 nm [37].

However, in this force detection method, the force acting on the semiconductor tip is affected by the surface potential change due to the optical near-field (namely, SPV) and also by a nonuniform work function of the surface. Therefore, technical improvement is required to isolate the surface potential change due to the optical near-field from the nonuniform work function of the surface and hence to achieve high lateral resolution.

Here, we propose a novel method to compensate for the work function variation on the surface using the Kelvin probe technique [38] and to measure only the force induced by the surface potential change due to the optical near-field [39].

5.2.1 Method of Measuring Optical Near-Field Using Force

Figure 5.15 shows the measurement principle of the optical near-field using force. An optical near (evanescent)-field is conventionally generated on a prism surface by the total reflection of an optical beam. The prism has a semitransparent metal electrode on its surface. Because of the semitransparency of the metal electrode, the optical near-field generated is not weakened very much. When a semiconductor AFM tip is in the optical near-field, electron–hole pairs are created near the surface of the AFM tip, and hence the surface potential $\delta\phi$ is changed. This surface potential change $\delta\phi$ is measured as electrostatic force acting on the tip when the bias voltage V is applied between the tip and the semitransparent metal electrode. This detection mechanism for the evanescent field in Fig. 5.15 can be easily understood by ideal metal–insulator–semiconductor (MIS) junctions. Note that the metal electrode on the prism surface can also shield the electrostatic field caused by residual charges under the prism surface and can avoid contact electrification between the tip and the prism surface [40].

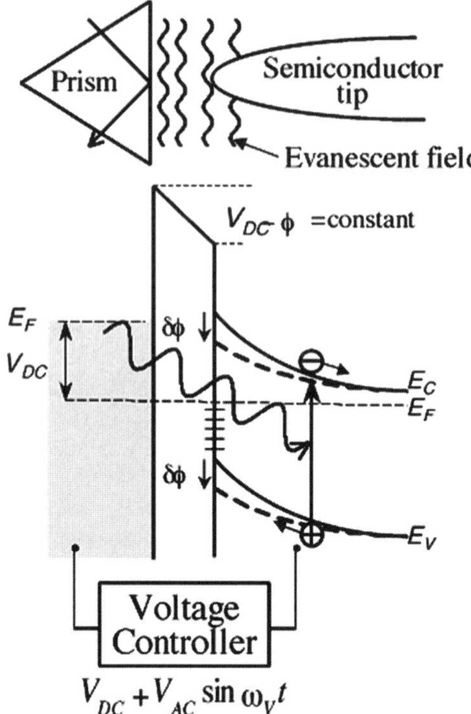

Fig. 5.15. Measurement principle of the optical near-field using force. To detect only the potential change $\delta\phi$ due to the optical near-field, the Kelvin probe technique was used to maintain a constant contact potential difference (CPD) between the tip and the sample. A sinusoidal voltage signal ($V = V_{\mathrm{dc}} + V_{\mathrm{ac}} \sin \omega_{\mathrm{v}} t$) was applied to the sample. $V_{\mathrm{dc}} - \phi$ becomes constant when the ω_{v} component of the total force is maintained at a constant level

The total force F_{total} acting on the semiconductor tip is expressed by following equation:

$$F_{\mathrm{total}} = F_{\mathrm{vdW}} - \frac{1}{2}\frac{\partial C}{\partial z}[(V - \phi) - \delta\phi]^2 \,. \tag{5.1}$$

Here, F_{vdW} is the force due to van der Waals (vdW) interaction between the tip and the surface. C and z are the capacitance and the distance between the tip and the metal electrode, respectively. ϕ is the contact potential difference (CPD) between the tip and the metal electrode without light.

When a sinusoidal voltage signal ($V = V_{\mathrm{dc}} + V_{\mathrm{ac}} \sin \omega_{\mathrm{v}} t$) is applied to the sample, if $\delta\phi$ is sufficiently small compared with $V_{\mathrm{dc}} - \phi$, the total force F_{total} acting on the tip is given approximately by the following equation:

$$F_{\text{total}} = F_{\text{vdW}} - \frac{1}{2}\frac{\partial C}{\partial z}\left[(V_{\text{dc}} - \phi)^2 + \frac{V_{\text{ac}}^2}{2} + 2V_{ac}(V_{\text{dc}} - \phi)\sin\omega_{\text{v}}t \right.$$
$$\left. - \frac{V_{\text{ac}}^2}{2}\cos 2\omega_{\text{v}}t\right] + \frac{\partial C}{\partial z}(V_{\text{dc}} - \phi)\delta\phi\,. \tag{5.2}$$

Here, the final term involves the surface potential change $\delta\phi$ due to the optical near-field. On the basis of the spherical tip model (tip radius R) [41], the distance dependence of the capacitance is given by $\partial C/\partial z = 2\pi\varepsilon R/z$ at $z < R$ and becomes nearly constant because the distance z is controlled as nearly constant by the vdW force F_{vdW}. As a result, the surface potential change $\delta\phi$ and the CPD ϕ between the tip and the surface affect the final term. In near-field imaging, the effect of the nonuniform work function of the sample on the prism surface should be extracted. So, the Kelvin probe technique [38] was used to maintain the electric potential $V_{\text{dc}} - \phi$ at a constant level. The bias voltage V_{dc} was controlled to keep the ω_{v} component of the total force constant. Thus, we can measure the surface potential change $\delta\phi$ on the semiconductor tip by extracting the final term. Actually, by modulating the laser beam to generate the optical near-field at a frequency ω, we can detect the surface potential change $\delta\phi$ from the ω component.

Figure 5.16 shows the experimental setup to detect the force due to the optical near-field. The force was measured in a high vacuum. The FM detection method [35] was used to detect the force acting on a semiconductor AFM tip. In this method, a piezoelectric tube scanner performed cantilever scanning and excitation at its mechanical resonant frequency. The vibratory amplitude of the oscillating cantilever was measured by an optical fiber interferometer. The positive feedback system with an automatic gain control (AGC) circuit was used to maintain the vibratory amplitude of the cantilever at a constant level. The FM demodulator detected the frequency shift $\Delta\nu$ of the oscillating cantilever. In the Kelvin probe technique, a sinusoidal voltage ($V_{\text{ac}}\sin\omega_{\text{v}}t$) was applied between the tip and the metal electrode on the prism surface. The ω_{v} component of the frequency shift $\Delta\nu$ (ω_{v}) was detected by a lock-in amplifier and fed to a voltage controller to maintain a constant potential difference $V_{\text{dc}} - \phi$. A laser beam was modulated at frequency ω, and the optical near-field was measured by detecting the ω component of the frequency shift $\Delta\nu$ (ω). The topography of the sample surface was measured using the dc component of the frequency shift $\Delta\nu$ (DC).

As a reference electrode, an Au thin film was sputtered on the prism surface. Its average thickness was about 15 nm. The Au film was also used as a sample surface. As a force sensor, we used an antimony (Sb)-doped n-type silicon cantilever. Its spring constant and mechanical resonant frequency were $k = 38$ N/m and $\nu_0 = 168$ kHz. The radius of curvature of the tip was less than 10 nm. An ac bias voltage with $V_{\text{ac}} = 0.55$ V and $\omega_{\text{v}}/2\pi = 800$ Hz was applied. The constant potential difference was set at $V_{\text{dc}} - \phi = -1$ V. An Ar laser beam ($I = 15$ mW, $\lambda = 488$ nm, s-polarization) was incident through a viewing port at a frequency of $\omega/2\pi = 400$ Hz.

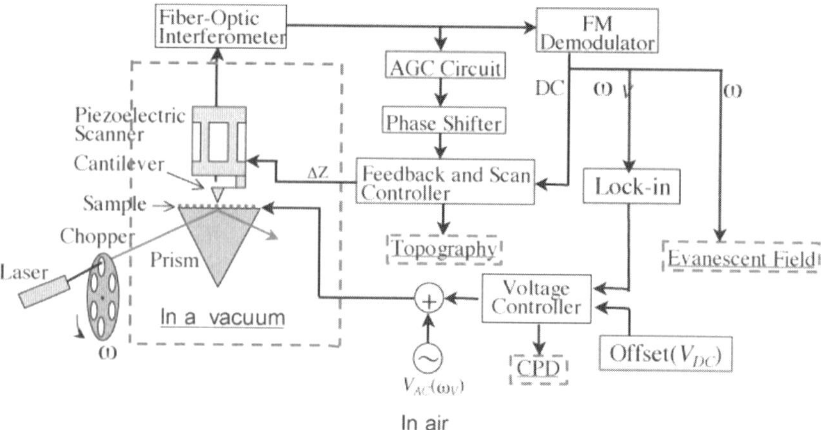

Fig. 5.16. Experimental setup to measure an optical near-field using force. The Kelvin probe technique was employed to extract the effect of the contact potential difference (CPD) between the tip and the surface. A sinusoidal voltage ($V_{\mathrm{ac}} \sin \omega_{\mathrm{v}} t$) was applied between the tip and the electrode on the prism surface. The ω_{v} component of the frequency shift $\Delta\nu$ (ω_{v}) was detected by a lock-in amplifier and fed to a voltage controller to maintain a constant potential difference $V_{\mathrm{dc}} - \phi$. A laser beam was modulated at the frequency ω, and the optical near field was measured by detecting the ω component of the frequency shift $\Delta\nu$ (ω). The topography of the sample surface was measured using the dc component of the frequency shift $\Delta\nu$ (dc)

5.2.2 Imaging Properties

Figure 5.17 shows the images of (a) the topography, (b) the CPD, and (c) the optical near-field measured on the Au film surface, respectively. The scan sizes were 800 × 800 nm. From the topography in Fig. 5.17a, we can see the grain structure with lateral diameters from 40 to 200 nm and corrugations from 3 to 30 nm. From the CPD image in Fig. 5.17b, the measured potential difference on the surface changed from 0.5 to 1.5 V. The minimum detectable value of the CPD was estimated at better than 30 mV from the noise level. The change of the CPD seems to result from contamination on the surface and/or the work function variation originating from the grain structure on the surface. From the optical near-field image in Fig. 5.17c, the frequency shift varied from 0.3 to 2.4 Hz.

In the topography (Fig. 5.17a), bright contrasts were observed in areas A and B. In the CPD image (Fig. 5.17b), dark contrasts were observed in both areas A and B. In the evanescent image dark contrast was observed in area A and bright contrast in area B. Although the contrast in area A was the same as that in area B in topography and in the CPD image, the contrast in area A differed from that in area B in the evanescent field image. From these results, the contrast in the optical near-field image in Fig. 5.17c was not due to either the topography or the CPD in Fig. 5.17c.

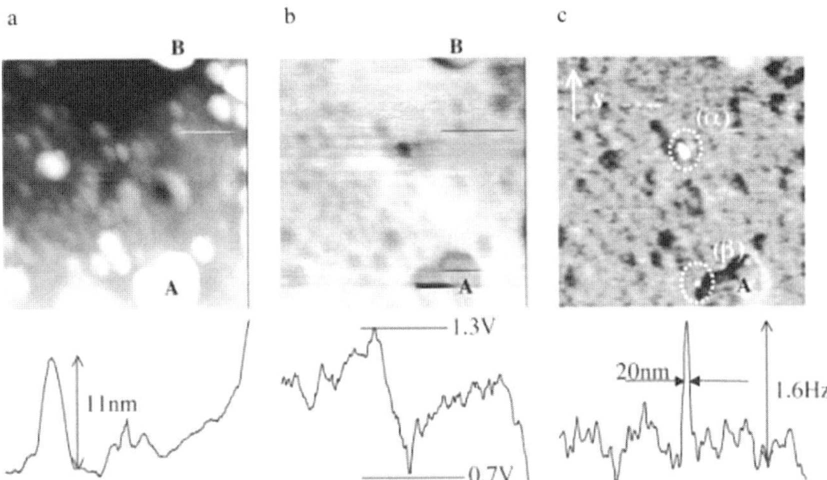

Fig. 5.17. a The topography, **b** the contact potential difference (CPD), and **c** the optical near-field measured for an Au film with a 15-nm average thickness. The scan area was 800 × 800 nm. $\Delta\nu$ (dc) = −4.6 Hz. $V_{ac} = 0.55$ V, $\omega_V/2\pi = 800$ Hz, and $V_{dc} - \phi = -1$ V. An Ar-ion laser beam ($I = 15$ mW, $\lambda = 488$ nm, s-polarization) was incident at a frequency of $\omega/2\pi = 400$ Hz

Interestingly, we observed anomalous bright contrasts in areas α and β circled by white dotted lines in the evanescent field image of Fig. 5.17c. In the topography of Fig. 5.17a or the CPD image of Fig. 5.17b, however, we did not observe such an anomalous contrast in both areas α and β. Here, for example, we show cross-sectional profiles across the bright contrast area α in Fig. 5.17 to compare the individual contrasts. We can recognize that, in area α, the contrast of the optical near-field image had less correlation with the topography and the CPD image. The peak values of the frequency shift in areas α and β of the evanescent field image were $\Delta\nu = 1.1$ and $\Delta\nu = 1.6$ Hz, respectively, with respect to the background. The half widths of these bright contrast areas α and β were estimated at 15 and 20 nm, respectively. This means that the lateral resolution of the optical near-field achieved was better than 15 nm ($\lambda/33$).

In summary, we presented a novel method for measuring the force due to the optical near-field using noncontact AFM. The Kelvin probe technique was employed to extract the effect of the contact potential difference (CPD) between the tip and the surface. The topography, the CPD, and the optical near-field of an Au film of 15-nm average thickness were imaged simultaneously. There was no obvious correlation among topography, the CPD, and the optical near-field. Interestingly, bright contrast areas were present in the evanescent field image. The lateral resolution of the evanescent field was better than 15 nm ($\lambda/33$).

5.3 High Efficiency Light Transmission Through a Nano-Waveguide

We will describe a new way to make an optical beam with a diameter much smaller than the wavelength in free space. The concept of low-dimensional optical waveguides are introduced and their transmission properties are discussed theoretically. By using low-dimensional optical waveguides, we will produce high efficiency light transmission through nanostructures.

5.3.1 Low-Dimensional Optical Wave and Negative Dielectric

The smallest diameter of an optical beam is of the order of a wavelength due to diffraction even in a waveguide. If there is a way to decrease the smallest beam diameter without a diffractive limit, it would contribute greatly to optical devices and optical measurements in the nanometer range. To achieve this, we proposed a low-dimensional optical wave [42]. Here, we will describe low-dimensional optical waveguides from basic ideas to applications.

First, we define the dimension of an optical wave for convenience. The definition is that the dimension is the number of real components in wave number vector $\boldsymbol{k}$. An optical wave is defined as three-dimensional (3-D) when it has three real components k_x, k_y, and k_z in the $\boldsymbol{k}$ vector. For example, optical waves in free space or a dielectric waveguide are 3-D optical waves. Similarly, an optical wave is defined as two-dimensional (2-D) and one-dimensional (1-D) when two and one components of $\boldsymbol{k}$ are real and the others are imaginary, respectively. When all of the components of $\boldsymbol{k}$ are imaginary, it is defined as zero-dimensional (0-D). We call 2-D, 1-D, and 0-D optical waves "low-dimensional optical waves."

Consider an optical wave at angular frequency ω propagating in a medium with refractive index n. The components of $\boldsymbol{k}$ must satisfy

$$k_x^2 + k_y^2 + k_z^2 = |\boldsymbol{k}|^2 = (\frac{2\pi n}{\lambda_0})^2 = \epsilon\mu_0\omega^2 , \tag{5.3}$$

where ϵ is the dielectric constant of the medium, λ_0 is the free-space wavelength, and μ_0 is the free-space permeability. In a 3-D optical wave, each component of $\boldsymbol{k}$ has an upper limit, i.e. $|k_i| \leq |\boldsymbol{k}| \equiv k$ $(i = x, y, z)$, since k_i are real in (5.3). This means that the maximum spread in k space, Δk_{m}, is $\Delta k_{\mathrm{m}} = 2k$. According to the uncertainty relation between the spread in k space Δk and in r space Δr, $\Delta r \Delta k \geq \pi$, the smallest beam diameter of a 3-D optical wave is limited to the effective wavelength as follows,

$$\Delta r \geq \frac{\pi}{\Delta k} \geq \frac{\pi}{\Delta k_{\mathrm{m}}} = \frac{\lambda_0}{4n} . \tag{5.4}$$

Equation (5.4) is the origin of the diffractive limit that is inevitable as long as an optical wave is 3-D.

If the dielectric constant is negative (i.e., $\epsilon < 0$), at least one of k_i must be imaginary to conserve (5.3). This means that an optical wave becomes low-dimensional in a negative dielectric (ND). k_i in a low-dimensional optical wave does not have an upper limit in contrast to a 3-D optical wave. This implies that the beam diameter (Δr) is not restricted by the minimum ($\lambda_0/4n$) according to the uncertainty relation. Since the dielectric constant of a metal is a complex number with a negative real part larger than the imaginary part in the visible and near-infrared range, a metal is considered a ND. Thus, it is possible to realize low-dimensional optical waves by using a metal as a ND.

5.3.2 One-Dimensional Optical Waveguides

Next, we propose waveguides of a low-dimensional optical wave. Simple examples of them are thin metal films and gaps. Two-dimensional optical waves can propagate in these structures as a coupled mode of a surface plasmon polariton (SPP) which is a surface wave at the interface of a metal and a dielectric [43,44]. Metal films and gaps are considered 2-D optical waveguides because of their $\boldsymbol{k}$ vector.

We proposed 1-D optical waveguides to make an optical beam with a diameter much smaller than λ_0. One-dimensional optical waveguides are formed by rolling thin ND films into thin cylindrical shapes. Figure 5.18 shows a schematic view of various 1-D optical waveguides. There are many types of 1-D waveguides, e.g., ND pins, holes, coaxial lines, tubes, parallel lines, and parallel holes.

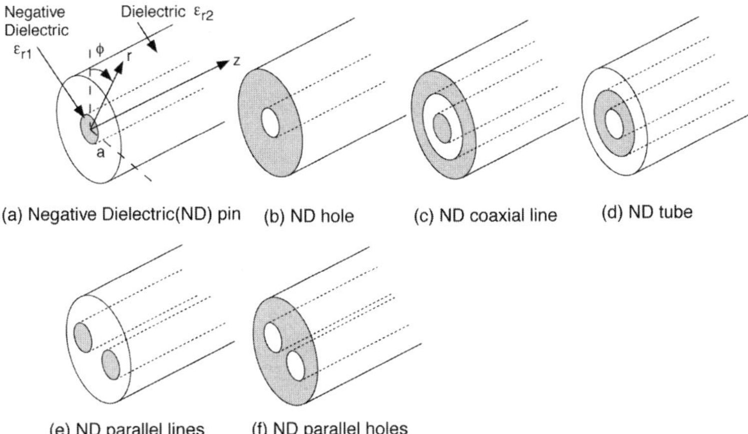

Fig. 5.18. Schematic view of various types of 1-D optical waveguides. Gray areas are NDs. The coordinates used in the analysis are shown in **a**

Many studies of dispersion relations of SPPs in metal cylinders have been done since the 1970s [45]. The recent progress of scanning near-field optical microscopy (SNOM) has attracted much interest in this system again [46]. Our proposal described here has been done with a point of view different from previous studies. In a previous paper, we pointed out that the beam radius of a 1-D optical wave in a ND pin can be shrunk to nanometer order by decreasing its core radius [42]. In this article, we focus particularly on ND pins, holes, and tubes.

5.3.3 Negative-Dielectric Pin and Hole

We will discuss here a propagation mode in a lossless ND pin and hole (see Fig. 5.18a,b), where there is only one dielectric/ND interface. We have calculated a phase constant for a 1-D optical wave by solving Maxwell's equations. The lowest (0th) order TM mode ($H_z = 0$) and hybrid mode ($H_z \neq 0$ and $E_z \neq 0$) are obtained as propagative modes. The TE mode ($E_z = 0$) is not a propagative mode in an 1-D optical wave. In the analysis, cylindrical coordinates $(r,\ \phi,\ z)$ are selected, as shown in Fig. 5.18a. An electromagnetic field propagating in the z direction in a ND pin and hole is assumed in the form $e^{i(\omega t - \beta z)}$, where β is a phase constant. The field in the TM mode in a core ($r < a$) with relative dielectric constant ϵ_{r1} is given as follows:

$$E_{z1} = AI_0(\gamma_1 r), E_{r1} = \frac{i\beta}{\gamma_1^2}\frac{\partial E_{z1}}{\partial r}, H_{\phi 1} = \frac{i\omega\epsilon_{r1}\epsilon_0}{\gamma_1^2}\frac{\partial E_{z1}}{\partial r}, \tag{5.5}$$

and when clad ($r > a$) with ϵ_{r2}, the field is

$$E_{z2} = BK_0(\gamma_2 r), E_{r2} = \frac{i\beta}{\gamma_2^2}\frac{\partial E_{z2}}{\partial r}, H_{\phi 2} = \frac{i\omega\epsilon_{r2}\epsilon_0}{\gamma_2^2}\frac{\partial E_{z2}}{\partial r}, \tag{5.6}$$

where A and B are any constants and a is the radius of the core. I_ν and K_ν are νth order modified Bessel functions. γ_j $(j = 1, 2)$ is defined as follows:

$$\gamma_j = \sqrt{\beta^2 - \epsilon_{rj}\epsilon_0\mu_0\omega^2}. \tag{5.7}$$

From boundary conditions of $E_{z1} = E_{z2}$ and $H_{\phi 1} = H_{\phi 2}$ at $r = a$, one can obtain the characteristic equation of the TM mode as

$$-\frac{\epsilon_{r1} I_1(\xi_1)}{\xi_1 I_0(\xi_1)} = \frac{\epsilon_{r2} K_1(\xi_2)}{\xi_2 K_0(\xi_2)}, \tag{5.8}$$

where $\xi_1 = \gamma_1 a, \xi_2 = \gamma_2 a$. Similarly, one can obtain the characteristic equation of the νth order hybrid mode as

$$\left[\frac{1}{\xi_1}\frac{I_\nu'(\xi_1)}{I_\nu(\xi_1)} - \frac{1}{\xi_2}\frac{K_\nu'(\xi_2)}{K_\nu(\xi_2)}\right]\left[\frac{\epsilon_{r1}}{\xi_1}\frac{I_\nu'(\xi_1)}{I_\nu(\xi_1)} - \frac{\epsilon_{r2}}{\xi_2}\frac{K_\nu'(\xi_2)}{K_\nu(\xi_2)}\right] \tag{5.9}$$

$$= \nu^2\left(\frac{\beta}{k_0}\right)^2\left(\frac{1}{\xi_1^2} - \frac{1}{\xi_2^2}\right)^2, \tag{5.10}$$

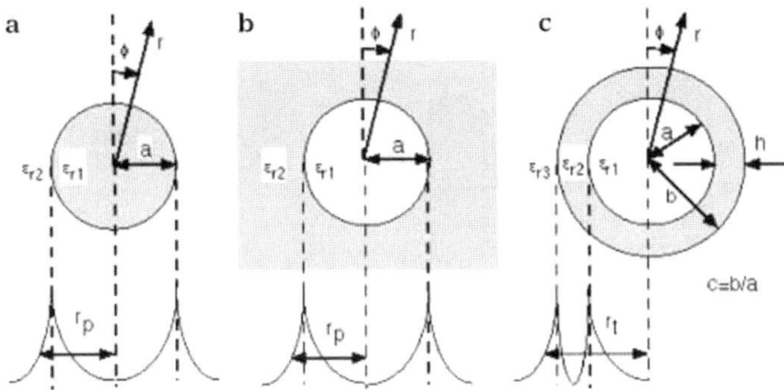

Fig. 5.19. Cross-sectional view and field distribution of $H_\phi(r)$ in (**a**) a ND pin, (**b**) a hole, and (**c**) a tube. Gray areas are ND

where k_0 is the free-space wave number. Substituting $\nu = 0$ in (5.9), one can derive (5.8) by using $I_0'(x) = I_1(x)$ and $K_0'(x) = -K_1(x)$. Equations (5.8) and (5.9) have forms similar to the equation characteristic of a step index dielectric optical fiber.

Electromagnetic fields in a ND pin and hole are localized at the interface and decay exponentially in both dielectric and ND, as shown schematically in Fig. 5.19a,b. Thus, we define the beam radius r_p of the TM mode by the equation $H_{\phi 2}(r_p) = \frac{1}{e} H_{\phi 2}(a)$ (see Fig. 5.19). This definition gives

$$K_1(\gamma_2 r_\mathrm{p}) = \frac{1}{e} K_1(\gamma_2 a) . \tag{5.11}$$

β and r_p as functions of a can be calculated by solving (5.7)–(5.11) numerically. In the numerical calculations for a ND pin, the dielectric constant of a metal core is taken as $\epsilon_{r1} = -19$ which is the experimental value for silver at $\lambda_0 = 633$ nm at 300 K, and the dielectric constant of a cladding is $\epsilon_{r2} = 4$ [47]. In a ND hole, the dielectric constant is $\epsilon_{r1} = 4$ and $\epsilon_{r2} = -19$. We note that propagative modes in a ND pin are only 1-D optical waves, whereas both 1-D and 3-D optical waves exist in a ND hole.

Figure 5.20a shows β with respect to a for the TM and hybrid modes of a ND pin. The r_p of the TM with respect to a is shown in Fig. 5.20b. From Fig. 5.20a, it is observed that β of the TM increases infinitely as a decreases, and the cutoff effect is never observed. We stress that such a dispersion relation of the TM mode is significant for the formation of an optical beam whose diameter is much smaller than λ_0. Actually, as shown in Fig. 5.20b, r_p decreases to zero as a decreases without a diffractive limit. As for hybrid modes, β of $\nu > 1$ shows a cutoff, whereas β of $\nu = 1$ asymptotically approaches 2 when a approaches zero, and the cutoff effect is never observed. Therefore, it is possible to make and propagate an optical beam of a nanometer-order diameter by using the TM mode of a ND pin.

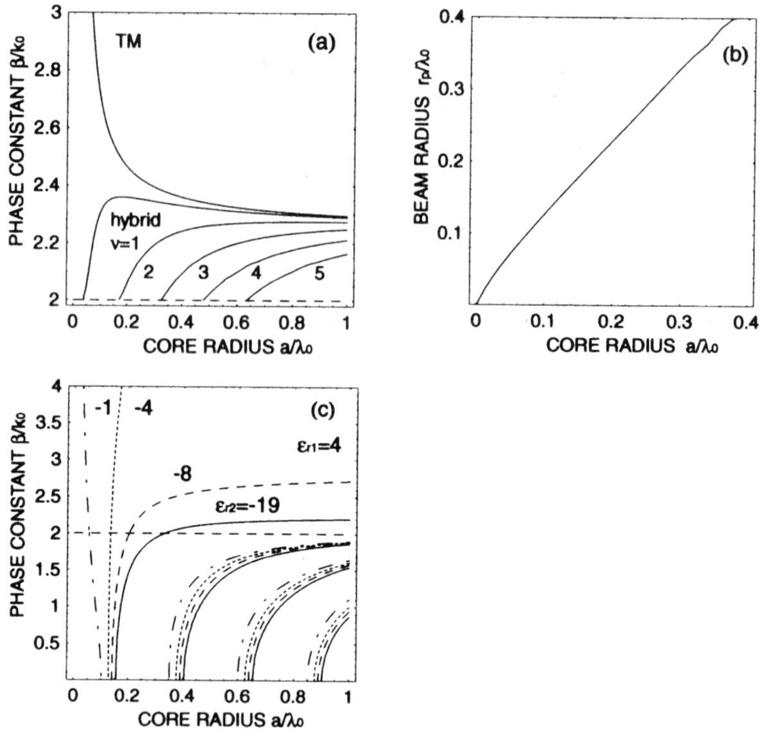

Fig. 5.20. **a** Phase constants of the TM and hybrid modes ($\nu = 1$–5) versus core radius in a ND pin. **b** Beam radius of the TM versus the core radius in a ND pin. **c** Phase constant of the TM in a ND hole versus the core radius for $\epsilon_{r2} = -19$ (*solid*), -8 (*dashed*), -4 (*dotted*), -1 (*dash-dotted*). The phase constant is normalized by k_0. The core and beam radius are normalized by λ_0. The dimension of an optical wave is 1-D above the horizontal dashed lines and 3-D below the lines

In contrast to the ND pin, the dispersion curves in a ND hole are complicated and sensitive to $|\epsilon_{r2}|$. β of the TM mode in a ND hole with respect to a is shown in Fig. 5.20c, where the modes for different ϵ_{r2} are plotted. The optical wave is 1-D above the horizontal dashed line ($\beta/k_0 > 2$) and 3-D below the line ($0 < \beta/k_0 < 2$). When $|\epsilon_{r2}| > \epsilon_{r1}$, β shows a cutoff at $a/\lambda_0 \sim 0.15$ in the 3-D optical wave region. When $|\epsilon_{r2}| < \epsilon_{r1}$, β increases infinitely, a approaches zero, and does not show a cutoff. Furthermore, we note that the group velocity is negative in this case. Thus, an optical beam can be transmitted through a ND hole of nanometer-order diameter when $|\epsilon_{r2}| < \epsilon_{r1}$.

5.3.4 Negative-Dielectric Tube

We will discuss a propagative mode in a lossless ND tube (see Fig. 5.18d), where there are two dielectric/ND interfaces [48]. The cross-sectional view of a ND tube is shown in Fig. 5.19c, where the relative dielectric constants of core, cladding and outer cladding are ϵ_{r1}, ϵ_{r2}, and ϵ_{r3}, respectively. Inner radius a, outer radius b, and the aspect ratio $c \equiv b/a$ are defined. We obtained the characteristic equations of the TM and hybrid modes by the same procedure as that for a ND pin and hole. The characteristic equation of the TM mode is given as follows:

$$\begin{aligned}
&\frac{\epsilon_{r1}\epsilon_{r2}}{\gamma_1\gamma_2} I_1(\xi_1)I_1(\xi_3)K_0(\xi_2)K_0(\xi_4) + \frac{\epsilon_{r2}^2}{\gamma_2^2} I_0(\xi_1)I_1(\xi_3)K_0(\xi_4)K_1(\xi_2) \\
&+ \frac{\epsilon_{r1}\epsilon_{r2}}{\gamma_1\gamma_2} I_0(\xi_2)I_1(\xi_1)I_0(\xi_4)I_1(\xi_3) - \frac{\epsilon_{r2}^2}{\gamma_2^2} I_0(\xi_1)I_1(\xi_2)K_0(\xi_4)K_1(\xi_3) \\
&+ \frac{\epsilon_{r1}\epsilon_{r3}}{\gamma_1\gamma_3} I_0(\xi_3)I_1(\xi_1)K_0(\xi_2)K_1(\xi_4) - \frac{\epsilon_{r1}\epsilon_{r3}}{\gamma_1\gamma_3} I_0(\xi_2)I_1(\xi_1)K_0(\xi_3)K_1(\xi_4) \\
&+ \frac{\epsilon_{r2}\epsilon_{r3}}{\gamma_2\gamma_3} I_0(\xi_1)I_1(\xi_2)K_0(\xi_3)K_1(\xi_4) \\
&+ \frac{\epsilon_{r2}\epsilon_{r3}}{\gamma_2\gamma_3} I_0(\xi_1)I_0(\xi_3)K_1(\xi_2)K_1(\xi_4) = 0\,,
\end{aligned} \tag{5.12}$$

where $\xi_1 = \gamma_1 a$, $\xi_2 = \gamma_2 a$, $\xi_3 = \gamma_2 b$, and $\xi_4 = \gamma_3 b$. γ_j $(j = 1, 2, 3)$ is defined by (5.7). The characteristic equation of the hybrid mode is the expansion of an 8×8 determinant, which is so complicated that we do not show it here.

The beam radius r_{t} of the TM mode in a ND tube is defined similarly to that of a ND pin (see Fig. 5.19c). r_{t} is defined by the equation $H_{\phi 3}(r_{\mathrm{t}}) = \frac{1}{\mathrm{e}} H_{\phi 3}(b)$. This definition gives

$$K_1(\gamma_3 r_{\mathrm{t}}) = \frac{1}{e} K_1(\gamma_3 b)\,. \tag{5.13}$$

Equations (5.12) and (5.13) and the characteristic equation of the hybrid mode can be solved numerically to obtain β and r_t. In the numerical calculations, the dielectric constant is taken as $\epsilon_{r1} = \epsilon_{r3} = 4$, $\epsilon_{r2} = -19$.

The dispersion relation for a ND tube is qualitatively different from those of a ND pin and hole. Figure 5.21a shows β of the TM and hybrid modes $(\nu = 1, 2, 3)$ with respect to a in the fixed ratio of $c = 1.1$. Figure 5.21b shows β of the TM with respect to a in the fixed outer radius of $b/\lambda_0 = 1$. It is observed that there are two solutions of a 1-D optical wave in each mode. They correspond to even (the upper curves) and odd (the lower curves) coupled modes of the surface wave in each interface. As shown in Fig. 5.21a, β of the even mode increases infinitely as a decreases, whereas the odd mode is cut off at $a/\lambda_0 = 0.4$–0.7. Figure 5.21b implies that two solutions are observed only when the metal is thin $(h \equiv b - a < 0.7\lambda_0)$, because they are attributed to the coupling effect of a surface wave. The example of field distribution E_z of the TM even mode at $a/\lambda_0 = 0.1$ is shown in Fig. 5.21c. r_{t} with respect

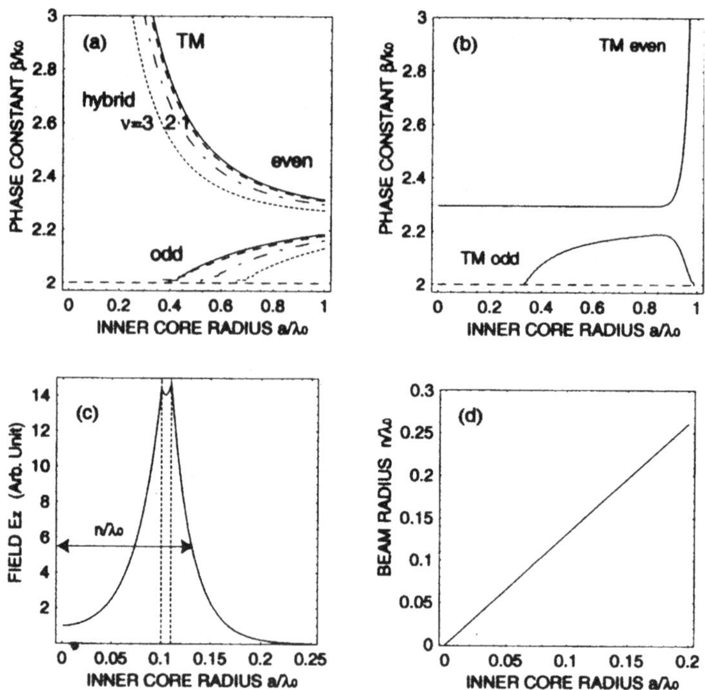

Fig. 5.21. **a** Phase constant versus core radius of TM (*solid*) and hybrid modes for $\nu = 1$ (*dashed*), 2 (*dash-dotted*), and 3 (*dotted*) in a ND tube with $c = 1.1$, and **b** of TM with $b/\lambda_0 = 1$. **c** Field distribution E_z of the TM even mode at $a/\lambda_0 = 0.1$, $b/\lambda_0 = 0.11$ ($c = 1.1$). **d** Beam radius of the TM even mode versus core radius. The dimension of the optical wave is 1-D above the horizontal dashed lines

to a is shown in Fig. 5.21d, where it is observed that r_t decreases linearly and approaches zero as a decreases. Therefore, we can conclude that the diameter of an optical beam in a ND tube can be shrunk in nanometer order by decreasing a and h as observed in the TM mode of a ND pin and hole.

5.3.5 Lossy Waveguides and Applications

In actual metals, a dielectric constant has a small imaginary part that induces a transmission loss, e.g., in silver $\epsilon_r = -19 - 0.53i$ at $\lambda_0 = 633$ nm [47]. We calculated the propagative constant ($\beta = \beta_R - i\beta_I$) for a lossy ND pin and tube to estimate the propagative length of a 1-D optical wave. Because we obtained a slight difference in β_R between lossy and lossless waveguides, we will only show the results for β_I here. Figure 5.22 shows β_I of the TM mode with respect to a in a ND pin (a) and a tube (b). In a ND pin, β_I is plotted for a different imaginary part of dielectric constant δ of a ND. In a ND tube, β_I is plotted for a different c. As shown in Fig. 5.22, it is observed that β_I

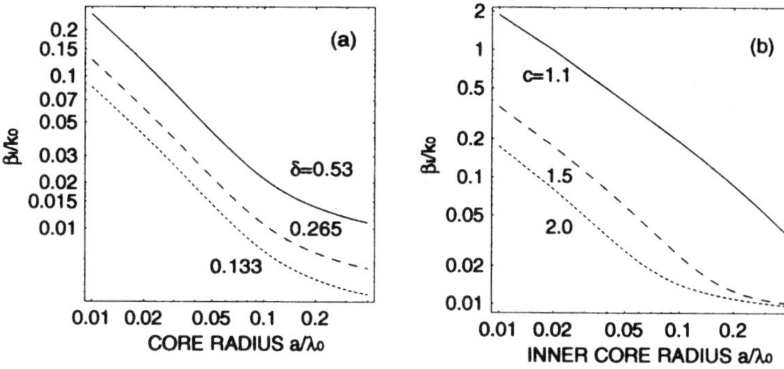

Fig. 5.22. Loss of ND pin and tube. **a** Imaginary part of β in a ND pin versus core radius at $\delta = 0.53$ (*solid*), 0.265 (*dashed*), 0.133 (*dotted*). **b** Imaginary part of β in a ND tube versus core radius at $c = 1.1$ (*solid*), 1.5 (*dashed*), 2.0 (*dotted*)

increases as a decreases in both a ND pin and tube. This indicates that a 1-D optical beam with a smaller diameter has a higher loss. In addition, the value of β_I depends on δ and c because loss is caused by the dissipation in a ND. Thus, one can reduce the loss and enhance the propagative length by reducing δ. As for the numerical example of a ND pin with a 20-nm diameter silver core ($\epsilon_{r1} = -19 - 0.53i$) and a dielectric cladding ($\epsilon_{r2} = 4$), the beam diameter is 33 nm and the transmission loss is 3 dB/410 nm at $\lambda_0 = 633$ nm.

Low-dimensional optical waves will play an important role in nanostructures with metal. One of the important applications of 1-D optical waveguides is a high efficiency optical head with high spatial resolution. It can be used for optical recording, manipulation, and a SNOM probe. There are many other attractive applications for "nanometer optical devices." One-dimensional optical waveguides contribute to optical integrated circuits or OEIC devices in the nanometer size. We stress that the transmission loss is not a very serious problem within a very short transmission ($< 10\lambda_0$) as far as application to optical devices in a nanometer size are considered.

References

1. T. R. Albrecht, S. Akamine, T. E. Carver, and C.F. Quate: J. Vac. Sci. Technol. **A8**, 3386 (1990)
2. L. C. Kong, B. G. Orr, and K. D. Wise: J. Vac. Sci. Technol. **B11**, 634 (1993)
3. S. Akamine, H. Kuwano, and H. Yamada: Appl. Phys. Lett. **68**, 579 (1996)
4. D. Drews, W. Ehrfeld, O. Haverbeck, M. Lacher, K. Mayr, W. Noel, and M. Stopka: *Tech. Dig. Int. Conf. Micro Opto Electro Mech. Syst.*, 1999, p. 100
5. C. Mihalcea, W. Scholz, S. Werner, S. Munster, E. Oesterschulze, and R. Kassing: Appl. Phys. Lett. **68**, 3531 (1996)
6. U. C. Fischer: Ultramicroscopy **42–44**, 393 (1992)
7. E. Betzig and J. K. Trautman: Science **257**, 189 (1992)

8. L. P. Ghislain, V. B. Elings, K. B. Crozier, S. R. Manalis, S. C. Minne, K. Wilder, G. S. Kino, and C. F. Quate: Appl. Phys. Lett. **74**, 501 (1999)
9. E. Betzig, J. K. Trautman, R. Wolfe, E. M. Gyorgy, P. L. Finn, M. H. Kryder, and C.-H. Chang: Appl. Phys. Lett. **61**, 142 (1992)
10. A. G. T. Ruter, M. H. P. Moers, N. F. van Hulst, and M. de Boer: J. Vac. Sci. Technol. **B 14**, 597 (1996)
11. D. Drews, W. Ehrfeld, M. Lacher, K. Mayr, W. Noell, S. Schmitt, and M. Abraham: Nanotechnology **10**, 61 (1999)
12. C. Mihalcea, W. Scholz, S. Werner, S. Munster, E. Oesterschulze, and R. Kassing: Appl. Phys. Lett. **68**, 3531 (1996)
13. P. N. Minh, T. Ono, and M. Esashi, *Nat. Conf. Phys. Sensors*, Tokyo, Japan, November 1998 (unpublished). The Cr-coated SiO_2 tip was placed close to a conducting substrate. A voltage was applied between the tip and the substrate. Due to a high electric field, Cr atoms around the tip were evaporated leaving an aperture.
14. P. N. Minh, T. Ono, and M. Esashi, *Proc. 11th IEEE Int. Conf. Micro Electro Mech. Syst.*, Orlando, 1999, Sens. Actuators A, **80**, 163 (2000)
15. P. N. Minh, T. Ono, and M. Esashi: Appl. Phys. Lett. **75**, 4076 (1999)
16. R. B. Maruss and T. T. Sheng: J. Electrochem. Soc. **129**, 1278 (1982)
17. R. B. Marcus, T. S. Ravi, T. Gmitter, K. Chin, D. Liu, W. J. Orvis, D. R. Ciarlo, C. E. Hunt, and J. Trujillo: Appl. Phys. Lett. **56**, 236 (1990)
18. S. Akamine and C.F. Quate: J. Vac. Sci. Technol. **B10**, 2307 (1992)
19. M. Ohtsu, ed.: *Near-Field Nano/Atom Optics and Technology* (Springer, Tokyo 1998)
20. P. N. Minh, T. Ono and M. Esashi: Rev. Sci. Instrum. **71**, 3111 (2000)
21. C. Mihalcea, W. Scholz, S. Werner, S. Munster, E. Oesterschulze, and R. Kassing; Appl. Phys. Lett. **68**, 3531 (1996)
22. G. A. Valascovic, M. Holton, and G. H. Morrison: Appl. Opt. **34**, 1215 (1995)
23. H. Zhou, A. Midha, L. Bruchhaus, G. Mills, L. Donaldson, and J. M. R. Weaver: J. Vac. Sci. Technol. **B17**, 1954 (1999)
24. M. B. Lee, N. Atoda, K. Tsutsui, and M. Ohtsu: J. Vac. Sci. Technol. **B17**, 2462 (1999)
25. U. C. Fischer and M. Zapletal: Ultramicroscopy **42–44**, 393 (1992)
26. T. Murashita: J. Vac. Sci. Technol. **B15**, 32 (1997)
27. D. W. van der Weide and P. Neuzil: J. Vac. Sci. Technol. **B14**, 4144 (1996)
28. A. Majumdar, J. Lai, M. Chandrachood, O. Nakabepu, Y. Wu, and Z. Shi: Rev. Sci. Instrum. **66**, 3584 (1995)
29. J. Mertz, M. Hipp, J. Mlynek, and O. Marti: Appl. Phys. Lett. **64**, 2338 (1994)
30. R. J. Hamers and K. Markert: Phys. Rev. Lett. **64**, 1051 (1990)
31. J. M. R. Weaver and D. W. Abraham: J. Vac. Sci. Technol. B **9**, 1559 (1991)
32. M. Abe, T. Uchihashi, M. Ohta, H. Ueyama, Y. Sugawara, and S. Morita: Opt. Rev. **4**, 232 (1997)
33. M. Abe, T. Uchihashi, M. Ohta, H. Ueyama, Y. Sugawara, and S. Morita: J. Vac. Sci. Technol. B **15**, 1512 (1997)
34. T. R. Albrecht, P. Grütter, D. Horne, and D. Rugar: J. Appl. Phys. **69**, 668 (1991)
35. Y. Sugawara, M. Ohta, H. Ueyama, and S. Morita: Science **270**, 1646 (1995)
36. Y. Sugawara, H. Ueyama, T. Uchihashi, M. Ohta, S. Morita, N. Suzuki, and S. Mishima: Appl. Surf. Sci. **113–114**, 364 (1997)

37. M. Abe, Y. Sugawara, Y. Hara, K. Sawada, and S. Morita: Jpn. J. Appl. Phys. **37**, L167 (1998)
38. M. Nonnenmacher, M. P. O'Boyle and H. K. Wickramasinghe: Appl. Phys. Lett. **58**, 2921 (1991)
39. M. Abe, Y. Sugawara, Y. Hara, K. Sawada, and S. Morita: Jpn. J. Appl. Phys. **37**, L1074 (1998)
40. H. P. Kleinknecht, J. R. Sandercock, and H. Meier: Scanning Microsc. **2**, 1839 (1988)
41. M. Abe, Y. Sugawara, K. Sawada, Y. Andoh, and S. Morita: Appl. Sur. Sci., **140**, 383 (1999)
42. J. Takahara, S. Yamagishi, H. Taki, A. Morimoto, and T. Kobayashi: Opt. Lett. B **22**, 475–477 (1997)
43. H. Raether: *Surface Plasmons on Smooth and Rough Surfaces and on Gratings* (Springer, Berlin 1988)
44. J. J. Burke, G. I. Stegeman, and T. Tamir: Phys. Rev. B **33**, 5186 (1986)
45. J. C. Ashley and L. C. Emerson: Surf. Sci. **41**, 615 (1974)
46. L. Novotny and C. Hafner: Phys. Rev. E **50**, 4094 (1994) and references therein
47. E. D. Palik: *Handbook of Optical Constants of Solids* (Academic Press, San Diego 1985)
48. J. Takahara, S. Yamagishi, and T. Kobayashi: Propagation of a nanometer diameter optical beam in negative dielectric tube, *Tech. Dig. 5th Int. Conf. Near Field Opt. and Relat. Tech.*, Shirahama, Japan, December, 1998, pp. 232–233

6 High-Density Optical Memory and Ultrafine Photofabrication

M. Irie

Optical recording uses focused or minimal size laser light to effect some optical property change in recording media, which is subsequently read back by the laser. Various approaches have been proposed to increase the recording density of optical memories: decreasing the recording mark size by using short-wavelength lasers, three-dimensional recordings, and near-field optical recording. Among them the most promising approach is a near-field optical recording. Recording density in conventional optical recording is limited by the diffractive limit of light and the numerical aperture of the lens. Therefore, the mark size cannot be reduced to less than the wavelength of light. In near-field optical recording, in contrast, the size depends only on the diameter of the probe tip aperture. Therefore, the recording density can be increased at will, in principle, if a small aperture tip is available.

So far, the memory media that have been used for near-field optical recording are magneto-optical or phase-change media [1–3]. These media are driven by a heat-mode recording method. The photon energy of the writing laser is converted to heat energy on the recording media and used for recording. The heat-mode recording requires a high-power recording laser, and the transfer rate is slow. Other candidates for near-field optical memory media are organic media, which are driven by a photon-mode recording method. Photon-mode photochemical recording offers advantages over heat-mode recording in resolution and writing speed. Photochromic media are one of the promising erasable photon–mode media. In the following, recent progress in photon-mode photochromic media and their application to near-field optical recording will be described.

Minimal size near-field light can be used for optical memory and also for producing fine patterns on substrate surfaces. So far, microfabrication technology using photoresists and conventional light sources, such as mercury lamps or excimer lasers, has been applied to producing integrated circuits. Light passed through photomasks with fine patterns induces photochemical degradation or crosslinking reactions of the spin-coated photoresist on silicon surfaces and wet development by chemicals produces fine patterns. The size of the patterns is limited to the wavelength of light because of the diffractive effect of conventional propagating light. This limitation can be eliminated by

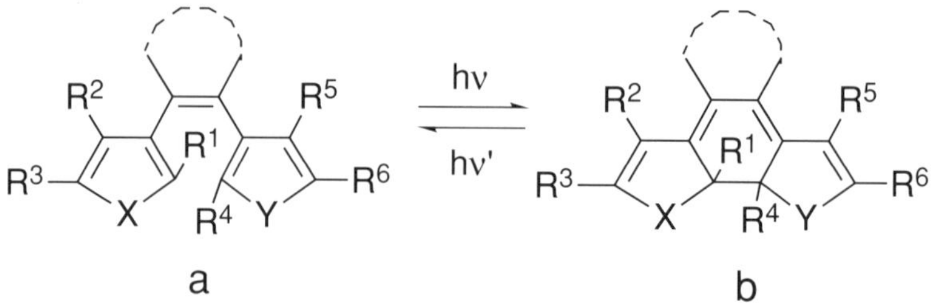

Fig. 6.1. Photochromism of diarylethene

employing minimal size near-field light as the light source for photochemical reactions. Ultrafine pattern formation by photoinduced chemical deposition and photochemical reactions will also be described.

6.1 Photochromic Memory Media

Photochromism is defined as a reversible transformation by photoirradiation between two forms of a chemical species that have different absorption spectra. When we apply photochromic media to photon-mode erasable optical recording, the following performance is required:

1. archival storage capability (thermal stability),
2. low fatigue (can be cycled many times without loss of performance),
3. high sensitivity and rapid response, and
4. nondestructive readout capability.

The more important requirements are thermal stability of both isomers and fatigue-resistant characters. Recently, a new class of photochromic compounds which fulfill these requirements has been developed. The compounds, named diarylethenes, undergo the following reversible photochromic reactions [4,5].

Typical examples of diarylethenes are shown in Table 6.1. Both isomers of diarylethenes, open-ring **a** and closed-ring **b** isomers, are thermally stable and never interconvert to each other at room temperature. For example, the temperature dependence of the thermal reaction rate from **2b** to **2a** was measured above 150°C and extrapolation of the temperature dependence indicated that the half-life of the colored closed-ring **2b** isomer is 1900 years at 30°C [6]. Diarylethenes that have thiophene or benzothiophene aryl groups undergo thermally irreversible photochromic reactions at ambient temperature.

The main disadvantage of organic materials for optoelectronic applications is their lack of durability. Although the durability problem has already been overcome for liquid crystals and organic photo-conductors, it still remains for photochromic materials. Photochromic reactions are always attended by rearrangement of chemical bonds. During the bond rearrangement,

Table 6.1. Photochromic diarylethenes

Open-ring Isomers (**a**)	Closed-ring Isomers (**b**)	Matrices
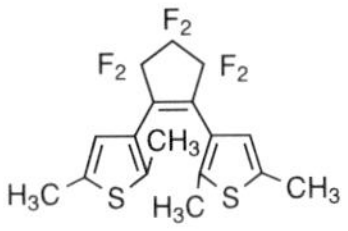 **1a**	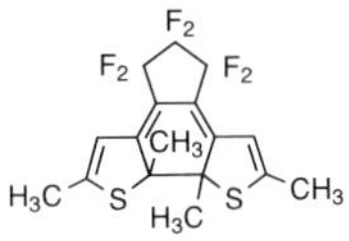 **1b**	in crystal
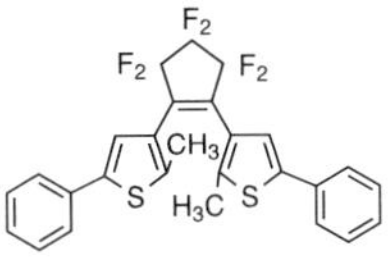 **2a**	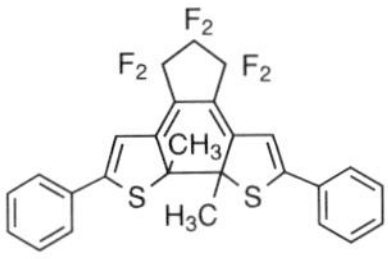 **2b**	in crystal
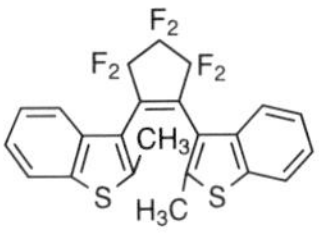 **3a**	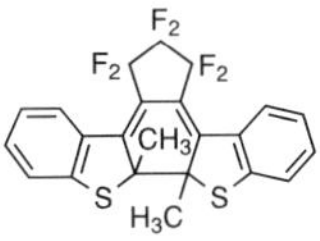 **3b**	in polymer film
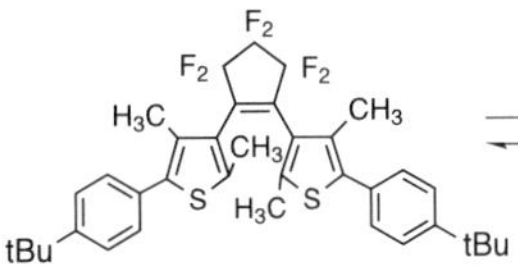 **4a**	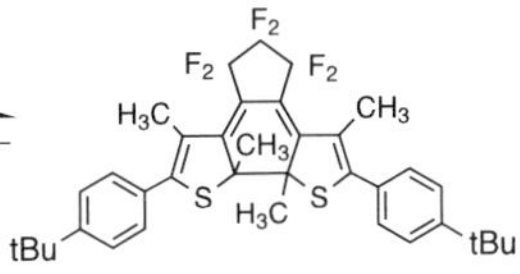 **4b**	in bulk amorphous solid
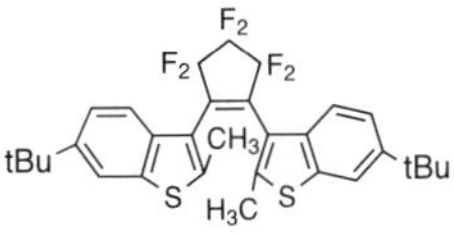 **5a**	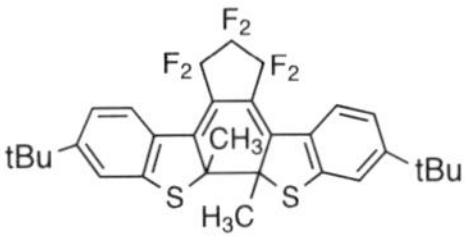 **5b**	in bulk amorphous solid

undesirable side reactions take place to some extent. This limits the durability of photochromic materials. Extensive examination of the fatigue resistance of various diarylethenes partly solved the durability problem. More than 10 diarylethenes that have benzothiophene aryl groups undergo fatigue-resistant photochromic reactions [4]. The photostimulated reversible coloration/decoloration cycles can be repeated more than 10^4 times.

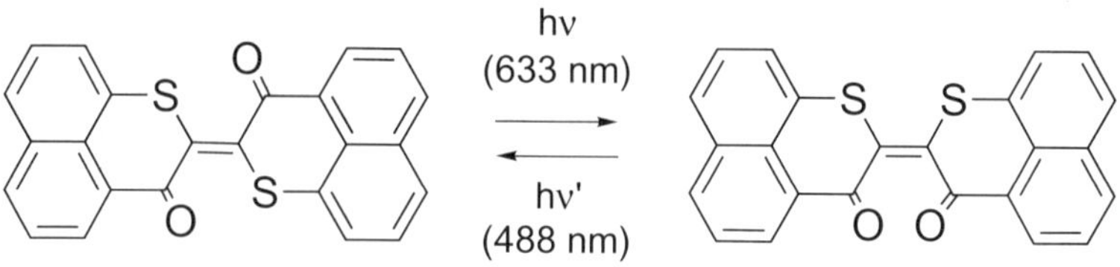

Fig. 6.2. Photochromism of perinaphthothioindigo

When we apply photochromic compounds to optical memory media, it is indispensable to disperse them in solid matrices, such as polymers. However, photochromic reactivity is suppressed in most dye/polymer systems, undesirable side reactions take place, and the sensitivity is low because of low solubility in polymer matrices. Thus, it is necessary to develop crystalline or bulk amorphous photochromic materials for practical applications. Table 6.1 summarizes these crystalline and amorphous photochromic diarylethenes. Although many photochromic compounds have been developed, the compounds shown in Table 6.1 are the only compounds that undergo thermally irreversible and fatigue-resistant photochromic reactions in crystalline and bulk amorphous states [6–9]. Photochromic molecules in crystals are protected from oxygen and other impurities. Therefore, intermolecular destructive reactions are suppressed. So far, the most robust photochromic compounds are crystalline diarylethenes.

Dynamic study by pico- and femtosecond laser photolysis revealed that both cyclization and cycloreversion reactions take place in less than a few picoseconds in solution as well as in the solid state [10–13]. The response time is very fast. The sensitivity depends on the photochemical quantum yields, the absorption coefficients of the colored closed-ring isomers, and the content of the compounds in the solid matrices. Photocyclization quantum yields of most diarylethenes are 0.2–0.5 [5]. The absorption coefficients are around $10^4 M^{-1} cm^{-1}$ [5]. These data show that the sensitivity is high enough when the content of the compounds is high, as in crystalline and bulk amorphous states.

Photochromic compounds change absorption spectra and also fluorescence. One of the typical fluorescent photochromic compounds is thioindigo. Various types of thioindigos were prepared, and it was found that naphthothioindigo (NTI) has fatigue resistance and efficient photochromic reactions even in polymer films [14,15]. In the following, near-field optical recording on photochromic memory media is described.

6.2 Near-Field Optical Memory

6.2.1 Diarylethenes

Two kinds of amorphous films containing diarylethene were prepared. One was polystyrene film containing **3a** (dye/polymer = 20 wt%, film thickness = 1.5 μm). For optical recording, the film was first colored homogeneously by irradiation with UV light ($300 < \lambda < 400$ nm), and then a small spot mark was written by an Ar-ion laser (529 nm, input laser power = 5 mW) from a micropipet whose diameter was 100 nm [16]. The written mark's diameter was around 1 μm. In such a thick medium, light scattering resulted in a large mark size. To obtain smaller spot marks, the medium thickness should be reduced to the mark size demanded. However, in dye/polymer systems, it is impossible to reduce the film thickness below 1.5 mm because sufficient contrast cannot be obtained. Therefore, a bulk amorphous film of **4a** was employed as the medium [9].

The film was prepared by spin-coating the hexane solution of dye **4a** (film thickness = 400 nm). The amorphous film showed a glass transition temperature at 68°C. The film was first colored by irradiation with UV light ($300 < \lambda < 400$ nm), and the recording was carried out with a He–Ne laser (633 nm, input laser power = 10 mW). For optical recording, the He–Ne laser was coupled to the other end of an optical fiber tip, and the tip was placed close to the medium surface by using a shear-force method. The diameter of the tip aperture was 80 nm. Figure 6.3 shows the first and second recorded marks. The recorded marks were detected with 633 nm light by the transmittance change. The mark diameter was as small as 80 nm, which was similar to the aperture diameter. The marks were stable and never disappeared in the dark but could be erased by irradiation with UV light. The write/read/erase cycle could be repeated more than 10^3 times.

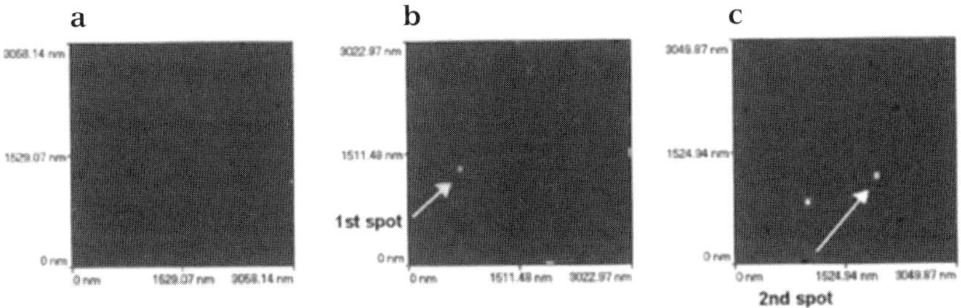

Fig. 6.3. Near-field optical recording on bulk amorphous thin film of **4**: **a** before recording, **b** the first spot mark, and **c** the second spot mark. The film thickness was 400 nm. Recording laser: He–Ne (633 nm, 10 mW)

6.2.2 Perinaphthothioindigo

In the above diarylethene memory media the transmittance change by photoirradiation was used as the basis of optical recording. In the transmittance-detection method, the sensitivity depends on the concentration of the dye and the film thickness. Therefore, it is not easy to obtain high sensitivity from thin films. Another detection method is measuring fluorescent intensity change by photoirradiation. The fluorescent-detection method is superior to the transmittance-detection method in the signal-to-noise ratio (SNR) at a high system bandwidth [17,18]. Theoretical calculation indicated that for a transmittance change of 0.9–1.0 as a result of recording, a readout-light power of 100 nW, and a system bandwidth of 1 MHz, the fluorescent-detection method could produce a sufficient SNR (higher than 26 dB), whereas the transmittance-detection method could not attain such a high SNR (below 20 dB).

NTI **6** is known to change fluorescent intensity by trans/cis photoisomerization [15]. The trans-form emits fluorescence ($\Phi = 0.07$), whereas the cisform does not fluoresce. The compounds were dispersed in spin-coated polystyrene films (film thickness = 300 nm), and recording was carried out with a He–Ne laser (633 nm, input laser power = 10 mW) [19]. The recorded marks were detected with 633-nm light by the fluorescent intensity change. Upon irradiation with 633-nm light for 60 s, trans NTI molecules in the irradiated areas are converted to cis forms, and the irradiated areas become nonfluorescent, or dark spots (dark spot recording). The spot mark size was estimated at approximately 110 nm, which is slightly larger than the aperture size.

Bright spot recording was carried out as follows. The whole area of the polystyrene film containing trans NTI was irradiated for 10 min with a 10-mW He–Ne laser. Upon irradiation with 633-nm light, the fluorescence of the film decreased and almost ceased after 10 min of irradiation. The nonfluorescent film was used for bright spot recording. The probe tip coupled with the Ar-ion laser (488 nm, input laser power = 7 mW) at the other end of the tip was placed close to the medium surface for 90 s. The recording process was repeated twice at different positions, and the recorded spots were read with the scanning probe tip coupled with the He–Ne laser (633 nm, 5 mW).

Figure 6.4 shows the two recorded bright spot marks. Upon irradiation with 488-nm light the nonfluorescent cis NTI molecules were converted to fluorescent trans forms. Therefore, the bright spots are due to the fluorescent trans NTI molecules. The spot marks were erased by irradiation with 633-nm light, as shown in Fig. 6.4b. The half width of the mark size was as small as 50 nm. This is smaller than the aperture size. The bright spot recording yielded a smaller spot mark than the dark spot recording. Although the difference in the spot mark size is apparent, based on theoretical calculations, the bright spot recording method, it was found, gives a higher SNR in com-

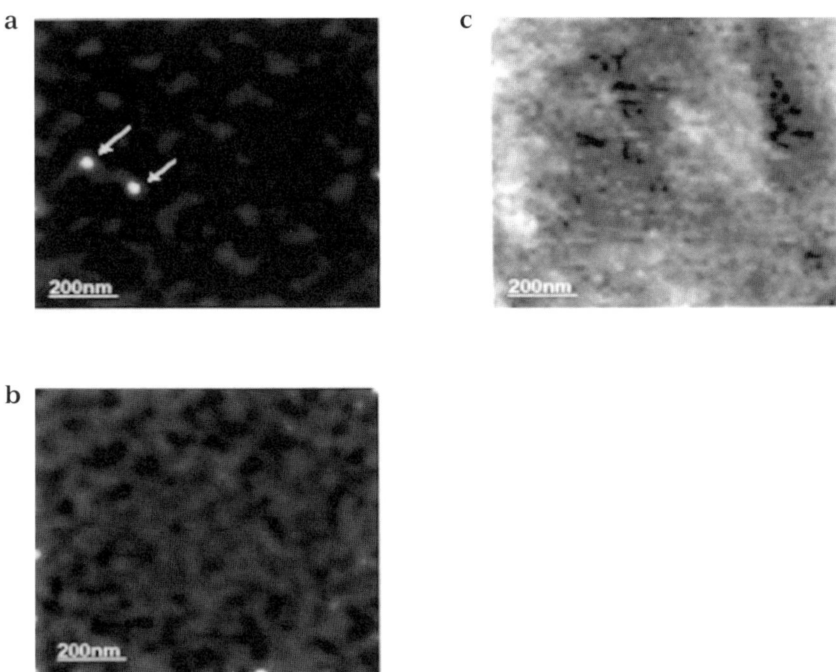

Fig. 6.4. a Two bright spot marks record recorded on polystyrene film containing 5×10^{-4}mol/kg NTI. Recording laser: Ar ion (488 nm, 7 mW), recording time: 90 s. **b** The same area, scanned after the marks were erased by irradiation with a He–Ne laser. **c** Topo image of the same area

parison with the dark spot recording method, when the writing laser power is low and the bandwidth is high [18].

Both the dark and bright spot recording methods shown above require long irradiation time to write spots. This is due to the low output light power from the small probe tip. To shorten the writing time and avoid damage to the probe tip, we employed a pulse Nd^{3+}:YAG laser (532 nm, 10 mJ, 20 ns) as the writing light source. Polystyrene film containing nonfluorescent cis NTI molecules was used as the recording medium. When the tip was as large as 200 nm, a single laser pulse could write a spot mark. The bright spot was erased by irradiation with a He–Ne laser, and no physical deformation was discerned. This result indicates that the recording time is 20 ns, which corresponds to a transfer rate of 50 Mbps. The rate may be increased by using a shorter laser pulse and increasing the throughput of the fiber probe.

6.3 Future Prospects for Near-Field Optical Memory

As described above, photochromic materials are promising memory media for near-field optical recording. Theoretical simulation was carried out to reveal the limiting performance of media. The most important measure of performance is the recording memory density. The density limit was theoretically studied by using Shannon's theory [20]. Shot noise and material noise were taken into account in the analysis of the signal-to-noise ratio. The conventional recording density limit, which is defined by the inverse of the detectable minimum recorded mark size, was 10^{11}–10^{12} bits/cm^2, whereas Shannon's calculated recording density limit was more than 10^{12} bits/cm^2 at a bandwidth 10^8 Hz. According to the theoretical simulations, the maximum memory density of a photochromic near-field optical memory is around 10^{12} bits/cm^2.

Data transfer rates are another important measure of performance. When memory density or capacity increases, it is indispensable to increase data transfer rates. Otherwise the memory media cannot be used practically. The rate in the recording process was calculated for the medium with sensitivity $\varepsilon = 10^4 \mathrm{M}^{-1}\mathrm{cm}^{-1}$ and $\Phi = 0.5$ in the recording process. The rate was about 10 kbps (bit per second) at a light power density of 10^2 W/cm^2, but it increased to more than 10 Mbps at 10^5 W/cm^2, which corresponds to a light power of around 10 μW from a 100-nm diameter tip aperture [21]. The rate in the readout process was also calculated. The rate at a light power of 10^{-6} W was around 1–10 Mbps.

Although the limiting recording memory density is around 10^3 times larger than conventional optical memories, the transfer rates are rather slow and remain at the levels of the systems currently used. For practical application an increase in transfer rates by increasing the output light power or the throughput of the fiber probe is required.

6.4 Nanofabrication: Chemical Vapor Deposition

Since the optical near field energy can be concentrated within a nanometric dimension smaller than the wavelength of light [22–24], it enables us to deposit various materials in nanometric dimensions by photodecomposing chemical gases. The combination of optical near-field technology with the photoenhanced chemical deposition (PE-CVD) process appears to be best technology for integrating nanometer-scale elements because it allows us to fabricate nanostructures and also has the advantage of in situ measurement of the optical properties of the fabricated nanostructures.

PE–CVD combined with optical near-field microscopy was carried out by using the optical near-field generated from a subwavelength aperture at the tip of a fiber probe introduced into a vacuum chamber. The process consists of two steps. In the first step, a vacuum chamber is filled with a

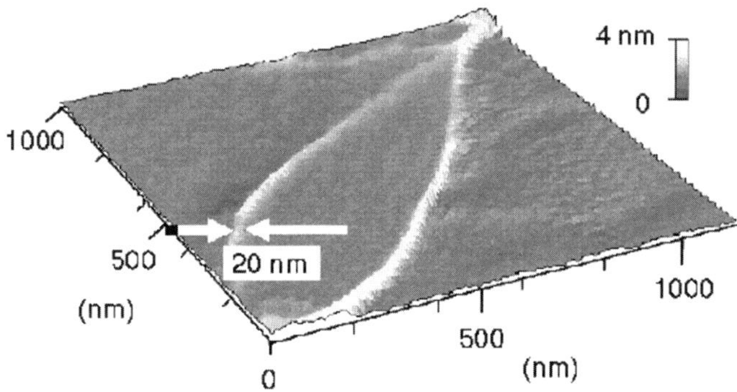

Fig. 6.5. Shear-force image of a loop-shaped pattern of Zn deposited on a glass substrate by the prenucleation method

metallo-organic gas for a few minutes and then evacuated, which leaves a few adsorbed monolayers on the substrate surface. By decomposing the adsorbed molecules with near-field light, nuclei for growth are formed. In the second step, conventional propagating light is directed onto the prenucleated area in the presence of a parent gas, and then the decomposed atoms are selectively deposited on the preexisting nuclei. This method has the advantage of being free from deposition at the probe tip.

Figure 6.5 shows a shear-force image of the loop-shaped Zn pattern on a glass substrate produced by the prenucleation method [25]. The vacuum chamber was evacuated to less than 10^{-5} torr prior to the prenuclei fabrication stage, then filled with about 10 torr of diethylzinc (DEZ) gas, maintaining the pressure for 20 min. Next, the chamber was reevacuated to the pressure of 10^{-5} torr, which leaves a few adsorbed monolayers on the substrate surface. Prenucleation was performed with a probe by using the second-harmonic (SH) light of an Ar-ion laser (244 nm) on the substrate covered with adsorbed molecules. Nuclei of Zn were formed by decomposing the DEZ gas adsorbed on the substrate with the optical near-field at the probe tip. In the growth stage after nuclei fabrication, the chamber was refilled with a few torr of DEZ gas, and the unfocused ArF excimer laser (193 nm, 10 mJ) irradiated the prenucleated substrate directly. Then, growth proceeded only on the preexisting nuclei.

As seen in Fig. 6.5, the minimum width of the pattern is as little as 20 nm. The width achieved here is two orders smaller than the minimum width reported so far by conventional PE–CVD using a far-field light [26]. Since the measured width includes the resolution of a vacuum shear-force microscope (VSFM) [27], depending on the shape of the probe used, the intrinsic width can be smaller than the value estimated from Fig. 6.5.

Another deposition method is direct gas-phase photodissociation. Metallo-organic gas was directly photodissociated in the gas phase, and the atoms

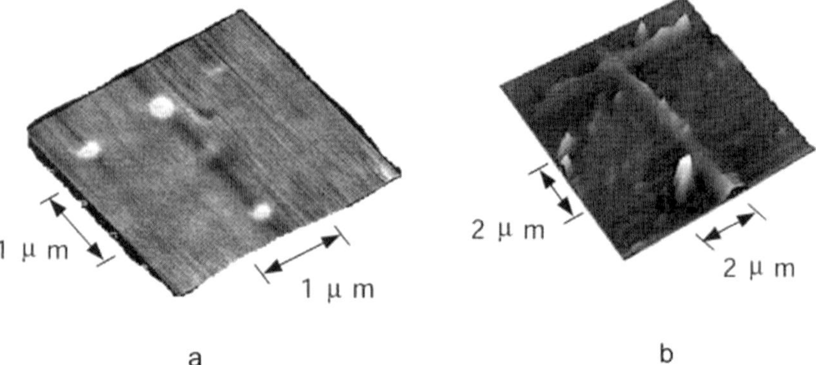

Fig. 6.6. Shear-force image of Zn on a glass substrate deposited by direct gas-phase photodissociation. **a** Dots. **b** A. T-shaped pattern

were deposited on the substrate. Figure 6.6a,b shows the shear-force image of deposited Zn dots and a T-shaped pattern. The gas pressure and input power of SH light (244 nm) were 1 mtorr and 10 mW, respectively. In the fabrication of the dots, the optical near-field on the probe tip was illuminated for a few second over five spots at an interval of 800 nm on the substrate. As shown in Fig. 6.6a, the dots are spaced by 800 nm in excellent agreement with the spacings of the illuminated points, which establishes the high controllability of positioning in fabricating nanostructures by this technique. The T-shaped pattern was prepared by scanning the substrate at a speed of 10–50 nm/s. A glance at Fig. 6.6b is sufficient to explain that the method makes it possible to fabricate subwavelength-scale structures with control of their size and position. The technique also allows us to fabricate nanostructure of oxides, insulators, and semiconductors, as well as metals containing Zn, Al, Cr, and W.

One of the most attractive features of this technique is its high spatial resolution. The lateral size of a fabricated pattern depends on the spatial distribution of the optical near-field energy, and its reproducibility also depends on the reproducibility of fabricating probes. Figure 6.7a,b shows the shear-force image of dots deposited by using a probe with an aperture diameter of 60 nm and the cross-sectional profile along the dashed line, respectively. Two dots with the diameter of 60 nm and 70 nm (full width at the half maximum of the cross-sectional profile) were fabricated at the very close distance of 100 nm. The diameter of the dots was comparable with the aperture diameter of the probe used, which suggests that the smaller dots can be fabricated by using a fiber probe with a smaller aperture.

One of the advantages of these CVD methods is that there is no limitation to substrate and deposited materials. Another technique in which CVD is combined with scanning tunneling microscopy (STM) has reportedly

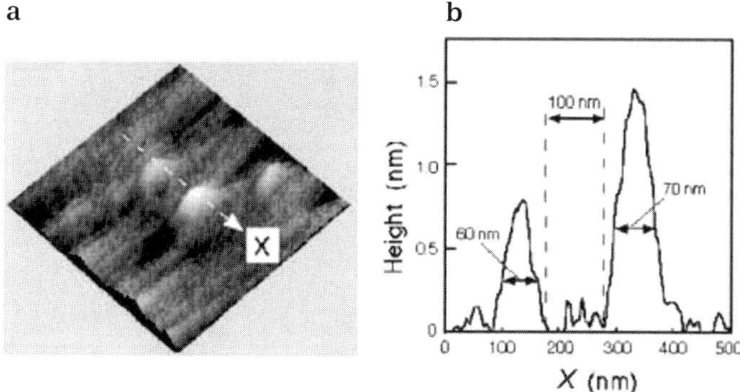

Fig. 6.7. Shear-force image of Zn dots fabricated by using a spliced fiber probe with an aperture diameter of 600 nm. **a** Three-dimension profile of the dots. **b** Cross senctional profile along the dashed line in **a**

produced nanostructures with dimensions close to the atomic level [28]. However, the main drawback comes from the impossibility of using nonconductive substrates and the growth patterns of nonconductive materials. However, this limitation is eliminated by using the previous optical near-field CVD method.

6.5 Nanofabrication: Organic Film

The recording of evanescent field distributions was first demonstrated in photolithographic applications [29,30]. A photomask with fine patterns was placed directly on a photoresist, and it was illuminated with light. If the distance between the photomask and photoresist was small enough, or the distance was smaller than the decay length of the evanescent field, it was possible to copy the fine structures of the photomask on the photoresist film. Contact recording of evanescent field distributions can be applied to near-field optical microscopy (NFOM) [31]. In the application, the evanescent field distribution near a specimen was recorded as a surface topography of a photosensitive film, and the topography was read out with an atomic force microscope (AFM). Since the system does not require a small aperture for illumination or detection of light, a higher signal-to-noise ratio is expected, and very fast phenomena can be observed.

The following urethane-urea copolymer has good prospects for recording intensity distribution as a highly resolving surface modulation.

The recording mechanism is based on trans–cis photoisomerization of pendant azo dyes [31,32]. The consecutive photoisomerization of the trans-to-cis flip-flop followed by the cis-to-trans flip-flop induces mass transport of copolymer chains due to pressure gradients corresponding to the intensity distribution [32].

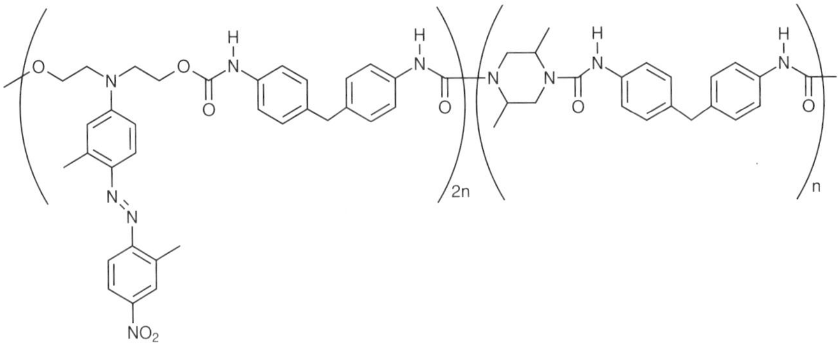

Fig. 6.8. Photoresponsive urethane-urea copolymer

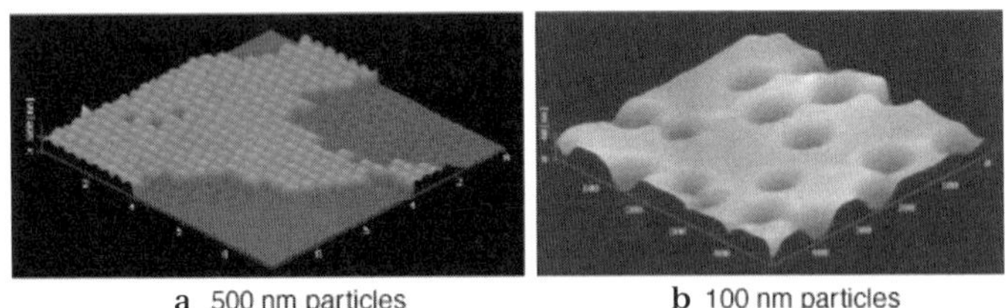

Fig. 6.9. Photofabricated patterns of particles: **a** 500 nm and **b** 100 nm diameter

Particles of 500 nm and 100 nm diameter were prepared as specimens. The 500-nm diameter particles were dispersed in water, and they were arranged in a hexagonal structure by using the self-organization process [31]. An air-cooled Ar-ion laser (488 nm, 30 mW) was used as a light source. The laser beam irradiated the particles and the films. The intensity distribution of the light localized near the surface of the particles modulated the topography of the film. After illumination, the film was immersed in water, and the particles were removed from the film.

Figure 6.9a shows the experimental result of the observation of particles of 500-nm diameter. The film topography was measured with an AFM. The structure of the particles can be clearly observed. The holes made by light illumination reflect the arrangement of particles in a hexagonal structure. Some residual particles which were not removed by washing were observed. It is possible to observe some "lattice defects" in the arrangements of the residual particles and holes.

Figure 6.9b shows the result with 100-nm diameter particles. The scan area of Fig. 6.9b was 500×500 nm. The particles were also clearly imaged. Dips of about 100 nm diameter were formed on the film by field enhancement at the particles. The pattern has a few tens of nanometers resolution. This is one of the promising methods for producing fine patterns on substrate surfaces.

References

1. D. G. Crowe: Appl. Opt. **30**, 4480 (1991)
2. E. Bezig, J. K. Trautman, R. Wolfe, E. M. Gyorgy, P. L. Finn, M. H. Kryder, and C.-H. Chang: Appl. Phys. Lett. **61**, 142 (1992)
3. S. Hosaka, T. Shintani, M. Miyamoto, A. Horotsune, M. Terao, K. Fujita, and S. Kammer: Thin Solid Films **273**, 122 (1996)
4. M. Irie and K. Uchida: Bull. Chem. Soc. Jpn. **71**, 985 (1998)
5. M. Irie: Chem. Rev. **100**, 1685 (2000)
6. M. Irie, T. Lifka, S. Kobatake, and N. Kato: J. Am. Chem. Soc. **122**, 4871 (2000)
7. S. Kobatake, T. Yamada, K. Uchida, N. Kato, and M. Irie: J. Am. Chem. Soc. **121**, 2380 (1999)
8. T. Yamada, S. Kobatake, K. Muto, and M. Irie: J. Am. Chem. Soc. **122**, 1589 (2000)
9. T. Kawai, N. Fukuda, D. Gröschl, S. Kobatake, and M. Irie: Jpn. J. Appl. Phys. **38**, L1194 (1999)
10. H. Miyasaka, S. Arai, A. Tabata, T. Nobuto, N. Mataga, and M. Irie: Chem. Phys. Lett. **230**, 249 (1994)
11. N. Tamai, T. Saika, T. Shimidzu, and M. Irie: J. Phys, Chem. **100**, 4689 (1996)
12. J. Ern, A. T. Bens, A. Bock, H.-D. Martin, and C. Kryschi: J. Luminescence **76/77**, 90 (1998)
13. H. Miyasaka, T. Nobuto, A. Itaya, N. Tamai, and M. Irie: Chem. Phys. Lett. **269**, 281 (1997)
14. J. Blanc and D. L. Ross: J. Phys. Chem. **72**, 2817 (1968)
15. G. M. Wyman and B. M. Zarnegar: J. Phys. Chem. **77**, 831 (1973)
16. M. Hamano and M. Irie: Jpn. J. Appl. Phys. **35**, 1764 (1996)
17. T. Tsujioka and M. Irie: Appl. Opt. **37**, 4419 (1998)
18. T. Tsujioka and M. Irie: Appl. Opt. **38**, 5066 (1999)
19. M. Irie, H. Ishida, and T. Tsujioka: Jpn. J. Appl. Phys. **38**, 6114 (1999)
20. T. Tsujioka and M. Irie: J. Opt. Soc. Am. B **15**, 1140 (1998)
21. T. Tsujioka and M. Irie: Jpn. J. Appl. Phys. **38**, 4100 (1999)
22. M. Ohtsu: *Near-Field Nano/Atom Optics and Technology* (Springer, Berlin, Tokyo, New York 1998)
23. M. Ohtsu: J. Lightwave Technol. **13**, 1200 (1995)
24. M. Ohtsu and H. Hori: *Near-Field Nano-Optics* (Kluwer Academic/Plenum, New York 1999)
25. V. V. Polonski, Y. Yamamoto, M. Kourogi, H. Fukuda, and M. Ohtsu: J. Microsc. **194**, 545 (1999)
26. D. Ehrlich, R. M. Osgood Jr., and T. F. Deutch: J. Vac. Sci. Technol. **21**, 23 (1982)
27. V. V. Polonski, Y. Yamamoto, J. D. White, M. Kourogi, and M. Ohtsu: Jpn. J. Appl. Phys. **38**, L826 (1999)
28. R. Wiesendanger: Appl. Surf. Sci. **54**, 271 (1992)
29. M. Fujihira, H. Monobe, H. Muramatsu, and T. Ataka: Ultramicroscopy **27**, 176 (1995)
30. O. J. F. Martin, N. B. Piller, H. Schmid, H. Biebuyck, and B. Michel: Opt. Express **3**, 280 (1998)

31. Y. Kawata, C. Egami, O. Nakamura, O. Sugihara, N. Okamoto, M. Tsuchimori, and O. Watanabe: Opt. Commun. **161**, 6 (1999)
32. C. Egami, Y. Kawata, Y. Aoshima, H. Takeyama, F. Iwata, O. Sugihara, M. Tsuchimori, O. Watanabe, H. Fujimura, and N. Okamoto: Opt. Commun. **157**, 150 (1998)

7 Near-Field Imaging of Molecules and Thin Films

M. Fujihira, S. Itoh, A. Takahara, O. Karthaus, S. Okazaki, and K. Kajikawa

7.1 Near-Field Imaging of Molecules and Thin Films

Characterization of materials on a nanometer scale becomes more important and indispensable year by year [1]. This is due to the development of new materials whose properties and functions are related to the structure and size of materials of the order of nanometers [2]. In this chapter, we will report our recent work on scanning near-field optical microscopy (SNOM [3]), or near-field scanning optical microscopy (NSOM [4]) in connection with chemistry and materials science. This work has been mainly carried out and developed in this project. In the first section, we describe some research into SNOM itself for studies of a variety of chemistries and materials.

7.1.1 Preparation of Organic Thin Films

As an example of materials with a fine structure on a nanometer scale, Langmuir–Blodgett (LB) films have been mainly used in the following [3,5]. Phase separation on a nanometer scale in mixed LB films of hydrocarbon (HC) and fluorocarbon (FC) amphiphiles was first studied with friction force microscopy (FFM) [6]. Details of the methods of preparing the LB films were already described in the literature [3,5].

7.1.2 Control of Tip–Sample Separation

The materials studied are sometimes soft and can be damaged readily during their characterization by SNOM. For this reason, we developed a new feedback mode for controlling tip–sample separation based on the same principle as that of atomic force microscopy (AFM) [3,7,8]. In most SNOM instruments, a shear-force tip feedback mode developed by two groups in 1992 [9,10] is now widely used. The controlling mechanism of the shear-force mode, however, has not been clear. For AFM control, optical fiber probes were bent, and the fibers themselves were used as cantilevers for force sensors based on the optical beam deflection method. To reduce the spring constant for a cylindrical optical fiber cantilever, the diameter of the optical fiber of 125 μm was further reduced by etching in a HF solution [11]. By using this slim optical fiber probe with a diameter of 30–40 μm, the spring constant can be reduced to the order of 1 Nm^{-1}, and the damage to sample surfaces can be reduced dramatically. The most standard control of tip–sample separation can be done by the tapping mode tip feedback.

7.1.3 Various Modes of Observations

A soft optical fiber probe and AFM control of tip–sample separation described above enabled us to develop various modes of simultaneous scanning probe microscopic (SPM) observations of sample surfaces by SNOM. The first type of simultaneous observations was SNOM with FFM [12]. In this method, contact mode AFM control was used for tip–sample separation, and topographic images were obtained simultaneously with frictional and fluorescent images without mechanical damage. To control the tip–sample separation in a pure noncontact mode, scanning Maxwell stress microscopy (SMM) [13] with SNOM was also attempted [14]. In this method, a $2w$ component of the Maxwell stress between the conductive tip and the sample substrate was used as the feedback signal. For Maxwell stress, ac voltage was applied between the tip and the substrate. Here, topographic, optical, and surface potential images were recorded simultaneously. Vapor-deposited metal films, such as gold and platinum, and indium-tin oxide (ITO) films on quartz or glass substrate can be used as the conductive and semitransparent substrate. Local photovoltaic measurements under pulsed illumination were also applicable, as well as SNOM observation [13].

7.1.4 Optical Recording on Organic Thin Films

Near-field illumination can be used for local photochemistry. An aperture of the SNOM probe tip or a photomask with smaller patterns than the wavelength of light was used as the near-field illumination. For small separation between the mask and photochemically active thin film samples, direct contact with the mask on the sample surface was also attempted [15]. Contact in the near-field region was more easily attained by evacuating air between them. Patterns made photochemically were observed by FFM [16] because the illuminated part became hydrophilic [17].

7.2 Two-Dimensional Morphology of Ultrathin Polymer Films

The scanning near-field optical microscope (SNOM) allows one to "see" optical images with high spatial resolution beyond the diffractive limit of light [18]. A variety of configurations of SNOM have been proposed and developed so far to visualize specific interactions in nanometer dimensions between an evanescent electric field and materials. Among them, we employed illumination mode SNOM [19,20]. Dye molecules in the specimen are excited by the evanescent field of light through a small aperture at the end of the raster-scanning optical fiber probe, and the fluorescence from the pinpoint is collected and detected with a photon counting apparatus as a function of the lateral coordinates. Besides recording the topography of the surface like an atomic force microscope (AFM), it can monitor both the morphology and the spectroscopic properties of materials simultaneously, even if the surface is quite smooth at the molecular level. In particular, the fluorescence from the local area pointed to by the fiber probe conveys valuable information on the molecules through photophysical and photochemical phenomena, including exciton dynamics [21].

As an attractive object of SNOM, we have been studying the morphology of two-dimensional polymer monolayers and the phase-separated structure of binary blends of amphiphilic polymers [22]. Polymer monolayers, which are prepared at the air/water interface, have been extensively investigated because the sequential deposition of the monolayers onto a solid substrate enables one to fabricate ultrathin polymer films in an arbitrary sequence [23,24], which are likely to provide new types of functional molecular assemblies for controlling fundamental photo and electronic processes in a nanostructure. We have used various optical techniques such as Brewster angle microscopy [25], energy transfer spectroscopy [26], and surface plasmon spectroscopy to study the properties of ultrathin polymer films [27]. SNOM is expected to provide novel insight into thin film morphology, using the versatile abilities associated with its high spatial resolution.

This report is concerned with a two-dimensional phase-separated polymer blend. The morphology of polymer blends has been widely investigated in three-dimensional bulk systems, but little in two dimensions [28,29]. The monolayer used here consists of a blend of poly(isobutyl methacrylate) (PiBMA) and poly(octadecyl methacrylate) (PODMA). These polymers in themselves are known to form stable monolayers at an air/water interface [30,31]. They were labeled with different fluorescent probes for SNOM. The samples deposited on a glass plate are extremely thin and flat, which also agrees with our purpose, i.e., to demonstrate the potential abilities of SNOM.

7.2.1 Materials, Preparation of Films, and Apparatus

Materials. PiBMA and PODMA were synthesized by radical polymerization of corresponding methacrylate monomers. Pyrene (Py) and perylene (Pe) chromophores were incorporated separately into PiBMA and PODMA by copolymerization with 1-pyrenylmethyl methacrylate or 3-perylenylmethyl methacrylate. The molecular weights of the polymers obtained were 39 000 for PiBMA-Py and 11 900 for PODMA-Pe, in which the mole fractions of chromophores were evaluated at about 2–3% by UV absorption and NMR measurements. Figure 7.1 shows the chemical structures of these copolymers.

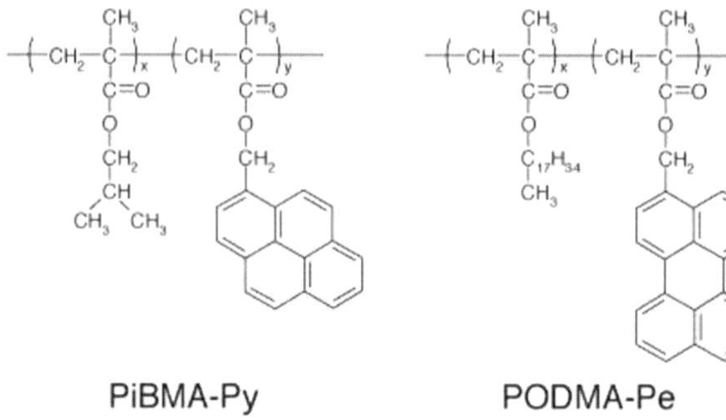

PiBMA-Py

PODMA-Pe

Fig. 7.1. Chemical structures of sample polymers

Preparation. A solution of PiBMA-Py and PODMA-Pe (1:1) was spread on the surface of pure water at 20°C to make a mixed monolayer. After evaporation of the solvent, the monolayer was transferred onto a glass plate or a silicon wafer by vertical dipping. To promote phase separation of the monolayer, the temperature of the subphase was raised to 40°C and kept constant for 60 min. The phase-separated monolayer cooled to 20°C was compressed and transferred onto a glass plate in a similar manner.

Apparatus. Illumination mode SNOM (SP-301, Unisoku) was combined with light sources and photon counting systems, as shown in Fig. 7.2. We employed the 325-nm and 442-nm lines of a He–Cd laser (1K5351R-D, Kimmon Electric) as a major light source for imaging and also the 395–440 nm pulsed second-harmonic light of a Ti:sapphire laser (Model 3950, Spectra

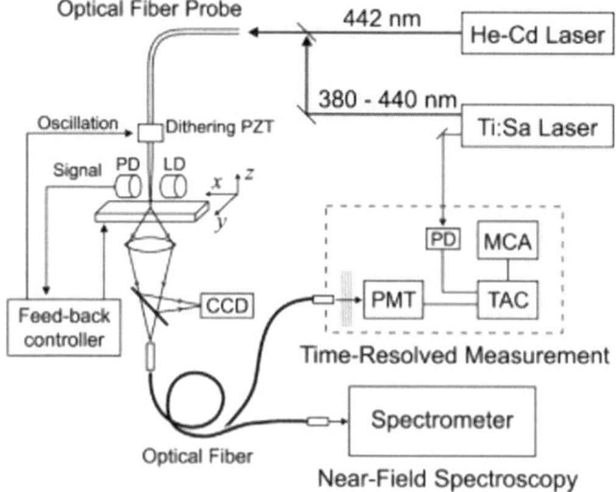

Fig. 7.2. Schematic illustration of the illumination mode SNOM used in this study

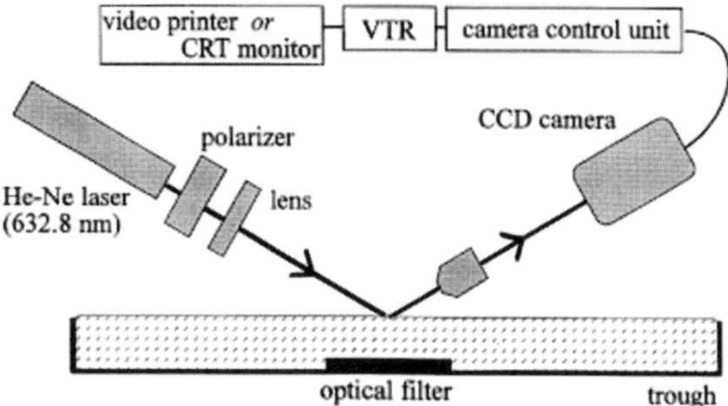

Fig. 7.3. Brewster angle microscope setup

Physics) for the time-resolving measurements in the local area specified by the probe. The excitatory light was coupled with an optical fiber probe from one end; the other end was controlled by a shear-force feed-back system to regulate the distance from the specimen surface to several nanometers. The fiber probe used in this study was fabricated from a pure silica optical fiber by a heating-and-pulling process (P-2000, Sutter Instrument) and then followed by etching in a buffered hydrogen fluoride solution. The use of the pure silica fiber is essential for the present experiments because commercially available fiber probes are not transparent to UV light around 300–400 nm and emit luminescence from the dopants of the glass [32,33]. The fluorescence from the sample was collected by a high NA objective lens and detected with a photon counting system equipped with a photomultiplier (R2949, Hamamatsu). The photon signal was fed into the pulse counter of a computer for imaging and into a constant fraction discriminator and a time-to-amplitude converter for time-resolving. For spectroscopic analysis, the fluorescence was detected through a monochromator (Unisoku). The lateral resolution of our system is typically 50–100 nm. Measurements were performed at room temperature, and no image processing was applied.

The Brewster angle microscope (BAM) is a powerful tool for observing the morphology and dynamic behavior of monolayers on a water surface [34,35]. Although the lateral resolution is limited to ca. 10 μm, it enables one to observe the in situ features on water with high contrast even for thin organic layers 1 nm thick. Figure 7.3 shows the setup of the BAM used in this study. p-Polarized light of a 50-mW He–Ne laser (GLG5800, NEC) at 632.8 nm through a polarizer (Glan-Thompson prism) impinged on the water surface. The angle of incidence was set at 53.1° which is the Brewster angle for the air/water interface. The optical image of the reflected light was magnified with an objective lens ($f = 50$ mm) and introduced in a CCD camera (Hitachi) equipped with a camera controller (C2400, Hamamatsu).

7.2.2 Observation of Two-Dimensional Morphology

Figure 7.4 shows a typical BAM image of a mixed monolayer of PiBMA and PODMA (1:1) at 20°C after the solution is spread on a water surface. On larger surface areas, it forms so-called island structures, i.e., many domains (islands) of the condensed monolayer appear on the surface. By compression to an appropriate pressure, the whole surface is covered with the mixed monolayer which is obviously inhomogeneous due to phase separation of the component polymers. The monolayer is composed mainly of three parts: a white part that consists of a thick PODMA monolayer, a gray part that is a mixture of PiBMA and PODMA but has a structure too fine to be seen with the lateral resolution of a BAM, and a dark part that is mainly the PiBMA monolayer.

The inner structure of the gray domain can be observed by AFM. The monolayer was deposited on a silicon wafer at an appropriate pressure, and the topography was recorded. As Fig. 7.5 shows, a network structure of higher (brighter) lines surrounds the lower (darker) domain. From the known thickness of the monolayers (3.0 nm for PODMA and 1.1 nm for PiBMA) [30,31] and from the heights observed by AFM, it can be safely said that the bright lines represent the PODMA monolayer and the dark domain corresponds to the PiBMA monolayer. This indicates that in the course of evaporation of the solvent, the cohesive PODMA chains gathered and made a network including the liquid-like PiBMA domains inside, and fixed the whole morphology because of the solid character of the PODMA monolayer.

The PODMA monolayer is formed by the strong cohesive force due to the long alkyl side chains that crystallize in film formation. Since the side-chain crystal melts around 35°C, the monolayer changes to a liquid phase at 40°C [36]. Figure 7.6 is again a BAM image for the same mixed monolayer but taken at 40°C. The bright and dark domains are visualized with high contrast, and they are completely separated from each other; phase separation

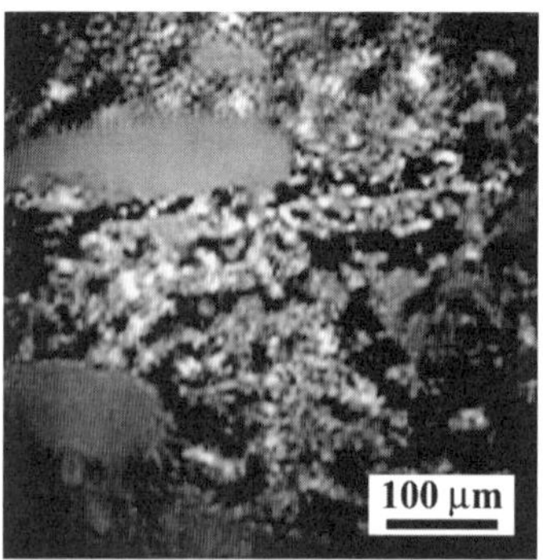

Fig. 7.4. Morphology of a mixed monolayer of PiBMA and PODMA observed by BAM at a surface pressure of 0 mNm^{-1} at 20°C

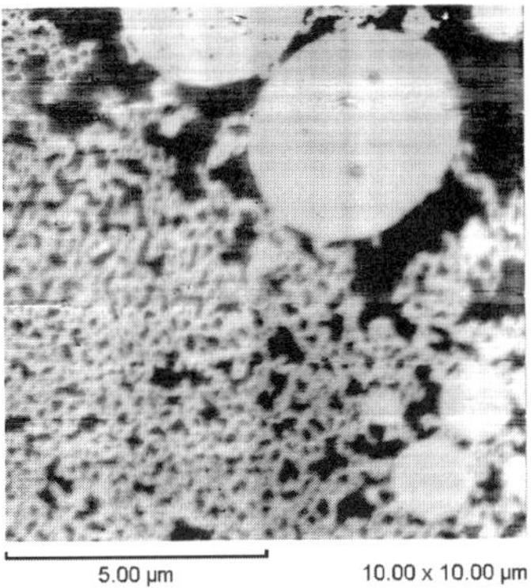

Fig. 7.5. AFM image of a mixed monolayer transferred to a silicon wafer

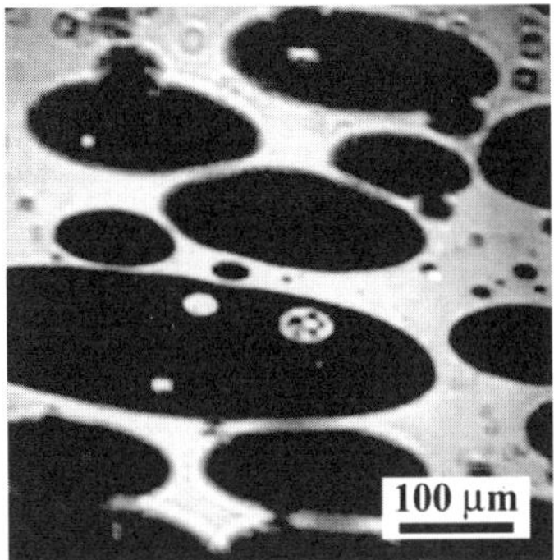

Fig. 7.6. BAM image of a phase-separated monolayer on a water surface after annealing at 40°C for 60 min

was complete. The AFM picture (not shown here) indicated that each domain had a very flat surface and a height difference of 2.0 nm at the boundary, which is in agreement with the thickness difference of these monolayers.

Since these polymers were labeled with fluorescent dyes, the phase-separated morphology could be visualized as a fluorescent image by the scanning of the fiber probe under the illumination of light at an appropriate wavelength. Figure 7.7 shows a pair of SNOM images which were taken by excitation at two different wavelengths; PiBMA-Py was selectively excited at 325 nm in Fig. 7.7a, PODMA-Pe was excited at 442 nm in Fig. 7.7b. Since the mixed monolayer was completely separated into individual domains by annealing at 40°C, we obtained complementary black-and-white pictures in Fig. 7.7a–b. The bright part indicates the domain of the monolayer containing labels selected by excitation, and the other black part is actually dark because no penetration of chains took place toward the counterpart. Every point in this area is bright in either Fig. 7.7a or b, showing that the

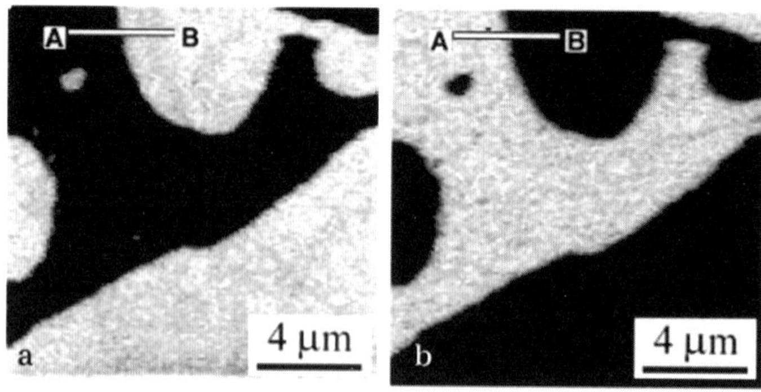

Fig. 7.7. Fluorescent SNOM images of a phase-separated monolayer: **a** taken by the fluorescence of pyrene excited at 325 nm, **b** taken by the fluorescence of perylene excited at 442 nm

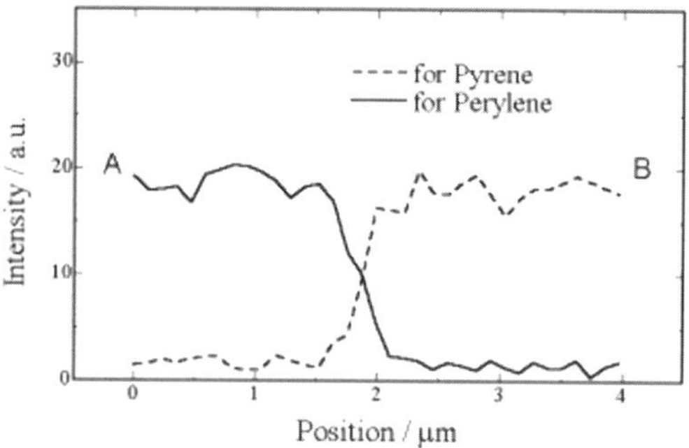

Fig. 7.8. Fluorescent intensity line profiles between the points A and B indicated in Fig. 7.7a,b

monolayer was successfully transferred onto the glass plate without any defect.

Figure 7.8 shows a line profile of the fluorescent intensity across the border of two domains. The intensity remained constant inside each domain, but at the border it suddenly dropped to the base level of dark counts. This again indicates the homogeneous distribution of polymer chains in each domain but separated from each other in the equilibrium state after heat treatment.

Before annealing, however, very complicated images were observed for the gray domain of the BAM image (see Fig. 7.4). Figure 7.9 was taken by excitation of perylene attached to PODMA. The network structure of PODMA was observed here as the bright parts in Fig. 7.9. However, we could find pyrene (PiBMA) emission in the PODMA network and perylene emission (PODMA) as a spot in the PiBMA domain. These facts show that

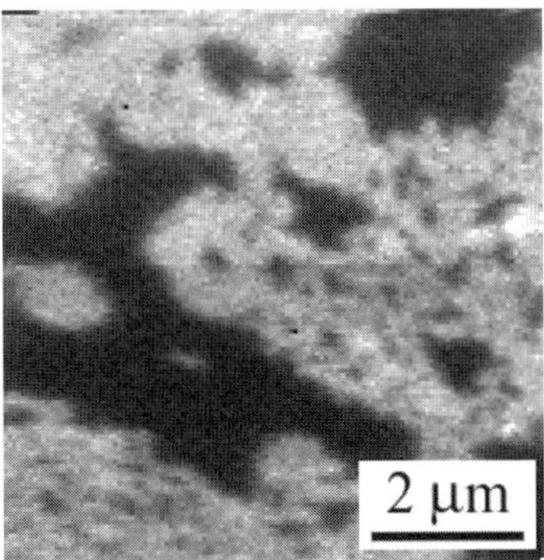

Fig. 7.9. Fluorescent SNOM image of a mixed monolayer before annealing, taken by perylene emission. The excitatory wavelength was 442 nm

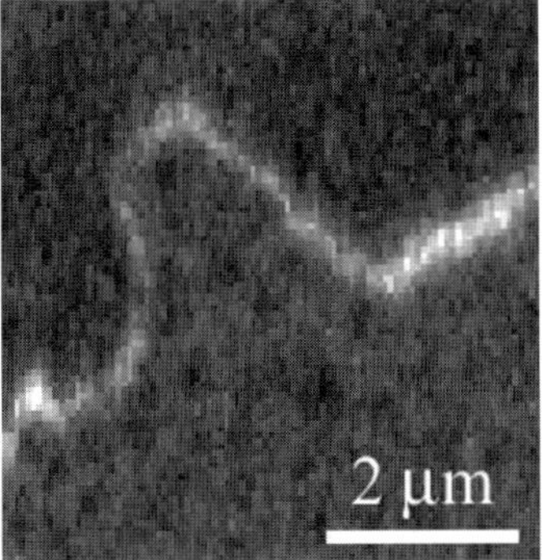

Fig. 7.10. Energy transfer image of a mixed monolayer after annealing. Pyrene was selectively excited at a wavelength of 325 nm, and the fluorescence from perylene was collected through an optical filter (Y-44, Hoya)

the separation is not complete and the other polymer coexists in each domain before annealing.

Next, the energy transfer method was applied to the SNOM measurements. Excitatory energy on a donor (Py) can transfer to an acceptor (Pe) when they are close to each other at a distance of a few nanometers. The critical distance of energy transfer is known to be 3.3 nm for the Py–Pe chromophores, and the transfer efficiency is markedly affected by alteration of the distance in the range of 1–10 nm. Therefore, energy transfer takes place only in the area where PiBMA and PODMA are mixed, and both chains contact within such a short distance. When Py is selectively excited and the Pe emission is recorded by raster-scanning of the fiber probe, the SNOM can map the excitation energy transfer efficiency on the monolayer plane. Figure 7.10 shows an energy transfer image for an annealed sample. This picture was taken by the fluorescence from Pe ($\lambda > 440$ nm) through the excitation of Py

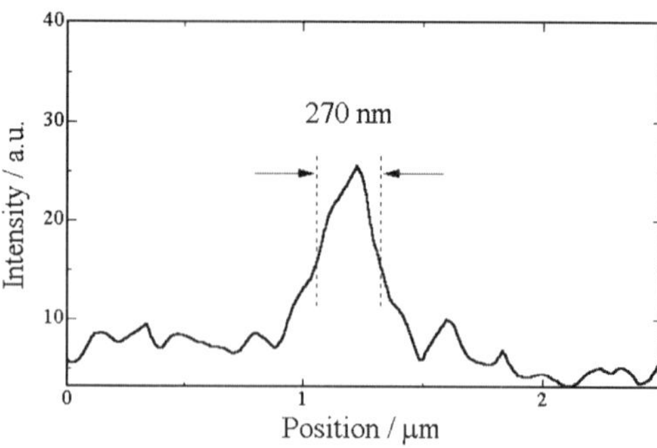

Fig. 7.11. A line profile of perylene fluorescent intensity through energy transfer from pyrene. The intensity increased steeply at the interface between the PiBMA and PODMA phases

at 325 nm. In this image, only the phase boundary appeared as a bright line where both polymers were close together at the molecular level. The other parts besides the line were quite dark because the emission detected was either weak Py fluorescence at longer wavelengths or Pe fluorescence excited directly at its small absorption band. This energy transfer measurement again revealed that phase separation was complete and no polymer chains of one component remained in the phase of the other component. At the interface, however, PiBMA and PODMA were mixed. Figure 7.11 shows the line profile of the fluorescent intensity across the interface. The typical value of the interface width was 200–400 nm, which was defined as the fwhm of the intensity profile. Considering the lateral resolution of the SNOM, the real width of the boundary was estimated at 100–300 nmm which is very large compared to the value observed for the three-dimensional boundaries of the bulk polymer blends. This value was independent of the experimental conditions, e.g., the surface pressure at the deposition and the annealing time, suggesting that the interfacial width is inherently determined by the two-dimensional miscibility of these polymers.

Finally, an interesting ability of SNOM is demonstrated by using a polymer monolayer of PiBMA–Py as a specimen. Since pyrene is apt to fade away in UV light, intense irradiation at a point on the surface makes a dark hole in the fluorescent image. This means that the trace of the fiber probe can be recorded on the monolayer. Figure 7.12 shows the "writing" literally with a fiber probe controlled by the mouse of a computer; the Japanese Kanji letter means "blue." The line width is about 100 nm, on average, which shows the convolution of the writing and reading processes through the evanescent field beneath the probe. Since the specimen is an ultrathin monolayer 1 nm thick, we can get rid of the blur in the depth direction. This indicates that SNOM is a leading candidate for high-density recording in combination with photosensitive thin films.

Fig. 7.12. SNOM image of a PiBMA–Py monolayer. A Japanese letter that means "blue" was written beforehand by 325-nm light from a fiber probe

7.2.3 Conclusion

SNOM studies on polymer monolayers revealed quite interesting morphological phenomena. The phase-separated thin films demonstrated the potential abilities of SNOM well. Both the morphology and the spectroscopic properties of materials can be monitored simultaneously, even if the surface is flat and smooth at the molecular level. In particular, the fluorescence from the local area provides molecular information through photophysical phenomena. The combination of SNOM with the fluorescent method is expected to be a novel and useful technique for exploring nanoscale structures, properties, and functions of materials.

7.3 Observation of Polyethylene (PE) Crystals

Since birefringence reflects the optical anisotropy of materials, it gives information on molecular or crystal orientation. Near-field scanning optical microscopy (NSOM) is a technique that overcomes the diffractive limit of far-field optical microscopy [37]. The structures of molecular aggregates in polyethylene (PE) single crystals and melt-crystallized films have been investigated by far-field light microscopy, transmission electron microscopy, and X-ray diffraction [38–40]. Since NSOM detects optical properties with high resolution, NSOM can reveal the structure of molecular aggregates which cannot be investigated by other techniques. In this study, polarizing NSOM was used to investigate molecular or crystal orientation in PE single crystals [41–43] and PE spherulites [44].

7.3.1 AFM and NSOM Observation of PE Single Crystals

The morphologies of polymer thin films demonstrate organization on a number of length scales from the chain folding of molecules to lamellae to the organization of spherulites or fibers. The crystalline lamella is a fundamental structural unit of crystalline polymers. Figure 7.13 shows a typical structural model of a polyethylene single crystal. Single crystals of linear polyethylene

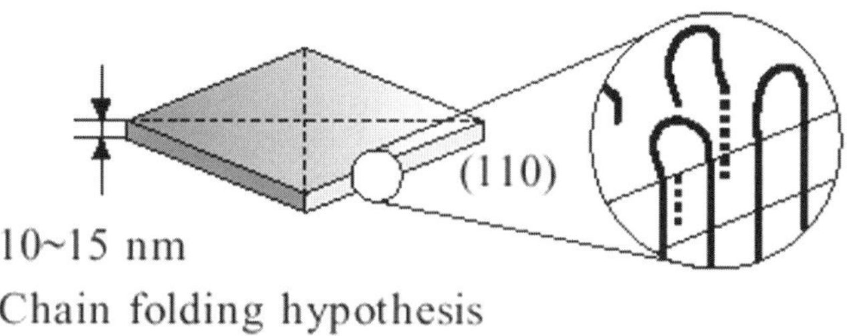

Fig. 7.13. Typical structural model of a polyethylene single crystal

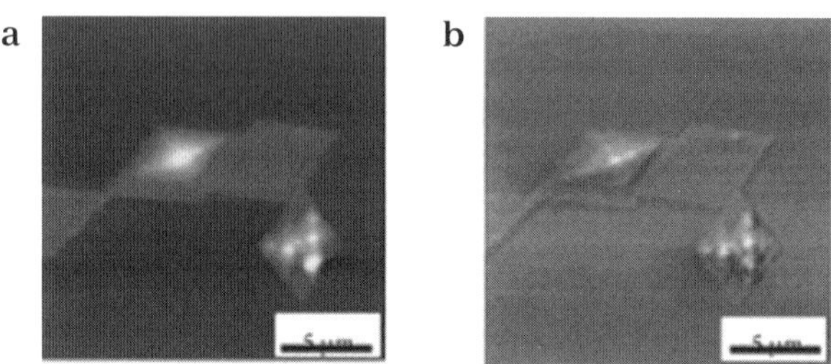

Fig. 7.14. The shear-force topographical **a** and polarized NSOM **b** images of the HDPE (Mw = 32 k) single crystal

have been prepared from dilute solutions in xylene and similar solvents. The c axis is oriented perpendicularly to the surface of a single crystal. The large surfaces that contain the chain folds are commonly referred to as fold surfaces. A PE single crystal was prepared from a dilute solution in *p*-xylene. The PE used was high-density polyethylene (HDPE) with a narrow molecular weight distribution. HDPE (weight average molecular weight, Mw = 32 k) was dissolved in p-xylene at 411 K to obtain a 0.01 wt% solution. The solution was kept at 339 K for 30 min. Then, the solution was heated again to 368 K and crystallized at 343 K for 48 hrs. A small amount of the solution with suspended single crystals was dropped onto a cover glass and allowed to air dry at room temperature. NSOM was used to observe of the HDPE single crystals. A beam from an Ar laser (488 nm) that passed through a one-quarter wave plate was launched into the optical fiber probe. The aperture system acts as a one-quarter wave plate. The observation was done under crossed Nicols. The measured extinction ratio for this setup without a specimen was between 20:1 and 12:1.

Figure 7.14 shows the shear-force topographical (a) and polarized NSOM (b) images of the HDPE (Mw=32k) single crystal. A characteristic lozenge-shaped lamellar crystal ca. 5 µm long was observed. Some of the single crystals showed spiral growth, which looked like a pyramidal shape. The NSOM image contrast seems to correspond to that of the topographic image. The intensity at the apex of the spiral growth was strongest in the NSOM image. The light transmittance from the part close to the apex of the spiral growth should be low, since the thickness of that area is larger than the periphery of the HDPE single crystal. The bright part in the NSOM image suggested the formation of a birefringent part during spiral growth or local deformation upon solvent removal. The HDPE crystal has a uniaxial refractive ellipsoid, and it is obvious that the PE chain is oriented normally to the PE single crystal surface. Therefore, the HDPE single crystal might be optically isotropic in an incident polarized light. The presence of the highly birefringent part might suggest tilting of the c axis from the perpendicular direction to the substrate.

7.3.2 AFM and NSOM Observation of Melt-Crystallized PE Thin Films

When polymer films are prepared by melt crystallization or by drying concentrated solutions, spherulites are a common morphological feature. The PE thin film was prepared by melt crystallization. The polyethylene used was HDPE (Marlex 9) with a weight average molecular weight, $M\mathrm{w}$ of 520 k. HDPE was dissolved in *p*-xylene at 413 K to prepare a 0.5 w/w% solution. The solution was placed on a cover glass slip which was heated to 423 K and the *p*-xylene was allowed to evaporate. Then, the HDPE spherulites were grown at the crystallization temperature of 373 K for 90 h. The spherulitic structure was observed by POM under crossed Nicols. The surface structure of PE spherulites was observed with an atomic force microscope (AFM: Seiko SPA300) under contact-mode operation. The local birefringence of the PE spherulites was observed by polarized NSOM.

Figure 7.15 shows the POM image of a PE spherulite. A large spherulite with a diameter of ca. 90 μm was observed. The boundaries of the spherulites are nonspherical but smooth after impingement. The straight boundary between the spherulites suggests that all of the spherulites were nucleated simultaneously. The extinction cross which corresponded to the directions of polarizer and analyzer was clearly observed. In addition, one can find light extinctions periodically along the circles. The regular variation in light intensity as a function of radial distance is due to a corresponding variation in the orientation of the refractive index ellipsoid. The sample appears isotropic where the axis of the refractive index ellipsoid is parallel to the incident light beam, i.e., optical axis. The maximum intensity of the transmitted light is observed at locations where the long axis of the refractive index is perpendicular to the incident light beam. Figure 7.16 shows the AFM image of the melt-crystallized PE thin film prepared under the same conditions as those of Fig. 7.15. AFM was operated in the contact mode. The bright region corresponds to the higher height region. The bright part observed in Fig. 7.16 corresponds to the nucleus of the spherulite. Peaks and valleys are observed in

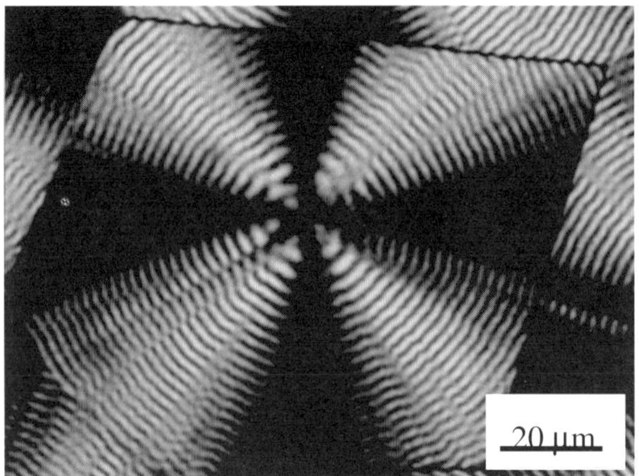

Fig. 7.15. POM crystal image of a melt-crystallized PE thin film

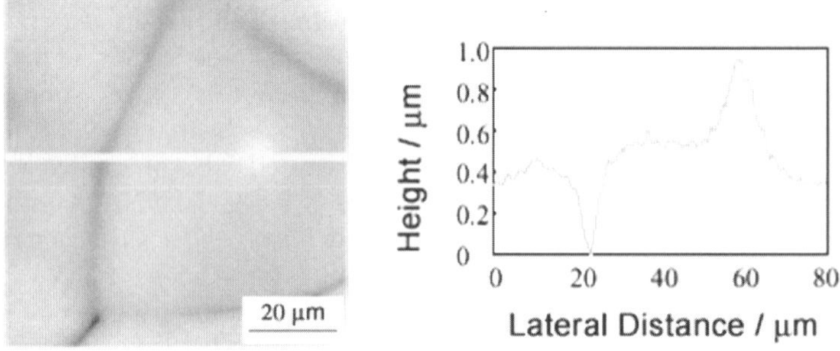

Fig. 7.16. AFM image and height profile of a melt-crystallized PE thin film

the line profile in the direction of the spherulite radius. The distance between the peaks in the line profile along the spherulite radius is in accordance with the extinction ring interval visible in the POM image shown in Fig. 7.16.

A polarizing NSOM observation was carried out for the melt-crystallized PE thin film prepared under the same condition as those of Figs. 7.15 and 7.16. Figure 7.17 shows the polarizing NSOM, the shear-force topographic images of the PE spherulite, and the line profile of the height and light intensities. The extinction cross that corresponded to the directions of polarizer and analyzer was also clearly observed. However, the part of the dark cross which corresponded to the vibration at direction of light at the exit of the NSOM aperture is not darker than that in the direction of the analyzer. This is due to the depolarization effect at the exit of the NSOM aperture. In addition, the NSOM image showed light extinctions periodically along the circles.

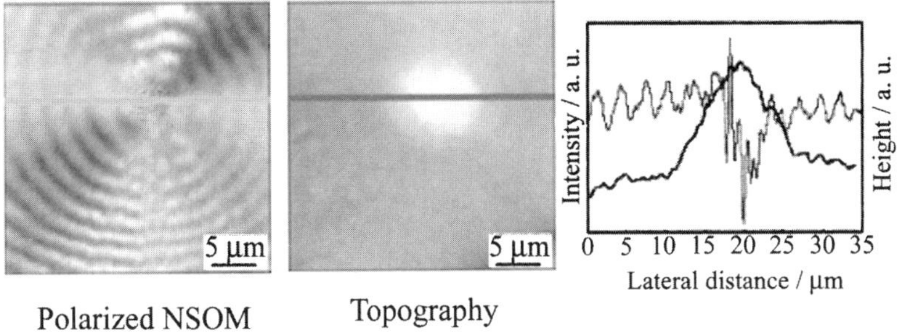

Fig. 7.17. Polarized NSOM, shear-force topographical images, and the line profile of light intensities and height for a melt-crystallized PE thin film

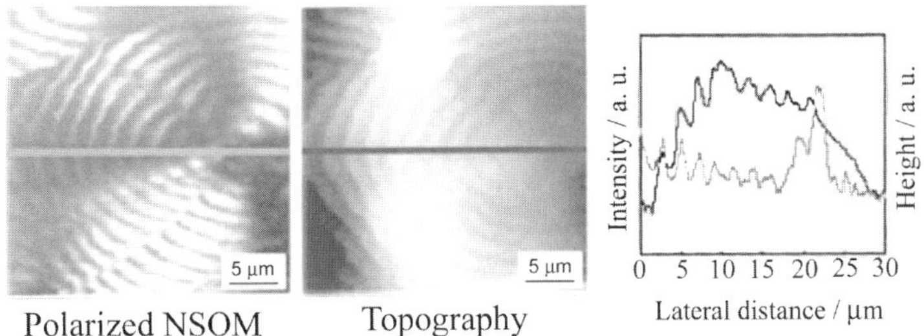

Fig. 7.18. Polarized NSOM, shear-force topographical images and the line profile of the light intensities and height of melt crystallized PE thin film after etching

This coincided with the extinction ring observed by POM. The comparison of the light intensity change and the height change in the radial direction of the spherulite revealed that the period of transmitted light intensity change agreed with that for the height change. These results apparently indicate that the higher part of the topographic image corresponds to the bright part in the NSOM image in which the c axis of the HDPE lamellar crystal is parallel to the film surface.

To reveal the internal structure of the spherulite, the melt-crystallized PE film was etched with a permanganic reagent [45] that consists of 0.8% weight of potassium permanganate in concentrated sulfuric acid. The melt-crystallized PE film on the glass substrate was immersed in the etching reagent at 293 K for 1.5 min. The surface was rinsed with dilute sulfuric acid, water, and acetone and dried. Figure 7.18 shows the polarizing NSOM, shear-force topographic images of the etched surface of the melt-crystallized PE and the line profile of the height and light intensities. A higher height region at the center of the spherulite was removed by etching. The surface corrugation also becomes clear after etching. It is well known that the etching rate of the amorphous phase is higher than that of the crystalline phase. This

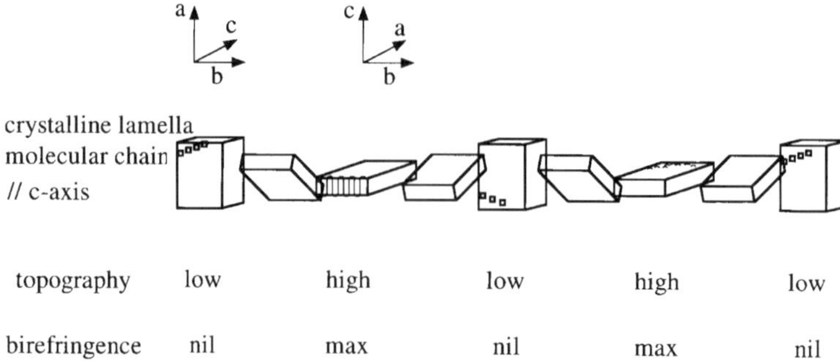

Fig. 7.19. Schematic representation of the relationship between birefringence and surface topography

indicates that the amorphous phase was selectively removed from the surface of the melt-crystallized PE film. A comparison of NSOM images before and after etching suggested that the influence of the surface amorphous phase was negligible in the NSOM image. A relationship between the NSOM image intensity and height change in the radial direction of the spherulite was carefully investigated before and after etching. It was revealed that the higher height region corresponded to the highly birefringent region. Figure 7.19 shows a schematic representation of a helicoidal lamella ribbon grown in the radial direction of the spherulite. The surface topography and local birefringence observed in each position in the radial direction of the spherulite are also shown in Fig. 7.19. The surface of the film is parallel to the plane of this figure. The period of the light intensity change corresponds exactly to the distance between peak and peak or valley and valley in the topographic image. The local birefringence measurement by polarizing NSOM also revealed that the bc plane is parallel to the film surface at the peak of the corrugation. On the other hand the ab plane is parallel in the surface in the valley of the corrugation.

The temperature dependence of the period of the banded structure is discussed on the basis of the morphological observation of PE film melt-crystallized at different temperatures. Figure 7.20 shows the shear-force topographic and polarizing NSOM images of the etched surface of the PE melt-crystallized at 375 and 377 K. Even though the difference in crystallization temperature is 2 K, the difference in the period of the banded structure is clearly observed in the NSOM image. A plot of the period of the banded structure observed by three different methods versus crystallization temperature is shown in Fig. 7.21. The period of the banded structure increased with an increase in crystallization temperature. The results obtained from POM, AFM, and polarizing NSOM agreed well. It has been reported that the lamellar thickness increases with an increase in crystallization temperature. An increase in the period might be related to the difficulty of the twist-

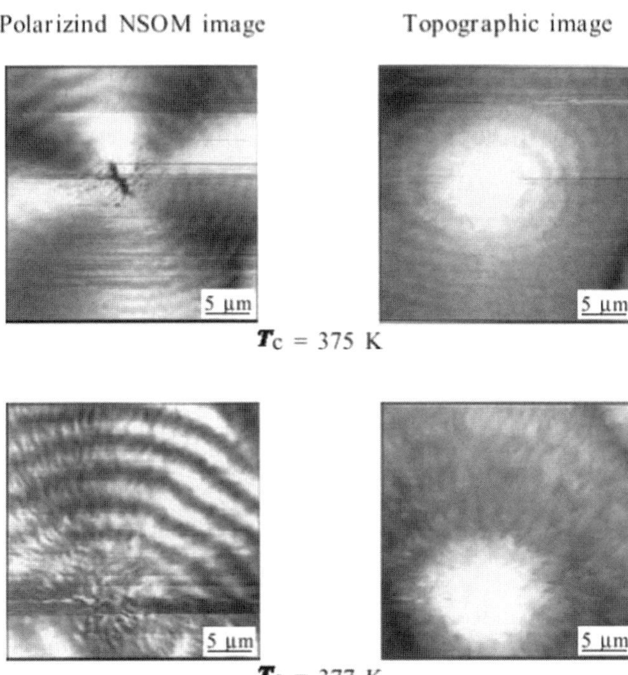

Fig. 7.20. Polarized NSOM and shear-force images of PE melt-crystallized at 375 and 377 K

ing of the lamella due to the increase in lamellar thickness with increasing crystallization temperature [46]. The peak-to-valley height difference in the periodic concentric circles also increased with increasing T^c. At $T^c > 385$ K, no period was observed in the AFM images. This might be related to the lamellar thickening and the lower effectiveness of the space filling in the interior region surrounded by twisted lamellae. In polyethylene spherulites, multilayered crystalline lamellae are oriented in the radial direction and fill the interior space by branching. It has been reported that the degree of the lamellar branching decreases with increasing T^c, which means that the morphology changes from the spherulitic to the axial structure [47]. Therefore, it was suggested that the morphology of melt-crystallized PE thin films changed from the spherulitic to the axial structure in the crystallization temperature region of 371 K $< T^c <$ 385 K.

7.3.3 Conclusions

Polarizing near-field optical microscopy (NSOM) was applied to the morphological observation of polyethylene single crystals and melt-crystallized

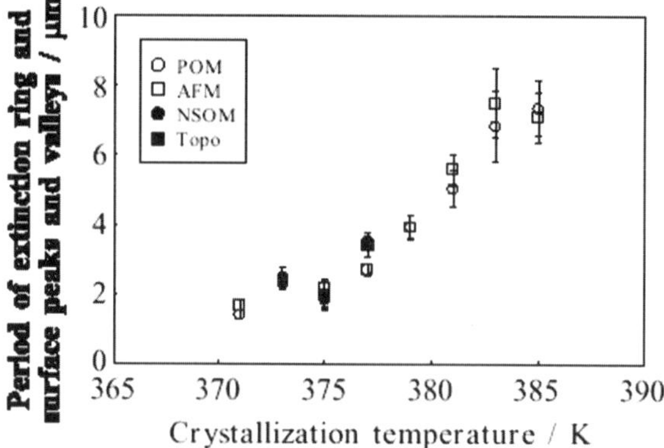

Fig. 7.21. Crystallization temperature dependence of the period of the banded structure of PE spherulite

PE thin films. Even though the birefringence of PE crystals was small, the local structure of the lamellae was clearly observed in detail by polarizing NSOM. The period of the banded structure observed in the thin film by polarizing NSOM agreed well with that obtained by far-field polarizing optical microscopy and AFM observation.

7.4 Preparation of Micrometer-Sized Chromophore Aggregates

7.4.1 Control of Aggregation

Chromophores may show a drastic change in electronic absorption spectra and/or the emission spectra depending on their environment (e.g., solvatochromism). The aggregation of dyes can lead to strong shifts in absorption and emission spectra. Usually the aggregation is controlled by modifying the intermolecular van der Waals and dipole interactions. Lateral substitution of the dyes with bulky groups can weaken the interaction and has the tendency to reduce aggregation. In the same way, incorporation of dyes into a matrix can be used to control aggregation. The first approach shows its effect in solution as well as in bulk and in thin films, whereas the second approach can be used only in solvent-free systems, i.e., in bulk or in thin films. Organization of dyes, may it be in solution, bulk, or in thin films, incorporates the molecules in a continuous medium. Thus, the rotational orientation and/or the spatial "addressability" is poorly controllable. Furthermore, diffusion through the medium (solution, bulk, or thin film) makes it difficult to control the exact size of the aggregates and the interaggregate spacing. Here, we present our work on the self-organization of dyes into micron-size structures and the effect of the aggregate state on the size of the structures.

7.4.2 Mesoscopic Patterns

We can already show that by using a dewetting process during the casting of dilute polymer solutions on various substrates, a variety of mesoscopic patterns can be produced. The patterns can consist of dot or line structures, in which each structure is typically between 5 and 100 nm high, between 10 and 100 µm wide, and has a separation of 1–100 µm. The size distribution in each sample is very narrow and the aggregates form highly ordered two-dimensional patterns of equidistantly arranged dots or lines (Fig. 7.22). The preparation of the patterns is achieved by evaporating a solution droplet on a substrate or by a dip-coating process, in which a substrate is pulled out of a polymer solution.

7.4.3 Mechanism of Pattern Formation

The mechanism of pattern formation involves hydrodynamic instabilities during the casting or dipping process. Microscopic in situ observation during the casting process reveals that the edge of the solution droplet develops a fingering instability, in which the normally straight contact line of the liquid with the substrate shows a regular undulation. Fluorescent microscopy of a solution of a fluorescent-labeled polymer shows that each "finger" contains the polymer in a higher concentration than the interfinger space and the homogeneous bulk in the center of the solution (Fig. 7.23) [48]. Fingering instability is well known in fluid dynamics and can be caused by a temperature gradient perpendicular to the contact line [49] or a concentration gradient in the solution [50,51]. The temperature gradient – or alternatively the concentration fluctuations – are caused by the evaporation of the solvent, which occurs fast compared to the diffusion of the molecules, so that a nonequilibrium state is created and maintained in the solution during solvent evaporation. The

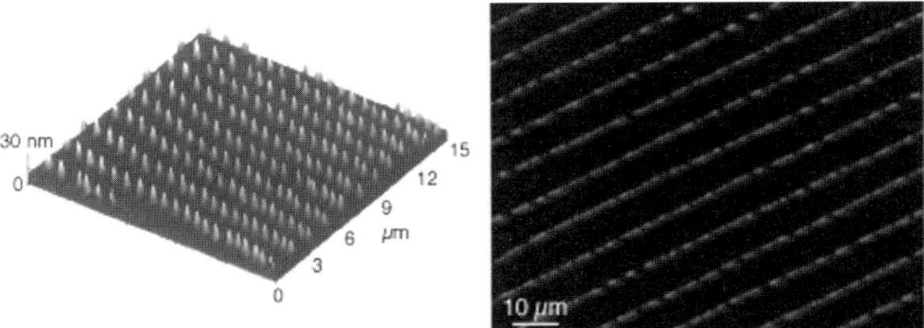

Fig. 7.22. AFM image of a polystyrene dot pattern on mica and a line pattern of a polyion complex

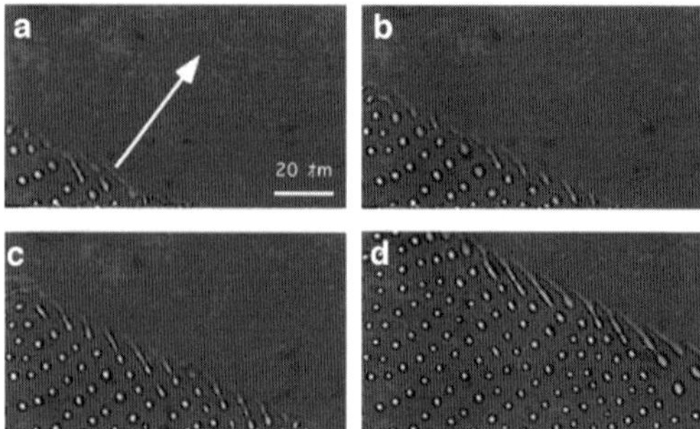

Fig. 7.23. Series of video frames of a microscopic view of the edge of an evaporating benzene solution of polystyrene. The white arrow indicated the receding direction of the three-phase line. The time difference between frame **a** and frame **d** is 0.8 s

polymer is deposited onto the substrate because evaporation of the solvent reduces the volume of the solution, and thus leads to receding of the contact line (Fig. 7.23), where the areas between the fingers recede faster than the fingers themselves. During the receding, a "bottleneck" formation at each finger leads to the generation of a dot pattern, whereas smooth receding of the contact line leads to the deposition of the polymer in a continuous line originating from each finger.

Since the pattern is caused primarily by hydrodynamic instability, the chemical nature of the solute plays only a minor role in determining the pattern. Recently, we showed that a wide variety of compounds form mesoscopic patterns [48,52–62].

7.4.4 Chromophore-Containing Mesoscopic Patterns

In the following, we show that the confinement of molecules in (sub)micrometer-size droplets can also be used to control the aggregative state of chromophores. Besides the two more classical methods of chemical modification and matrix incorporation (see Chap. 7.4.1), the mesoscopic size effect is a third way of controlling chromophore aggregation, which will lead to a new and deeper understanding of chromophore aggregation and novel aggregate structures not observed in continuous films.

Two different samples are chosen to demonstrate the incorporation of polymeric chromophores into two different mesoscopic patterns:

1. a dot pattern of an azobenzene-containing polyion complex and the aggregate structure dependence on the mesoscopic dot-size and
2. a line pattern of poly(hexylthiophene) and its near-field optical characterization.

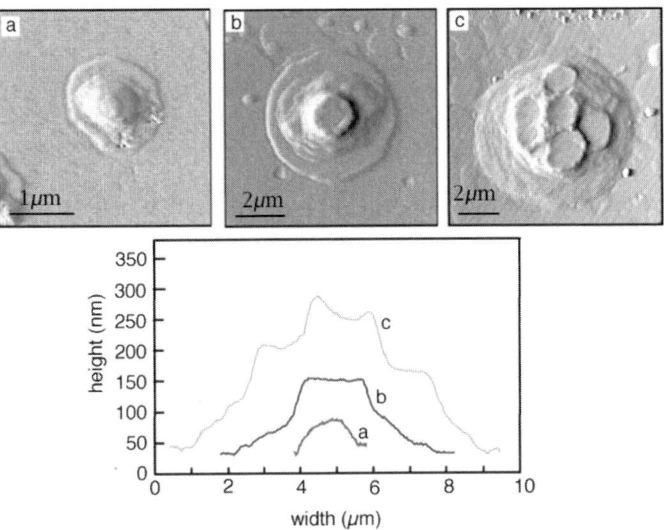

Fig. 7.24. Atomic force microscope images of three different azobenzene droplets. The graph below the pictures shows the cross section of each droplet

7.4.5 Azobenzene-Containing Polyion Complex

Droplet samples were prepared by casting a dilute polymer solution of the polyion complex onto a solid substrate, with subsequent annealing at 130°C for two hours. The sample that contained chromophore droplets of various sizes was observed by optical microscopy and AFM. Above a critical size of ca. 2 μm, the droplets show birefringence and liquid crystal domain formation that was confirmed by polarizing microscopy. Atomic force microscopy investigation of droplets of various sizes showed that, below a critical height of 50 nm and a diameter of 2 μm, the shape of the droplet is more or less spherical. In larger droplets with an intermediate height range around 100 nm, a central single domain with a flat surface was observed. Optical microscopy confirmed that it is an optically homogeneous monodomain of an azobenzene aggregate. Above a height of 200 nm, the droplets contain multiple domains, and each domain is optically homogeneous. The edge of each droplet consists of an amorphous and unstructured rim 50–100 nm high that shows no birefringence in the polarizing microscope.

It is already known from observing of thick continuous cast films that the polyion complex shows a temperature- and humidity-dependent bistable aggregation. A parallel arrangement of the chromophores (so-called H-aggregation) is accompanied by a blue shift of the absorption, whereas a tilted arrangement of the chromophores lead to a red shift [63–66]. Based on these results, we investigated the influence of the mesoscopic droplet size on the aggregation and the bistable switching between H and tilted aggregates. Since the size in which a change of aggregation state should be observed is in the micrometer range, either far-field microscopic or near-field techniques have to be employed. A far-field microscope approach was chosen for the azobenzene chromophores because the droplets are well separated and the 5-μm lateral resolution of the microscope spectrometer is sufficient for the spectroscopy of single droplets.

The domain formation in the mesoscopic droplet structures concurs with the aggregation of the azobenzene into tilted aggregates, as observed in the UV-vis spectra of single droplets obtained by a microscope spectrometer. Before annealing, the UV-vis absorption spectrum shows an absorption maximum at 330 nm. After annealing, a shift to a longer wavelength of the 345-nm aggregation occurs. Treatment with humidified air via a nozzle leads to a decrease of the 345-nm absorption and to a new absorption maximum at 313 nm, which is attributed to H-aggregation (Fig. 7.25). By comparing the course of the spectral change of several drops, a dependence of the final UV-vis spectra on the drop size was observed. For large drops, with a large absorbance in the tilted aggregate, a complete rearrangement into H-aggregates was found (Fig. 7.25a). In smaller drops, with a diameter comparable to the threshold size for domain formation, an incomplete rearrangement was predominant, and the characteristic sharp absorption maximum at 313 nm was not observed, even though the absorption of the tilted aggregate diminished (Fig. 7.25b). Figure 7.25c shows the time-dependent relative absorbance of the two absorption regions at 345 nm and 313 nm. Several drops were investigated, and it was found that drops with an initial absorbance maximum below 0.45, i.e. small drops, show an incomplete H-aggregation, i.e., the final ratio between the absorption at 342 nm and 312 remains larger than unity. Larger droplets show an absorption coefficient smaller than unity and a clear sharp H-aggregate in the spectra. This results indicate that indeed a size effect in the humidity-induced aggregation switching is observed.

The above results show that a distinct size dependence of the morphological and optical properties of a chromophore containing droplets exists on the mesoscopic scale. Since the size of the droplet, below which no chromophore aggregation occurs, is very large compared to the molecular size, the disruption of aggregation is most likely due to surface effects of the substrate or due to the surface tension of the droplet. The AFM images indicate that the first few layers of molecules on the substrate are in an amorphous state. Only when the size of the droplet is sufficiently large, can the chromophores that are more distant from the substrate form aggregates. Even though the switching of the chromophores from the H- to a tilted aggregation is controlled by direct intermolecular interactions on a nanometer scale, a mesoscopic size effect can be obtained by the fact that the chromophores show this peculiar aggregative behavior.

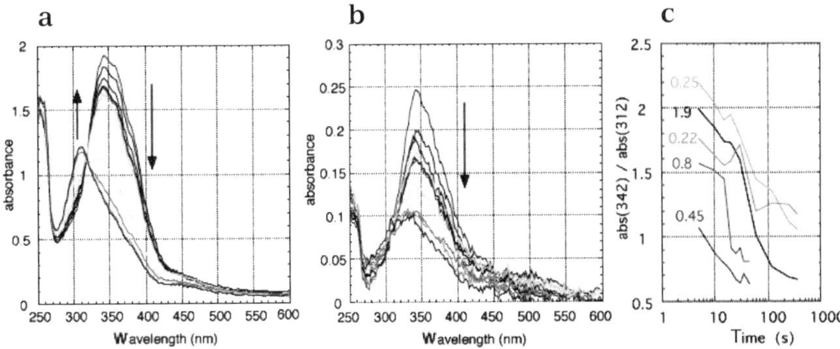

Fig. 7.25. A series of UV-vis spectra of single azobenzene droplets. Spectra were taken before and during treatment with humidified air (90% r.h.) at 5 s intervals. **a** A droplet with a diameter of more than 5 μm. **b** A droplet with a diameter of 3 μm. **c** Course of the spectral changes. The numbers at the left side of each line are the initial absorbance at 342 nm

7.4.6 Mesoscopic Line Pattern of Poly(hexylthiophene)

Patterning of conductive polymers is of interest for electronics and optoelectronics. We can already show that mesoscopic line patterns of poly(hexylthiophene) (PHT) can be produced and that the electric conductivity and photoconductivity can be measured by contact with micromanipulator-controlled microelectrodes [57,61]. Since the conductivity depends very much on the orientation of the polymer backbone and on the domain structure of the polymer in its mesophase, the homogeneity of the line structure has to be confirmed. The small size of the lines, approximately 1 μm wide and a few tens to hundreds of nanometers high, renders optical microscopy unsuitable for investigation. Instead, near-field scanning microscopy is the method of choice. Regioregular PHT shows a broad absorption maximum between 400 and 600 nm. Upon excitation with actinic light between 510 and 550 nm, a narrow fluorescent emission band can be seen at 640 nm.

The measurement system consisted of an Ar-ion laser with a 488-nm wavelength for excitation, coupled to a Lumina (Topometrix) scanning near-field optical microscope that was mounted on a Nikon TE 300 microscope. The fluorescence was detected by a photomultiplier after passing through a cut-off filter (> 550 nm). Figure 7.26a shows the SNOM trace of a single PHT line. The fluorescent intensity is homogeneous along the line and shows no polarization. The grain structure in the line is most probably due to noise because it has the same characteristics (grain size and "roughness") as the background level in the upper left corner. The plot of the fluorescent intensity profile of the line is shown in Fig. 7.26b. The width of the line at half intensity is 1.5 μm, which is only half of the simultaneously measured topographic width (data not shown). The discrepancy can be explained by the

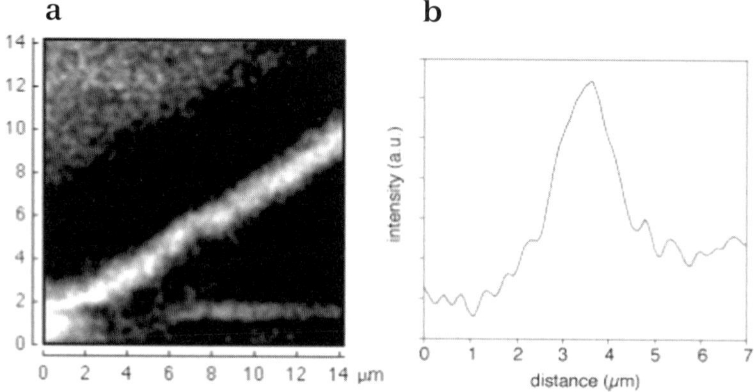

Fig. 7.26. **a** Near-field scanning optical micrograph of a mesoscopic line of poly(hexylthiophene). **b** Plot of the fluorescent intensity across the line

convolution of the topographic image with the blunt SNOM probe and also by the fact that the rims of the line emit much less fluorescent light, since they are much thinner than the center part.

With the present results, we can show that far-field microscope-spectroscopy and near-field imaging provide important insight into the morphology of mesoscopic chromophore droplets, into their aggregate structure, and into the bistable switching of tilted aggregates to H-aggregates.

7.5 Application to Electrochemical Research

7.5.1 Fabrication of an Aluminum Nanoelectrode SNOM Probe to Stimulate Electroluminescent (EL) Polymers

Recently there has been growing interest in EL polymers in the international scientific community [67]. A number of orginal applications of these films are of extreme interest for several reasons. EL polymers can emit light by simple electric stimulation. The mechanism of light emission is, in principle, similar to that of LED devices where high mobility charge carriers are recombined in a process that induces photoemission. The great advantage of EL polymers is that these materials have all of the mechanical characteristics of polymers: they are flexible and easy to produce in any possible forms. Application as flexible display screens, luminous paint, and others are already foreseen by many research groups around the world [68,69].

The structural organization of the polymer was studied until now in a bulk condition with conventional optical diffraction-limited devices. For the first time in our laboratories we are studying the emission pattern of these materials with a near-field optical device modified to act as a nanoelectrode on the surface of the polymer. The light emitted by the film is collected

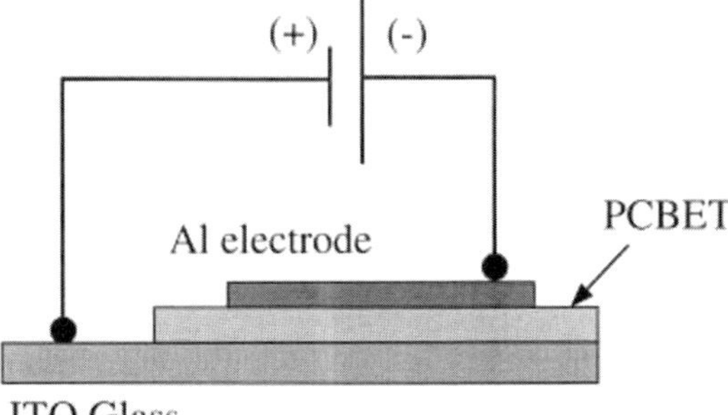

Fig. 7.27. Typical configuration for studying the optical emission of an EL polymer

directly by a sensor placed under the tip, in the same way as a conventional SNOM illumination-mode setup. To measure the in situ emission properties of the EL sample, we created an original SNOM head that hosts a conductive optical fiber (Jasco corp.). The fiber was electrically connected to a power supply to locally polarize the polymer and induce electroluminescence. We intended to use the localization power of SNOM systems to study the emission properties of these polymeric materials. We used a highly efficient EL polymer developed by Prof. Haiwoon Lee of Hanyang University. For optical emission, it is necessary to induce polarization on the surface by two electrodes on the film (called PCBET). Usually this procedure has been performed by depositing a polymeric film on conductive ITO glass that acts as the first electrode. A macroscopic aluminum layer is deposited on the film itself as the second electrode (Fig. 7.27). In this way, applying a suitable voltage between the electrodes induces a current flux in the film, and then electroluminescent phenomena occur.

The direct measurement of the optical emission is possible, in this case, only on a macroscopic scale in the part of the film enclosed between the two electrodes.

The original new procedure we developed is based on a SNOM setup for examining the optical emission of the EL polymer in any area of the film. The film is simply deposited on ITO glass without the necessity for using any other electrode. We insert the sample in a conventional SNOM sample holder. The SNOM tip is metal coated and connected through a thin wire to an external power supply. The ITO glass is connected to the other pole of the power as shown in Fig. 7.28. We established a conventional shear-force feedback condition to produce a localized nanometric scale contact between the polymeric sample and the probe. In this condition optical electroluminescence is induced, and the emission is highly localized in a nanometric area. We observed a weak optical EL signal that was mapped in a standard SNOM mode (Fig. 7.29b). The signal was taken by a photomultiplier detector located underneath the sample. The emission can be investigated in any desired place

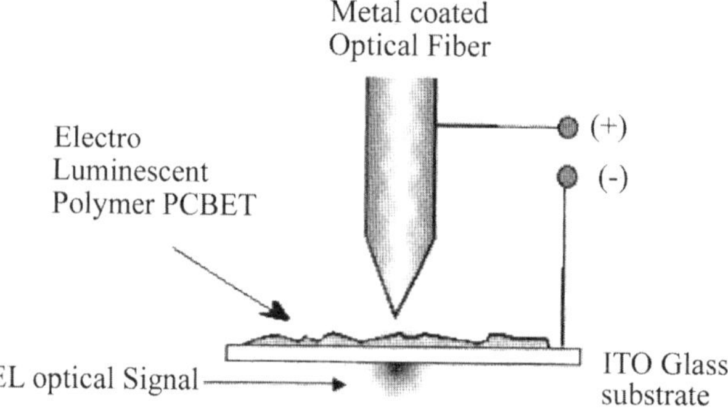

Fig. 7.28. Scheme of the core of our SNOM for electroluminescence. The probe is electrically connected to a 12-volt power supply. EL optical emission is then collected from the bottom of the sample

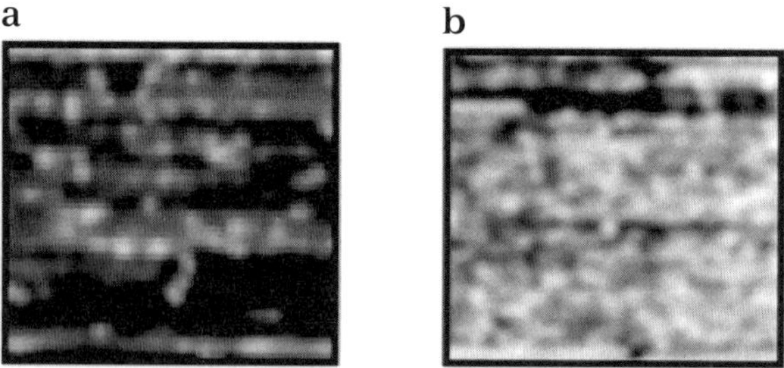

Fig. 7.29. **a** Shear-force image of PCBET surface. **b** SNOM picture of its local EL emission. Images are both 5×5 µm, potential is 10 volts

on the sample, and real time electrical research can study the emission and spectra properties of EL in terms of electrical parameters. We believe that this new approach to the study of EL materials is extremely promising for the analysis of their properties on a nanometric scale.

7.5.2 Integration of STM with SNOM Microscopy by Fabricating Original Chemically Etched Conducting Hybrid Probes

We developed a new type of Hybrid SNOM/STM system. The apparatus is designed to host particular fabricated metal-coated optical fibers that can act both as STM and SNOM probes. With this system, we can investigate the structure of different kinds of conductive samples with the possibility of performing a simultaneous SNOM experiment. The hybrid SNOM-STM probe can also be used to stimulate optical structural modification on particular photosensitive materials and analyze in real time with the STM the effects

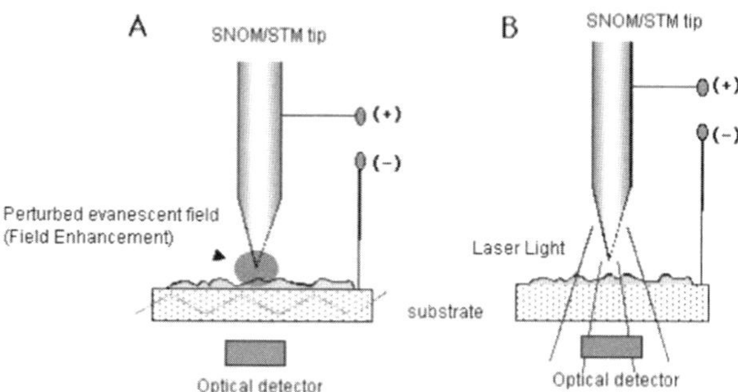

Fig. 7.30. Two optical configurations of the system

on the surface at nanometric resolution. In Fig. 7.30, we show two typical configuratiosn in which the system is set up to operate.

To investigate a sample surface in the STM mode and in the near-field mode, it is necessary to realize a hybrid probe that can act both as a conductive STM tip and as an optically-transparent SNOM tip. We used commercially available 3% doped glass fibers. We chemically etched the probes by using an NH4F:HF:H20 (ratio 10:1:1) solution and a film of silicon oil at the air/solution interface. The glassy fiber is inserted in this solution and etched for 18 hours. The "meniscus" formed at the oil/acid interface creates a sharp pencil-shaped probe with a cone angle of about 20°. After this procedure the tip is metal coated by sputtering it with gold. The fiber is rotated along its axis by 180° and coated again to obtain complete and uniform coverage. The very top of the tip has a small aperture where light can pass through for near-field optical detection. The tip itself is conductive for the STM experiment. The fiber holder is originally constructed to allow mechanical stability and good electrical conduction. A schematic of the holder is shown in Fig. 7.31.

The conductive metal-coated glass fiber is inserted in the probe holder, and tunnel current is induced between the external metal coating and the sample which is deposited on a conductive substrate. Feedback conditions were realized easily, and high resolution imaging has been achieved. In Fig. 7.32, we show two STM mappings of a gold surface obtained with this original glass tip. An optical imaging system and optical detector electronics are under development. With this system, we want to investigate the morphological and optical properties of several photochemical materials. The hybrid STM/SNOM tip can be used as a probing element and also as a stimulus to write nanoscale data on a photosensitive material for a high-density optical storage experiment. The relation between morphological structure and optical interaction can be researched deeply with this instrument.

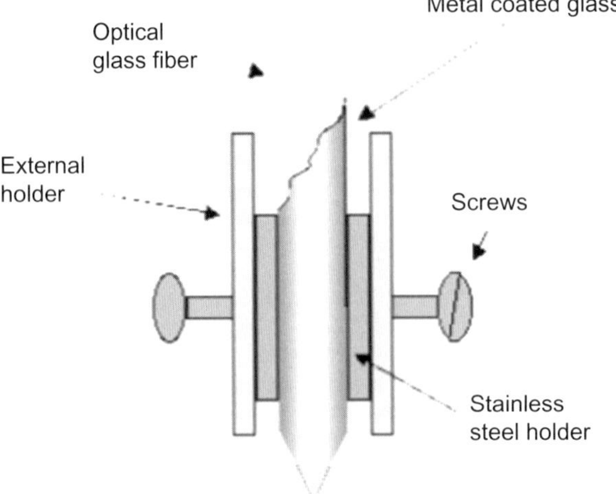

Fig. 7.31. A schematic of the probe holder

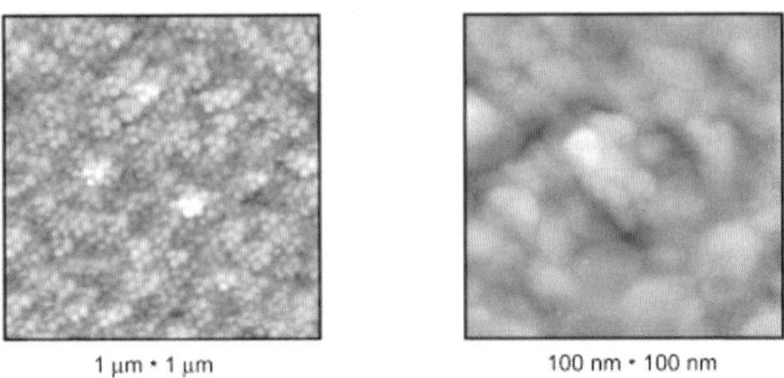

Fig. 7.32. STM images of a gold surface taken with a hybrid STM/SNOM optical fiber

7.5.3 Development of a New Type of AFM/SNOM Integrated System

Surface analysis was performed recently with several high-resolution probing methods like STM or AFM. These techniques alone cannot realize a direct study of the optical properties of the sample itself. Scanning near-field optical microscopy allows those investigations on the sample, but lateral resolution is still limited in the 10-nm range at best. To overcame this limit, we developed an original setup where we integrated an AFM system with SNOM in a new way.

The core of the system is an optically transparent silicon cantilever, coupled with a flat optical fiber. The center of the cantilever, where a sharp pyramidal tip is placed as a probing element, is transparent to light. There-

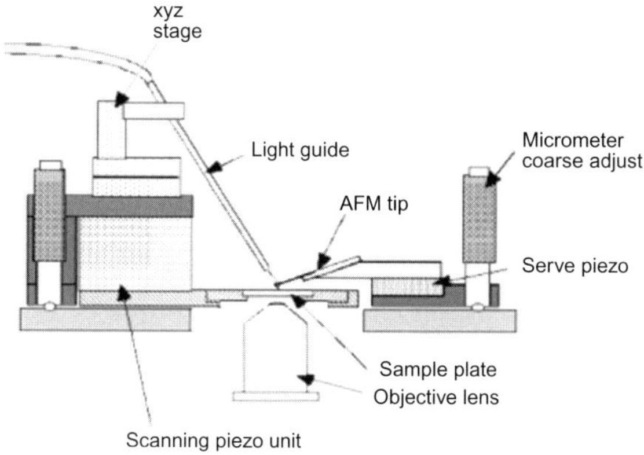

Fig. 7.33. Schematic of the AFM/SNOM integrated system

fore, the tip itself can also be used as an optical probe that is in direct contact with the sample, illuminated by laser light in the total internal reflection condition. This configuration differs from other typical SNOM configurations by the fact that the probe is actually in contact with the sample, so the optical coupling is strongly dependent on the near-field properties of the sample. The pyramidal tip is coated by metal to screen the far-field component of the light and has a small aperture at the very end. The light coupled to the tip is then collected by a conventional optical fiber that is placed 200 nm off the cantilever and is guided to a photon counting system or a spectrophotometer. In Fig. 7.33, we show a schematic of the core of the apparatus.

One of the keys of this probing method is the cantilever itself, which should be simultaneously a good AFM tip and an efficient SNOM probe. We avoided the use of an external laser to probe cantilever bending to simplify the optical detection system. Instead, we used a sensing piezo film on the cantilever body to detect movement, with the advantage of a direct electric signal proportional to the deflection and without the necessity for any beam alignment (Fig. 7.34).

To perform the first calibration tests we used a DVD ROM as a standard sample. In Figs. 7.35 and 7.36, we show the simultaneous scanning of an area of the sample and the corresponding AFM and SNOM maps.

The system is in its infancy, yet at this stage it can be extremely useful in examining samples where resolution at the atomic level range is not particularly necessary. Now, we aim to use our new SNOM/AFM system to image and analyze the dynamic properties of live biological samples with a high degree of sensitivity.

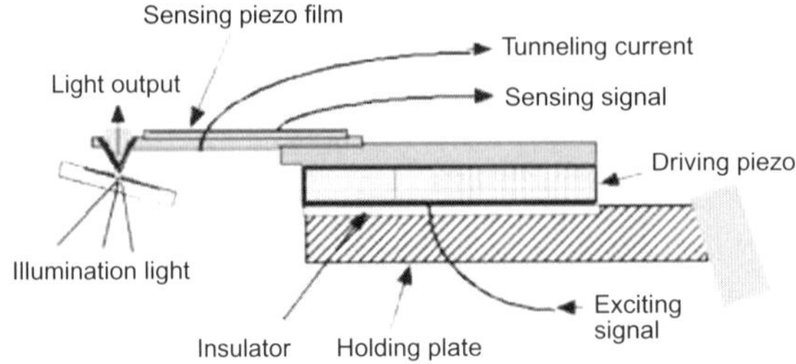

Fig. 7.34. Construction of AFM-SNOM tip with piezoelectric film

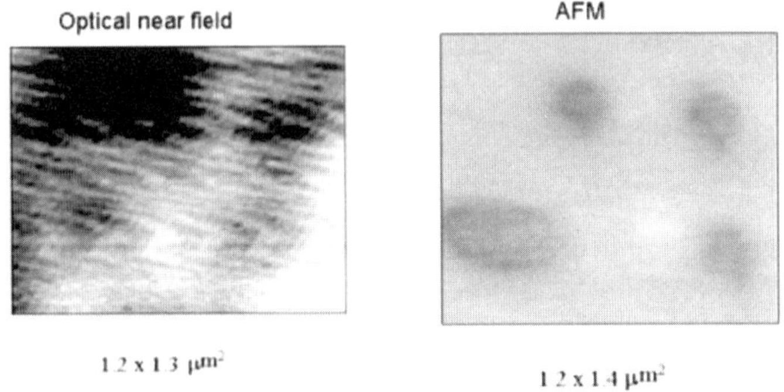

Fig. 7.35. DVD Dots seen simultaneously by AFM and AFM-SNOM

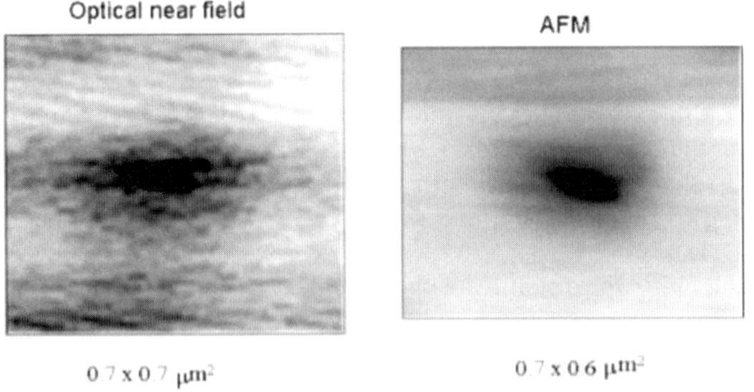

Fig. 7.36. DVD Dots seen simultaneously by AFM and AFM-SNOM

7.5.4 Biological Applications

Biological applications of SNOM systems are of extreme importance. Biological microscopy has led to great advances in understanding cell physiology, this in turn has led to significant advances in cell engineering and biomedical sciences. However, to date, live specimens have not truly been imaged on a nanometric scale. Some systems have come close to it. AFM microscope systems in the tapping mode have certainly been used to produce three-dimensional topographic images of live cells immersed in liquid media on a nanometric scale. However, these images are to some extent compromised by the artifacts that arise from the interactions between soft samples and the probe during the scanning procedure. Biologists need a system that can allow biological imaging on a nanometric scale in a purely noncontact mode. This can be provided by using a SNOM system.

We have already carried out a series of experiments testing a preliminary custom-made SNOM setup using a range of cultured cell types. Initially, we examined the dynamics of live liquid-immersed cardiac myocytes in culture. In vitro, cardiac myocytes contract spontaneously and rythmically [70] such that the membrane moves vertically about 1 μm. Actively beating rat cardiac tissues were examined using a SNOM system that scanned in the free-running mode without any feedback regulating the tip position, such that the probe tip was positioned approximately 1 μm above the sample. In this crude mode of operation the myocyte cell surfaces moved into and out of the near field during measurement, which resulted in images of lower lateral resolution (Fig. 7.27). However, this mode of operation allowed us to monitor the contraction profiles of the cells by stopping the scanning probe at specific positions. It was found that by scanning pixel by pixel over the contracting myocytes, contraction profiles could be measured with a subnanometric vertical sensitivity and these contraction profiles varied dramatically within adjacent 800-nm areas. Thus, even in the crudest mode of operation, the SNOM system, it was shown, has subnanometric vertical sensitivity and can detect submicron lateral features [71,72] in living cells. Then we imaged fixed rat hippocampal cells. The hippocampal cells were isolated and cultured using a technique described elsewhere [73], immersed in HBSS (Hanks balanced salt solution, GIBCO), and examined with a SNOM setup. In these experiments the SNOM system was used in the total internal reflection mode. The scanning probe was positioned in the near-field above the samples, and images were obtained by scanning pixel by pixel across the surface in the free-running mode. The maximum theoretical lateral resolution of this system was limited to approximately 100–200 nm by the size of the aperture of the scanning probe; see Fig. 7.38 for a sketch of the system. This work is as yet unpublished; however, it was found that in this simple mode of operation, the SNOM system clearly resolved the details of the cells, even to the extent of resolving internal features (Fig. 7.39). This is a feature of SNOM systems. SNOM images reflect the optical properties of the sample, so that internal features locally affecting the refractive index of the sample can be easily identified. It is important to note here that the neurons examined with the SNOM system were fixed; however, because of the noncontact mode of operation live cells may be examined as easily as fixed dead cells

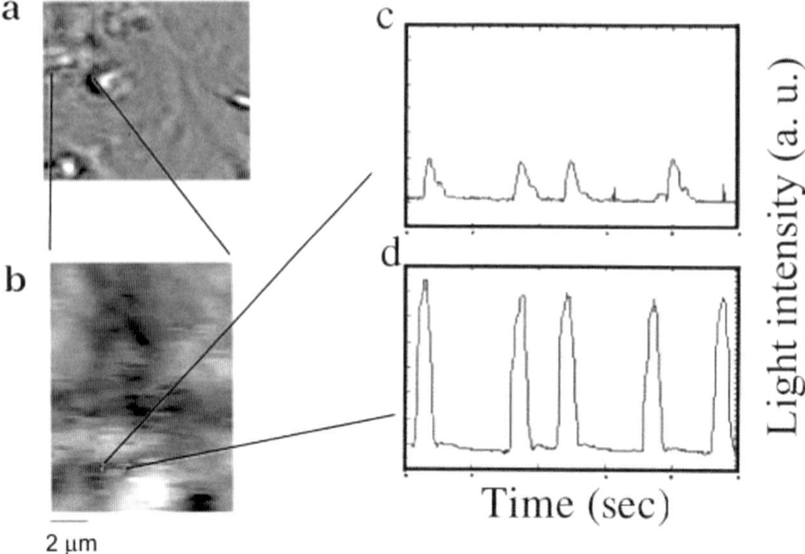

Fig. 7.37. Typical SNOM images of live cardiac myocytes in culture when the collecting probe is positioned approximately 1 µ m from the surface. **a** (scale bar = 10 µm) shows a low magnification SNOM image of cardiac myocytes. It is difficult to resolve details. The reason for this can be seen in **b**, which is a higher magnification scan of the boxed area in **a**. The clearly discernible horizontal bright features in **b** (scale bar = 2 µm) result from movement of the cell membrane in and out of the near field as the cell spontaneously contracts during the scan. By stopping the probe at specific points, the contraction profiles of the myocytes at those points could be examined (**c**,**d**). The recorded contraction profiles depend very much on the positioning of the tip, so that at points less than 1 µm apart, distinctly different contraction profiles can be recorded (**c**,**d**). This demonstrates that even in this crude mode of operation, the SNOM system can detect lateral features in fluid-immersed living cells of less than a micron

with near-field optics. A logical progression of our work is the development of SNOM systems for fluorescent imaging. There is no reason why SNOM systems could not used for the same purpose, and if they were, fluorescent imaging could be taken to much higher vertical and lateral resolutions in live cells. In initial experiments, we mounted a piezoelectric actuator on a standard fluorescent microscope (Olympus 1X70). In this configuration, the sample was illuminated via fluorescent optics, and the resulting fluorescence was collected with a SNOM tip and passed through a suitable filter before being fed into the optical detector. These first experiments in fluorescent imaging with a SNOM resulted in clear fluorescent SNOM images at a reso-

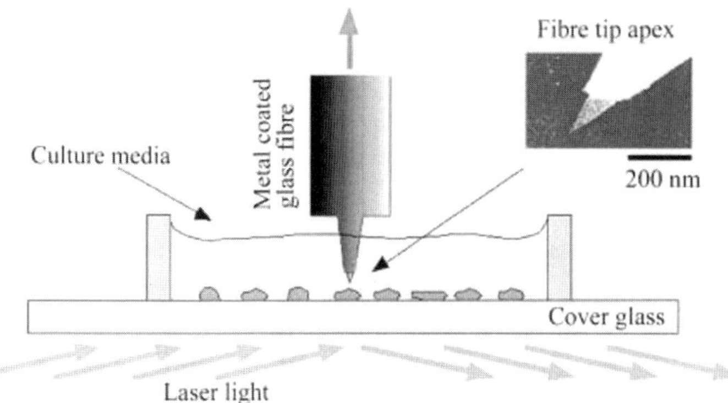

Fig. 7.38. A schematic of the SNOM configuration used to examine hippocampal neurons in a liquid medium. Laser light was applied at a critical angle to the underside of the sample. A sharpened optical fiber with a tip diameter of 100–200 nm was dipped into the evanescent field and used as a collecting probe. To work in liquid, the sample was surrounded by a Teflon ring filled with Hanks balanced salt solution. A typical SEM picture of the tip of an optical fiber probe is shown in the inset (scale bar = 200 nm)

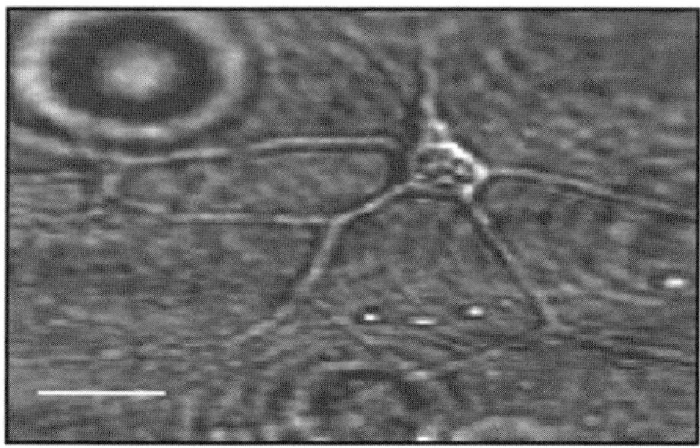

Fig. 7.39. The first SNOM image of a cultured cell in a liquid. The morphology of the hippocampal neuron can be easily resolved with the SNOM system (scale bar = 12 μm). Images obtained with a SNOM depend on the optical properties of the sample directly under the scanning probe. In this SNOM image, internal features that alter the refractive index of the sample can be resolved in the cell body. Unlike AFM systems, there are no artifacts associated with direct contact between the scanning probe and the sample, so SNOM systems can be used to examine soft samples such as live cultured cells. (Courtesy of M. Denyer, Bradford University)

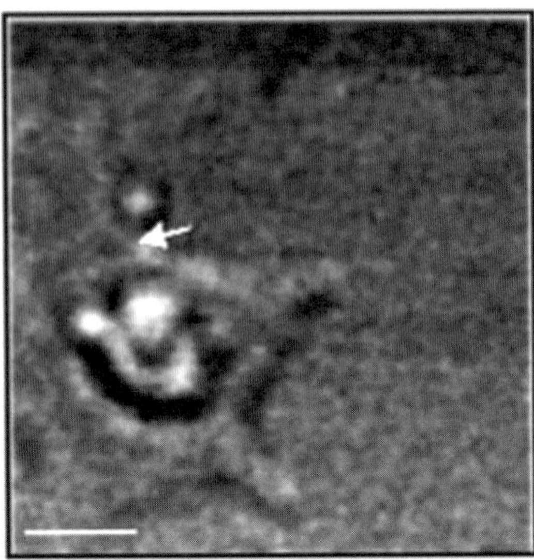

Fig. 7.40. The first fluorescent image of a tetramethyl-rhodamine isothiocyamate (TRITC) labeled phalloidin-stained cardiac cell. The fluorescent SNOM image has a resolution sufficient to allow resolving actin fibers (*arrow*) (scale bar = 7 μm). (Courtesy of M. Denyer, Bradford University)

lution that allowed identifying internal features (Fig. 7.40). These data clearly demonstrate that SNOM systems can be used to image fluorescent-labeled biological samples in a liquid media.

With the optimization of tip fabrication and tip positioning in relation to the surface, the SNOM has the potential of becoming a very powerful biological tool. The next step is to optimize SNOM tip fabrication and positioning in liquid media to produce images of live cells and fluorescent-labeled cells at much higher nanometric resolution, so that internal and external cellular features and mechanisms can be observed in real time. This may be easily achieved using our SNOM/AFM setup.

7.6 Second-Harmonic Generation in Near-Field Optics

When a high-power laser field **E** is incident in a nonlinear medium, the polarization **P** can be expanded in the form

$$\mathbf{P} = \epsilon_0(\chi^{(1)} : \mathbf{E} + \chi^{(2)} : \mathbf{EE} + \chi^{(3)} : \mathbf{EEE} + \cdots) , \qquad (7.1)$$

where $\chi^{(1)}$ is the linear susceptibility and $\chi^{(n)}$ is the nth nonlinear susceptibility. The second term is the origin of second-harmonic generation (SHG),

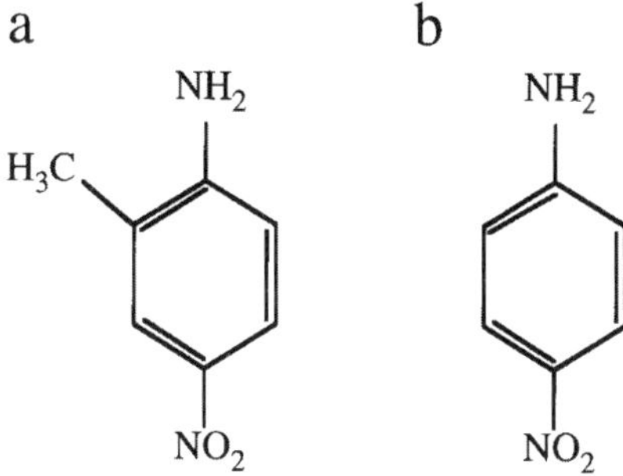

Fig. 7.41. Chemical structures of the materials used in this present study. **a** MNA **b** pNA

which is one of the second-order nonlinear optical phenomena. SHG is a two-photon coupling process, i.e., one coherent 2ω photon is generated by the coupling of two photons of frequency ω in a nonlinear medium. It is often applied to frequency conversion of high-power laser light. Since SHG is one of the second-order nonlinear optical effects, it is inhibited in a medium that has inversion symmetry under electric dipole approximation. Thus, SHG active crystal structure is limited.

There are few reports that observe SHG in near-field optical microscopy [74][75], compared with fluorescence because the conversion efficiency of SHG is much lower than that of fluorescence. However SHG provides interesting information that cannot be obtained from other optical effects: only the system (crystals, domains, films etc.) without inversion symmetry is observable. In this article, two microcrystal systems are examined: one is 2-methyl-4-nitroaniline (MNA), and the other is *p*-nitroaniline (pNA), whose chemical structures are depicted in Fig. 7.41. It is known that MNA is one of the most promising materials for second-order nonlinear optics because it forms noncentrosymmetric crystals with a second-order nonlinear susceptibility $\chi_{11} = 500$ pmV^{-1}. Although pNA has a chemical structure similar to that of MNA, it shows the absence of SHG activity, i.e., $\chi = 0$. However, as discussed below, weak SHG was observed from the microcrystal in our experiment.

To observe SHG, a high-power laser source is necessary as fundamental light. The laser source needed for measurements depends on the samples because the resonant frequency is individual for each sample. A mode-locked Ti:sapphire laser with a high repetition rate is the most realistic solution for observing the SHG signal from MNA and pNA microcrystals, for the following reasons:

1. The doubled frequency 2ω is in the near resonant region of both compounds: we can use the resonant enhancement of the material, which gives a signal intensity of about 100–1000 times higher.

2. To avoid the damage due to the high-power light incident on the sample and the tip, a laser with a high repetition rate and a short pulse width is suitable. If we use a nanosecond-laser light, the peak power of the pulse is more than 10^6 W, which may be too high for a tip. Using a typical mode-locked femtosecond-laser with a high repetition rate of about 100 MHz, we always obtain more than 10^5 photons s^{-1} in the far-field from SHG active monomolecular layers, and the peak power is less than 10^5 W.

7.6.1 Materials and Apparatus

The materials used in this study, MNA and pNA (Fig. 7.41), were purchased from Tokyo Kasei Co. Ltd. and were used as received. Microcrystalline samples were deposited by casting a chloroform solution on a prism surface. Absorption spectroscopy provides the average thickness of the material, which was about 300 nm.

The optical setup used in the current study is illustrated in Fig. 7.42. We used a mode-locked Ti:sapphire laser (Tsunami: Spectra Physics) operating with a pulse width of 130 fs and a repetition rate of 82 MHz ($\lambda = 760$ nm, 500 mW). The light was incident in total reflection geometry to generate SH light from the MNA microcrystals. The tip of a single-mode optical fiber, which was not sharpened for high sensitivity, was approached by using a piezoelectronic apparatus to monitor a shear force. It was fixed in the near-field region ($d \ll \lambda$) during the SHG measurement. The fundamental light was blocked by an interference filter in front of the photomultiplier tube (PMT). To increase the S/N ratio, we used an SR-830 lock-in amplifier (Stanford Research).

7.6.2 SHG Observation

When we approached the sample surface of the MNA microcrystals with the tip, a strong signal appeared on the lock-in amplifier. The typical signal

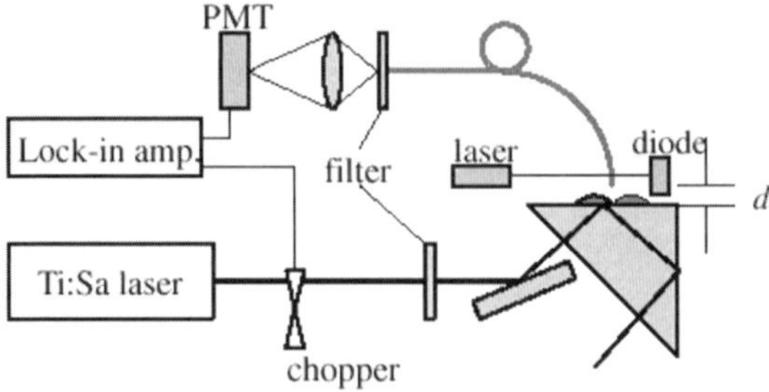

Fig. 7.42. Optical setup used in this present study

intensity was about the order of 10^3 photon s^{-1}. Bozhevolnyi and Geisler [74] also observed 600 SHG photon s^{-1} from a 2-docosylamino-5-nitropyridine (DCANP) Langmuir–Blodgett film. That signal intensity is compatible with ours. To check the origin of the signal, the square root of the SHG light intensity from the MNA microcrystals was plotted against the intensity of the fundamental light. Then the quadratic relation between them was obtained. In addition, the signal disappeared when the mode-locking is off. These facts safely led us to conclude that the origin of the observed signal is due just to the SHG process.

Although pNA crystals show the absence of SHG activity, weak SHG signals from the microcrystals were observed when the tip approached. The signal was 100 times weaker than that from the MNA microcrystals. The following three reasons can be considered:

1. SHG active pNA microcrystals were produced when we deposited the pNA microcrystals. It is reported that pNA has a SHG active crystal structure, although it is unstable [76]. The SHG efficiency is reportedly three times larger than that of MNA [76].
2. The interaction between the fiber tip and the microcrytals may break the symmetrical condition.
3. At the domain boundary, inversion symmetry is broken. SHG from the domain boundary is effectively detected because the tip is very close to the sample.

The first is not the case because the SHG from the pNA microcrystals was much weaker than that from MNA microcrystals in the present results. The second and third seem to be realistic; but it is difficult to draw a conclusion from the present results. More detailed study is required.

7.6.3 Conclusion

The SHG signal from MNA microcrystals was stable and had a high S/N ratio with an averaging time of 300 ms. Since the signal intensity by more than a factor of 100 is expected from improvements in the optical geometry, it will be easy to measure of each point in less than 100 ms. The procedure is in process. The short measurement time will make it realistic to image a SHG response by scanning a tip in the near-field region.

Acknowledgement. I thank Prof. Y. Ouchi of Nagoya University for allowing me to use the Ti:sapphire laser and for helpful discussions.

References

1. R. J. Colton, A. Engel, J .E. Frommer, H. E. Gaub, A. A. Gewirth, R. Guckenberger, J. Rabe, W. M. Heckl, and B. Parkinson, eds.: *Procedures in Scanning Probe Microscopies* (John Wiley & Sons, New York 1998)

2. *Nanotechnology Research Directions: IWGN (Interagency Working Group on Nanoscience, Engineering and Technology) Workshop Report*, M. C. Roco, R. S. Williams, and P. Alivisatos, eds. (1999)
3. M. Fujihira: in *Optics at the Nanometer Scale*, M. Nieto-Vesperinas and N. Garcia, eds., Nato ASI Series, E319, (Kluwer Academic, Dordrecht 1996)
4. R. C. Dunn: Chem. Rev. **99**, 2891 (1999)
5. M. Fujihira: in *Functionality of Molecular Systems* 2, K. Honda, ed. (Springer, Tokyo 1999)
6. R. M. Overney, E. Meyer, J. Frommer, D. Brodbeck, R. Luthi, L. Howald, H.-J. Güntherodt, M. Fujihira, H. Takano, and Y. Gotoh: Nature **359**, 133 (1992)
7. H. Muramatsu, N. Chiba, K. Homma, K. Nakajima, T. Ataka, S. Ohta, A. Kusumi, and M. Fujihira: Appl. Phys. Lett. **66**, 3245 (1995)
8. A. Lewis, K. Lieberman, N. K. Ben-Ami, G. Fish, E. Khachatryan, A. Strinkovski, S. Shalom, S. Druckmann, M. Ottolenghi, and U. Ben-Ami: Isr. J. Chem. **36**, 89 (1996)
9. E. Betzig, P.L. Finn, and J.S. Weiner: Appl. Phys. Lett. **60**, 2484 (1992)
10. R. Toledo-Crow, P.C. Yang, Y. Chen, and M. Vaez-Iravani: Appl. Phys. Lett. **60**, 2957 (1992)
11. H. Muramatsu, N. Chiba, N. Yamamoto, K. Homma, T. Ataka, M. Shigeno, H. Monobe, and M. Fujihira: Ultramicroscopy **71**, 73 (1998)
12. M. Fujihira, H. Monobe, A. Koike, G. R. Ivanov, H. Muramatsu, N. Chiba, N. Yamamoto, and T. Ataka: Ultramicroscopy **71**, 269 (1998)
13. M. Fujihira: Annu. Rev. Mater. Sci. **29**, 353 (1999)
14. Y. Horiuchi, K. Yagi, T. Hosokawa, N. Yamamoto, H. Muramatsu, and M. Fujihira: J. Microsc. **194**, 467 (1999)
15. M. Fujihira, H. Monobe, H. Muramatsu, and T. Ataka: Ultramicroscopy **57**, 176 (1995)
16. K. J. Kwak, T. Hosokawa, N. Yamamoto, H. Muramatsu, and M. Fujihira: J. Microsc. (submitted)
17. M. Fujihira, D. Aoki, Y. Okabe, H. Takano, H. Hokari, J. Frommer, Y. Nagatani, and F. Sakai: Chem. Lett. 499 (1996)
18. M. A. Paesler and P. J. Moyer: *Near-Field Optics: Theory, Instrumentation, and Applications* (John Wiley & Sons, New York 1996)
19. E. Betzig, J. K. Trautman, T. D. Harris, J. S. Weiner, and R. L. Kostelak: Science **251**, 1468 (1991)
20. E. Betzig and J. K. Trautman: Science **257**, 189 (1992)
21. M. Yamamoto, S. Ito, and M. Ohoka: in *New Functionality Materials*, Vol. C, T. Tsuruta, M. Doyama, and M. Seno, eds (Elsevier, Amsterdam 1993)
22. H. Aoki, Y. Sakurai, S. Ito, and T. Nakagawa: J. Phys. Chem. B **103**, 10553 (1999)
23. S. Ito, H. Okubo, S. Ohmori, and M. Yamamoto: Thin Solid Films **179**, 445 (1989)
24. S. Ito, K. Kawano, M. Yamamoto, H. Hasegawa, and T. Hashimoto: Macromolecules **28**, 3736 (1995)
25. N. Sato, S. Ito, and M. Yamamoto: Polym. J. **28**, 784 (1996)
26. T. Hayashi, T. Okuyama, S. Ito, and M. Yamamoto: Macromolecules **27**, 2270 (1994)
27. T. Hayashi, M. Mabuchi, M. Mitsuishi, S. Ito, M. Yamamoto, and W. Knoll: Macromolecules **28**, 2537 (1995)

28. G. J. Gaines: *Insoluble Monolayers at Liquid-Gas Interfaces* (Wiley Interscience, New York 1966)
29. M. Kawaguchi: in *Ordering in Macromolecular Systems*, A. Teramoto, M. Kobayashi, and T. Norisuje, edss (Springer, Heidelberg 1994)
30. K. Naito: J. Colloid Interface Sci. **131**, 218 (1989)
31. S. J. Mumby, J. D. Swalen, and J. F. Rabolt: Macromolecules **19**, 1054 (1986)
32. S. Mononobe, T. Saiki, T. Suzuki, S. Koshihara, and M. Ohtsu: Opt. Commun. **146**, 45 (1998)
33. M. Ohtsu, ed.: *Near-Field Nano/Atom Optics and Technology* (Springer, Tokyo 1998)
34. S. Henon and J. Meunier: Rev. Sci. Instrum. **62**, 936 (1991)
35. D. Honig and D. Mobius: J. Phys. Chem. **95**, 4590 (1991)
36. Y. Sakurai, N. Sato, S. Ito, and M. Yamamoto: Kobunshi Ronbunshu **56**, 850 (1999)
37. E. Betzig and J. K. Trautman: Science **257**, 189 (1992)
38. A. Keller: Phil. Mag **2**, 1171 (1957)
39. V. Peck and W. Kayne: J. Appl. Phys **25**, 1465 (1954)
40. Y. Fujiwara: J. Appl. Phys. **4**, 10 (1960)
41. T. Kasjiyama, I. Ohki, and A. Takahara: Macromolecules **28**, 4668 (1995)
42. T. Fujii, A. Takahara, and T. Kajiyama: *ACS Symposium Series, Microstructure and Tribology of Polymer Surfaces* (2000), p. 336
43. T. Fujii, K. Kojio, A. Takahara, and T. Kajiyama: Rept. Progr. Polym. Phys. Jpn. **41**, 277 (1998)
44. Y. Sakaki, T. Fujii, S. Sasaki, A. Takahara, and T. Kajiyama: Rept. Progr. Polym. Phys. Jpn. **42**, 195 (1999)
45. K. L. Naylor and P. J. Phillips: J. Polym. Sci. Polym. Phys. **21**, 2011 (1983)
46. L. Mandelkern, A. S. Posner, A. F. Diorio, and D. E. Roberts: J. Appl. Phys. **32**, 1509 (1961)
47. J. D. Hoffman, L. J. Frolen, G. S. Ross, and J. I. Lauritzen: J. Res. Natl. Bur. Stand. **A 79**, 671 (1975)
48. N. Maruyama, T. Koito, T. Sawadaishi, O. Karthaus, K. Ijiro, N. Nishi, S. Tokura, S. Nishimura, and M. Shimomura: Supramolecular Sci. **5(3/4)**, 331 (1998)
49. A. M. Cazabat, F. Heslot, S. M. Troian, and P. Carles: Nature **346**, 824 (1990)
50. R. Vuilleumier, V. Ego, L. Neltner, and A. M. Cazabat: Langmuir **11**, 4117 (1995)
51. X. Fanton and A. M. Cazabat: Langmuir **14**, 2554 (1998)
52. O. Karthaus, K. Ijiro, and M. Shimomura: Chem. Lett. **9** 821 (1996)
53. M. Shimomura, O. Karthaus, N. Maruyama, K. Ijiro, T. Sawadaishi, S. Tokura, and N. Nishi: Rep. Progr. Polym. Phys. Jpn. **40**, 523 (1997)
54. J. Hellmann, M. Hamano, O. Karthaus, K. Ijiro, M. Shimomura, and M. Irie: Jpn. J. Appl. Phys. **37**, L816 (1998)
55. O. Karthaus, L. Grasjo, N. Maruyama, and M. Shimomura: Thin Solid Films **327–329**, 829 (1998)
56. N. Maruyama, T. Koito, J. Nishida, X. Cieren, K. Ijiro, O. Karthaus, and M. Shimomura: Thin Solid Films **327–329**, 854 (1998)
57. M. Shimomura, T. Koito, N. Maruyama, K. Arai, J. Nishida, L. Grasjo, O. Karthaus, and K. Ijiro: Mol. Cryst. Liq. Cryst. **322**, 305 (1998)
58. O. Karthaus, L. Grasjo, N. Maruyama, and M. Shimomura: CHAOS **9(2)**, 308 (1999)

59. O. Karthaus, T. Koito, N. Maruyama, and M. Shimomura: Mol. Cryst. Liq. Cryst. **327**, 253 (1999)
60. M. Shimomura, J. Matsumoto, F. Nakamura, T. Ikeda, T. Fukasawa, K. Hasebe, T. Sawadaishi, O. Karthaus, and K. Ijiro: Polym. J. **31**, 1115 (1999)
61. O. Karthaus, T. Koito, and M. Shimomura: Mater. Sci. Eng. C **8–9**, 523 (1999)
62. O. Karthaus, X. Cieren, N. Maruyama, and M. Shimomura: Mater. Sci. Eng. C **10(1/2)**, 103 (1999)
63. K. Okuyama, M. Ikeda, S. Yokoyama, Y. Ochiai, Y. Hamada, and M. Shimomura: Chem. Lett. 1013 (1988)
64. M. Shimomura, Y. Hamada, N. Tajima, and K. Okuyama: J. Chem. Soc. Chem. Commun., 232 (1989)
65. M. Shimomura, N. Tajima, and K. Kasuga: J. Photopolym. Sci. Technol. **4**, 267 (1991)
66. M. Shimomura, S. Aiba, N. Tajima, N. Inoue, and K. Okuyama: Langmuir **11**, 969 (1995)
67. T. Chen, X. Wu, and R. D. Rieke: J. Am. Chem. Soc. **117**, 233 (1995)
68. R. H. Friend et al.: Nature **397**, (1999)
69. R. Friend, J. Burroughes, and T. Shimoda: Phys. World, June 1999
70. M. C. T. Denyer, M. Riehle, J. Hayashi, M. Scholl, C. Sproessler, S.T. Britland, A. Offenhaeusser, and W. Knoll: In Vitro Cell. and Dev. Biol. **35**, 352 (1999)
71. R. Micheletto, M.C. Denyer, M. Scholl, K. Nakajima, A. Offenhaeusser, M. Hara, and W. Knoll: J. Anat. **194**, 607 (1999)
72. R. Micheletto, M.C. Denyer, M. Scholl, K. Nakajima, A. Offenhaeusser, M. Hara, and W. Knoll: Opt. Rev. **6**, 268 (1999)
73. K. Goslin, H. Asmussen, and G. Banker: in *Culturing Nerve Cells*, G. Banker and K. Goslin, eds. (MIT Press, Cambridge, MA, London 1998) pp. 339–371
74. S. I. Bozhevolnyi and T. Geisler: J. Opt. Soc. Am. A **15**, 2156 (1998)
75. K. Kajikawa, K. Seki, and Y. Ouchi: in *Tech. Dig. 5th Int. Conf. Near Field Opt. Rel. Tech.*, Shirahama, Japan, 1998, pp. 2351–352
76. Y. Kamio, J. Watanabe, K. Kajikawa, H. Takezoe, A. Fukuda, T. Toyooka, and T. Ishii: Jpn. J. Appl. Phys. **30**, 1710 (1991)

8 Near-Field Microscopy for Biomolecular Systems

T. Yanagida, E. Tamiya, H. Muramatsu, P. Degenaar, Y. Ishii, Y. Sako, K. Saito, S. Ohta-Iino, S. Ogawa, G. Marriott, A. Kusumi, and H. Tatsumi

Direct visualization of functioning *biomolecular systems* (cells and subcellular organelles and molecules) has been one of the greatest dreams of biologists. Conventional techniques for high-resolution imaging of biological materials use high-energy electrons (e.g., electron microscope) or X rays (e.g., X-ray microscope). Since these high energy particles or waves significantly damage biolomecular systems, direct visualization of functioning *biomolecular systems* has been remained difficult. The recent development of near-field optics and related technology enable us to realize this dream partially. There are two ways to use near-field optical phenomena in biological applications; one is a scanning near-field optical microscope (SNOM) and the other is a total internal reflection fluorescence microscope. Technical details of these near-field optical microscopes are mentioned elsewhere (see related chapters in this book).

This chapter consists of three sections to report various applications of near-field optics and progress in molecular and cellular biology; *8.1 Near-Field Imaging of Human Chromosomes and Single DNA Molecules, 8.2 Imaging of Biological Molecules*, and *8.3 Cells and Cellular Functions*. An important application of the SNOM system is high-resolution imaging of biological molecules, e.g., DNA and choromosomes, intracellular cytoskeletons, the microstructure of cells to stabilize cell shapes and subcellular structures, organelles, vesicles for exocytosis and synapses. These advances in the application of SNOM are due to the development of optical probes made of glass fibers (see Chap. 3 *High-Resolution and High-Throughput Probes*) and feedback control systems. The highest resolution achieved by a glass fiber probe based SNOM is 15 nm for single fluorescent molecules and 4 nm for DNA samples [1]. These high-resolution probes are used in studies in this section. The observations by SNOM were made mainly on fixed molecules, cells, or tissues of biological specimens. However, the recent development of SNOM techniques in physiological solutions will allow imaging functioning molecules and living cells [2].

Total internal reflection fluorescence microscopy (TIRFM) [3] has been found useful in visualizing single biological molecules and/or cell membrane–substrate contact regions. In these techniques, the illumination of a water-covered glass at angles greater than the critical angle results in an "evanescent field" extending several hundred nanometers into the water. Specimens in the vicinity of the glass/water interface are illuminated by the evanescent field and thus can be observed. TIRFM is very sensitive to the distance between the cell membrane (or biological molecules) and the substrate, more so than even IRM (interference reflection microscope) [4]. The cell–substrate contact region is selectively illuminated, thus minimizing the exposure of the cell interior (or bulk solution) to the excitatory light. As a result, fluorescence from fluorophores both in the bulk solution and inside the cells is suppressed, thus reducing background fluorescence. Two types of applications have been developed: first, visualizing biological molecules with an evanescent field without background fluorescence for single molecule imaging (*the actin-myosin system, membrane receptors, and ATP synthase*). This application allows a new field of study, so-called "single molecule physiology" [5]. Another application of the evanescent field is visualization of functioning molecules and subcellular organelles, such as biological membranes and cell adhesive contact formation and synapses (*dynamics of cell membranes*).

8.1 Near-Field Imaging of Human Chromosomes and Single DNA Molecules

Conventional optics had a limit of resolution (about 250 nm) which was based on half of the wavelength of visible light, whereas near-field scanning optical microscopy (NSOM) demonstrated spatial resolution of only a few tens of nanometers and sensitivity down to the single molecule level [6–8]. This technique is analogous to scanning tunnel microscopy (STM) and atomic force microscopy (AFM). Various kinds of NSOM have developed with variations of the method for controlling the tip–sample separation such as utilizing STM [9], lateral shear force [6,7] and contact-mode AFM [10]. We have already developed a scanning near-field optical/atomic-force microscopy (SNOAM) in which a method of dynamic mode AFM was used to control the tip-to-sample separation[11,12]. An optical fiber with a sharpened tip was bent to use the probe as a cantilever for AFM, and the vibrational amplitude of the cantilever was held constant during scanning. SNOAM may be superior in biological observation to other NSOM systems because this system operates excellently in liquid [13]. It is safely applicable for observing soft samples with great variations in height, such as cultured cells. SNOAM may be also superior in liquids to other cyclic contact AFM (e.g., tapping mode); the latter uses flat type cantilevers [12,14,15], whereas the optical-fiber cantilever of SNOAM is round which helps to reduce the viscous resistance of liquids. Green fluorescent protein (GFP) originally isolated from the jellyfish *Aequorea victoria*, has become a useful reporter molecule for monitoring gene expression and protein localization *in vivo* and in real time [16,17].

The SNOAM system reported in this chapter provided us with simultaneous topographic and optical Images of human chromosomes and single plasmid DNA molecules.

8.1.1 SNOAM System

The SNOAM system is shown in Fig. 8.1. The optical-fiber cantilever is mounted on a bimorph and vibrated vertically against the specimen stage at the resonant frequency (typically 15–40 kHz). The vibratory voltage applied to the bimorph was between 0.1 and 5 ACV for an 0.11-nm/V bimorph. The vibratory amplitude is monitored by detecting the deflection of the laser beam, which is reflected on the ground surface of the optical-fiber cantilever. The probe–sample distance is controlled by decreasing the vibratory amplitude to an appropriate level when the distance between the probe and the

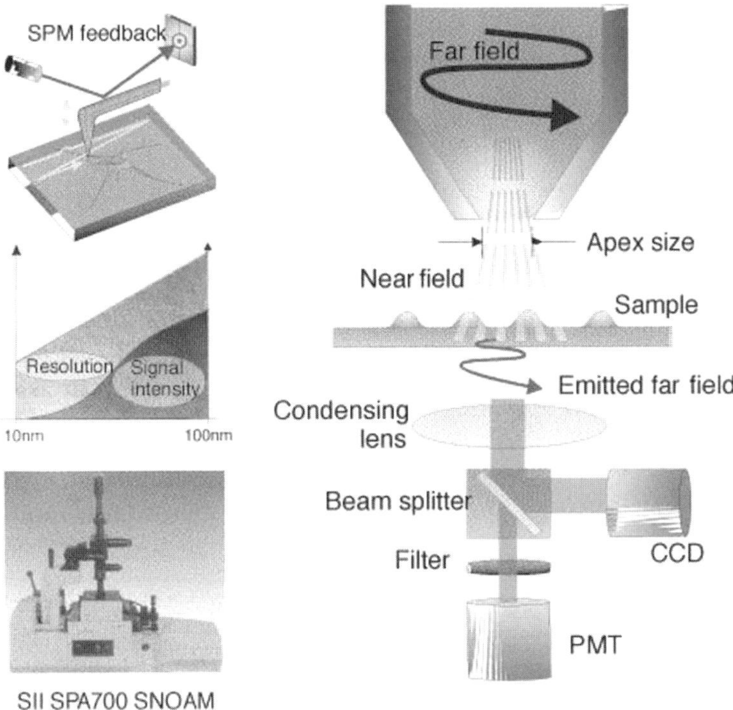

Fig. 8.1. An explanation of the SNOAM system. The optical configuration can be seen (*right*). The sample is raster-scanned relative to the probe to form the image (*top left*). The resolution of any SNOAM probe is determined by its aperture size (*left middle*). However, the smaller the aperture, the smaller the signal-to-noise ratio. Therefore, resolution is largely limited by detection efficiency. A picture of the SNOAM system (*bottom left*)

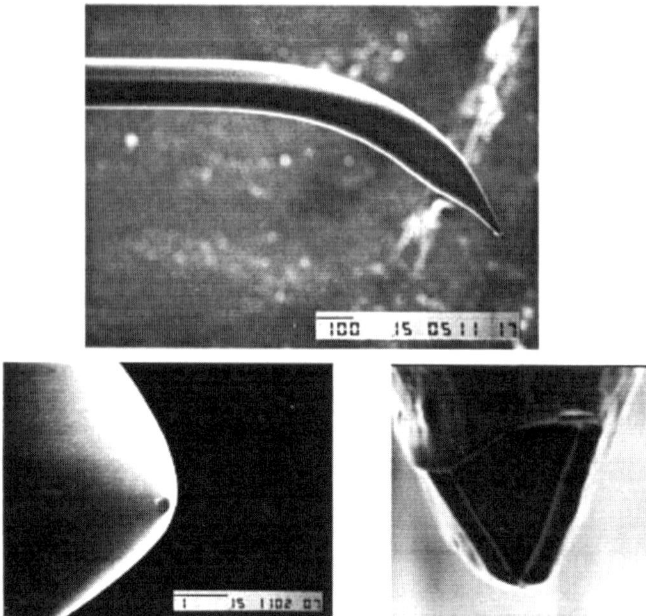

Fig. 8.2. Structure of the SNOAM probe

sample decreases (Fig. 8.2). This operation was controlled by a commercial AFM controller (model SPI 3700, Seiko Instruments Inc.). Laser beams (458, 488, 514.5 nm) from a multiline Ar-ion laser (max. 150 mW) were selected by a polychromatic AO modulator and coupled to the optical fiber on the other side of the optical-fiber probe. Signal light from the sample is corrected by an objective lens (typically: 100× oil immersion type) and separated by a dichroic mirror to the CCD camera and detectors. A photomultiplier and intensified-CCD (ICCD) camera with a spectrometer were connected as the detectors. A liquid chamber was designed for the present experiment. Water and cell culture media were held between the glass plate and an upper window. Both the probe and the sample were immersed in solution. The probe was prepared as described previously [18]. Briefly, an optical fiber was sharpened by chemical etching to make a tip, and then bent by the irradiation of a CO_2 laser. The probe was coated with a metal layer (aluminum or gold) 100–200 nm thick, and an aperture was made by the deposition of metal. Gold was especially used for a probe operated in liquid.

8.1.2 SNOAM Imaging of Human Chromosomes [19]

Chromosomes as genetic vehicles provide the basic material for a large proportion of genetic investigations from the construction of gene maps and models

Table 8.1.

Group 1–3 (A)	Large metacentric chromosomes readily distinguished from each other by size and centromeric position.
Group 4–5 (B)	Large submetacentric chromosomes which are difficult to distinguish from each other.
Group 6–12,X (C)	Medium sized metacentric chromosomes. The X chromosome resembles the longer chromosomes in this group. This large group presents a major difficulty in identifying individual chromosomes without banding techniques.
Group 13–15 (D)	Medium-sized acrocentric chromosomes with satellites.
Group 16–18 (E)	Relatively short metacentric chromosome (#16) or submetacentric chromosomes (#17 and 18).
Group 19–20 (F)	Short metacentric chromosomes.
Group 21–22,Y (G)	Short acrocentric chromosomes with satellites. The Y chromosome is similar to these chromosomes, but bears no satellites.

of chromosome organization to the investigation of gene function. The study of chromosomes has developed in parallel with other aspects of molecular genetics, beginning with the first preparations of chromosomes from animal cells through the development of banding techniques, which permit the unequivocal identification of each chromosome in a karyotype, to the present methods of analytical cytogenetics (Table 8.1).

A karyotype shows the metaphase chromosomes of an individual cell, arranged in pairs and sorted according to size. There are 22 pairs of autosomal chromosomes and a pair of sex chromosomes in humans. Karyotypes are a simple way to evaluate chromosomes. Many diseases and malformations are a direct result of missing, broken, or extra chromosomes. In a karyotype, scientists called cytogeneticists can recognize and identify many of these gross chromosomal abnormalities. The term "karyotype" can be defined as follows:

Characterization of the chromosome set of an individual or group, described in terms of number, length, centromeric position, and availability of satellites. The centromere is the point or region on a chromosome to which the spindle attaches during mitosis and meiosis. When chromosomes are stained by methods that do not produce bands, they can be arranged into seven readily distinguishable groups based on the descending order of size and the position of the centromere.

Basic karyotype analysis is based on the result of several conferences (ISCN – International System for Human Cytogenetic Nomenclature) established in Denver, USA, in 1960. Chromosome number and morphology were decided by banding or nonbanding techniques. When chromosomes are stained by methods that do not produce bands, they can be arranged into seven readily distinguishable groups (A–G) based on the descending order of size and the position of the centromere. Usually, a metacentric chromosome's centromere appears as a pronounced minimum in either the width or shape profile, whereas that of an acrocentric centromere is represented only by a smaller than usual gradient at one end of the profile.

Chromosmes are only a few microns long, and hence high-resolution microscopy is desired for karytoping. Confocal microscopy can achieve around 200 nm as mentioned earlier. A dedicated AFM can obtain resolutions higher than SNOAM, but it can only obtain topographic information. Using fluorescent SNOAM imaging, it is possible to verify topographic images and correct for any artifacts.

In this research, human metaphase chromosomes were imaged and identified by using SNOAM. The metaphase chromosomes were derived from human B cell lymphoblastoid line RPMI1788. The cells were grown in RPM11640 medium with 10% of fetal calf serum (FCS) at 37°C. Metaphase chromosomes were obtained after addition of colcemid (final concentration 0.05 mg/mL) which synchronized the cell cycle. Synchronized cells were harvested by centrifugation (500× g, 5 min). Chromosomes were prepared by the "surface-spreading whole-mount technique." To begin with, after centrifugation, collected cells were placed on a clean surface of distilled water with a clean platinum loop. Here, the cells burst due to osmotic pressure and then spread out rapidly over the water surface. The chromosomes were transferred to a glass coverslip by contact with the surface of the water. The sample was then dried.

The samples were then imaged using the Seiko SPA700 SNOAM system described earlier. The probe was operated in air using the cyclic-contact mode. SYBR green I (excitation at 497 nm, emission at 520 nm) was used as a fluorescent intercalater combined with DNA strands. The fluorescent images were detected at 520-nm emission with a 488-nm argon laser beam excitation. SYBR^{tm} Green I is the most sensitive stain available for detecting nucleic acids. Less than 20 pg of double-stranded DNA can be detected in a single band of a SYBR^{tm} Green I-stained gel using 254-nm epi-illumination.

Figure 8.3 shows a typical topographic image of human metaphase chromosomes that are recognized as three different types of chromosomes. In this case, chromosome (b) in Fig. 8.4 was imaged most clearly. From a morphological view of chromosome (b) in Fig. 8.4, its length was measured at approximately 7 nm, and a centromere was observed on one side of the chromosome. The candidate for chromosome (b) is #4 or #5 in group C. Similarly, chromosome (a) in Fig. 8.4 was also judged to be part of group E or F, and numbered 16, 19, or 20 because it was about 5–6 µm long and had a metacentric centromere. Chromosome (c) was designated a Y chromosome because of an acrometric centromere and its length of approximately 3–4 µm. Figure 8.5 shows the fluorescent chromosomes stained with SYBR^{tm} Green I

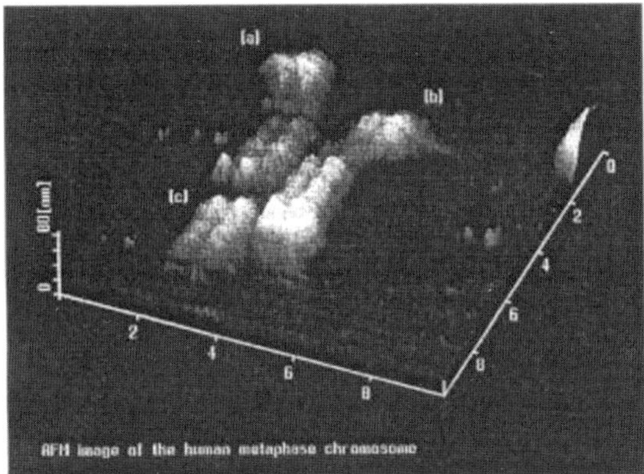

Fig. 8.3. Dynamic-mode AFM (DFM) image displaying the topography of a human metaphase chromosome on a glass coverslip in air. The image area is 10×10 μm. The dark-brown (umber) to white scale of the image is 80 nm. The height information in the cyclic-contact mode is calibrated just as for the height information in the conventional contact mode; by the voltages applied by the feedback amplifier to a piezoelectric scanner to keep the deflection signal constant

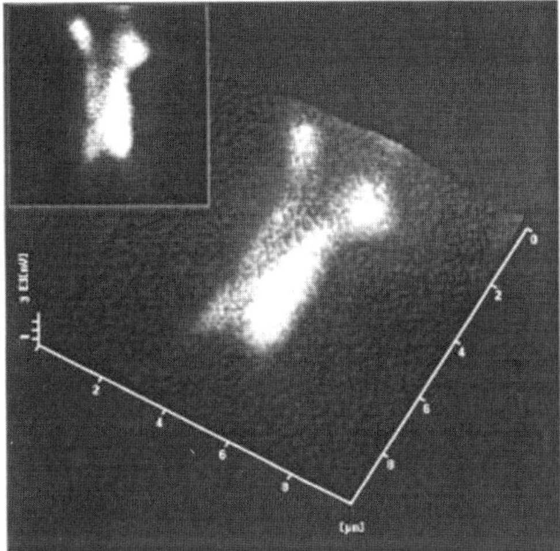

Fig. 8.4. Near-field fluorescent image of a human metaphase chromosome from Fig. 8.3. Fluorescence was obtained by stimulation with a 488-nm laser line from an Ar-ion laser. The maximum intensity of fluorescence from the chromosome was sevenfold higher than the background level

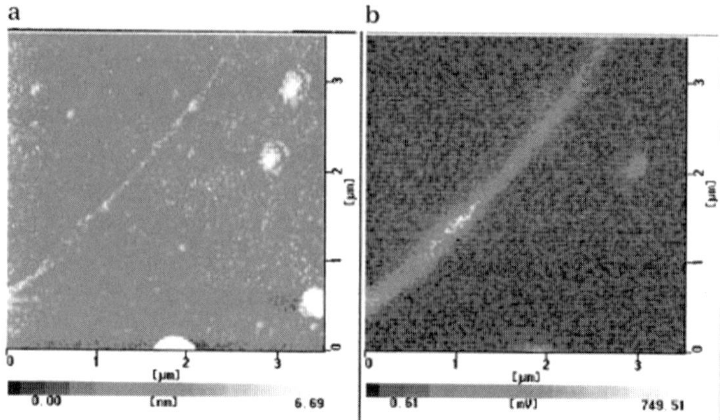

Fig. 8.5. Topography **a** and near-field fluorescent image **b** of DNA prepared from 5 μM DNA (base) with 5 μM YOYO-1. The scan area is 3.5 × 3.5 μm^2

corresponding to the topographic images in Fig. 8.4. The topographic images clearly indicated a duplicate structure on the metaphase chromosome, whereas the fluorescent images had a somewhat different shape. This probably results from the combination of SYBR^{tm} Green I and chromosomal DNA. SYBR^{tm} Green I, it is thought, binds noncovalently to the phosphate backbone of the DNA molecule.

8.1.3 SNOAM Imaging of a Single DNA Molecule [20]

Single DNA molecules, in which YOYO-1 was intercalated, were imaged and characterized using a SNOAM sysytem. Solutions of DNA, 5 μM (base concentration) with 5 μM and 0.5 μM YOYO-1, were prepared and cast on γ-APTES treated-cover slips. DNA immobilized on the cover slips aggregated in line. Figure 8.5a–b shows the topography and fluorescent images of the DNA on the coverslip prepared from 5 μ MYOYO-1 solution in 3.5 × 3.5 μm^2 areas. In the fluorescent image (Fig. 8.5b), the fluorescent intensity of the DNA is partially different. To characterize the difference of fluorescent intensity, areas of the cross section were measured. The areas of the topographic cross sections in each region were measured to check that the DNA was stained uniformly. Figure 8.6 shows the topography and fluorescent images of DNA prepared from a 500 nM YOYO-1 solution. In this case, the fluorescent intensities are not uniform, as shown in segments a and b in Fig. 8.6. In these segments, the topographic area of b (400 nm^2) is twice the size of a (220 nm^2), but the fluorescent intensity of b is three times lower than that of a. Similar results were consistently obtained in other areas of the sample. This indicates that YOYO-1 is intercalated in the DNA heterogeneously when the concentration of YOYO-1 is low. This result suggests that YOYO-1

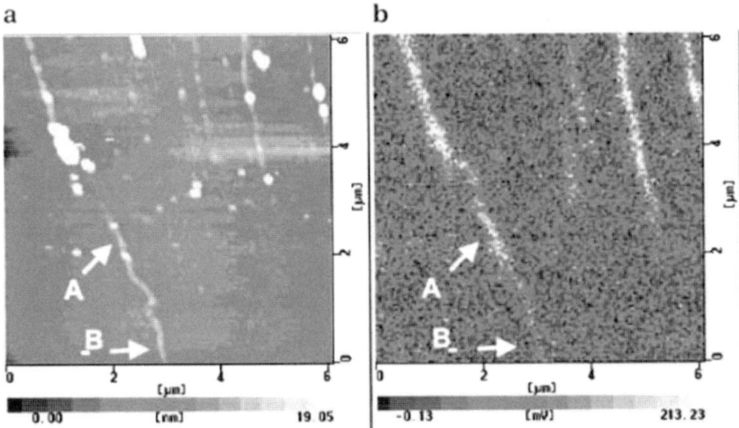

Fig. 8.6. Topography **a** and near-field fluorescent image **b** of DNA prepared from 5 μM DNA (base) with 500 nM YOYO-1. The scan area is 6×6 μm^2. The topographic area of segment **b** is twice that of **a**, but the fluorescent intensity of **b** is three times lower than that of **a**. This indicates that YOYO-1 intercalates in DNA heterogeneously when the concentration of YOYO-1 is low

binds to DNA partially cooperatively. To compare a conventional microscope and SNOAM, a fluorescent image was obtained with a conventional fluorescent microscope. Although a wide imaging area and a short imaging time are possible with a conventional microscope, the resolution is not better than 0.5 μm, and no topographic information is obtained. Therefore, a SNOAM is a useful tool for analyzing the aggregative structure and the intercalating state of DNA by comparing the fluorescent image and topography. This technique can be applied to direct observation of the hybridization phenomena of DNA and can be a useful tool for characterizing DNA at a molecular level. In these studies, topography is as important as high-resolution optical images in characterizing biological materials.

8.2 Imaging of Biological Molecules

8.2.1 Myosin-Actin Motors

Sliding Movement of Myosin and Actin. Actin and myosin are the major functional proteins involved in muscle contraction and cell motility. Muscle contraction occurs from the sliding movement between myosin and actin filaments [21]. The sliding model was first proposed to explain the changes in the striated pattern of muscle between relaxing and contracting states [22,23]. In nonmuscle cells, in which myosin and actin are less organized, the sliding motion of the actin filament over myosin drives such processes as motility,

a

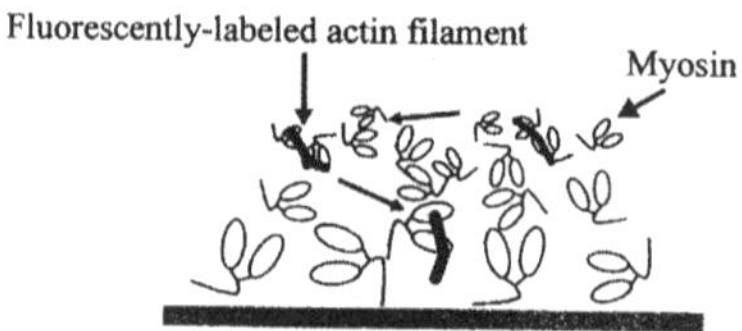

b

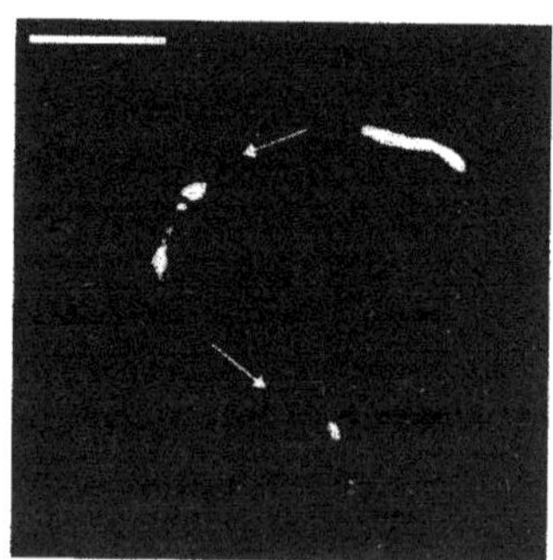

Fig. 8.7. Sliding movement of actin filaments. Actin filaments labeled with fluorescent phalloidin can be visualized under a fluorescent microscope. Their sliding movement occurs over myosin molecules adsorbed onto the glass surface in the presence of ATP (**a**). **b** Consecutive images of actin filaments taken at an interval of 2 s. Arrows show the direction of movement. The scale bar is 5 μm

cytoplasmic streaming, and changes in the shape of cells. In the 1980s, a single actin filament with fluorescent-labeled phalloidin was visualized, and the sliding motion of the actin filaments produced by myosin molecules adsorbed on a glass surface was observed in the presence of ATP under a microscope [24–26] (Fig. 8.7). The sliding velocity of the single actin filaments observed was similar to the velocity of the shortening of muscle, and the sliding motion was directional as in muscle. Thus, we recognized, for the first time, that a sliding motion occurs between individual isolated contractile proteins.

The sliding movement requires energy, which is supplied by the hydrolysis of ATP by myosin. The reaction comprises many steps including the binding of ATP, its hydrolysis to ADP and Pi, and release of the products. The interaction between actin and myosin depends on the states of the chemical reaction of ATPase, and the mechanical work results from the change in the interaction of actin and myosin. How is the energy supplied and used for mechanical work? How are these two processes coupled? To answer these questions experimentally, both the chemical and mechanical processes involved in the sliding movement need to be visualized. In the *in vitro* assay of the motility of actin filament described above, however, the interaction of myosin and actin cannot be directly observed, and it is not known where and how these molecules are involved in producing this sliding movement. To relate the sliding motion to the molecular interaction, we must be able to record the sliding movement of an actin filament produced by a single myosin molecule, or the sliding motion of single myosin molecules along an actin filament. Single molecule manipulation and imaging techniques are definitely necessary.

Measurement of Force Exerted by Single Myosin Molecules. In the *in vitro* motility assay, actin filaments slide on myosin molecules immobilized on the glass surface. When one of the ends of an actin filament was attached to a thin glass microneedle, the glass microneedle was pulled and deflected due to the interaction with myosin. By measuring the change in the displacement of the tip of the microneedle, whose spring constant is known, we can determine the force exerted on the actin [27,28]. To this end, the force generated by single myosin molecules can be determined by reducing the number of myosin molecules interacting with actin filaments [29,30]. Similar measurement was also done with a laser trap instead of a microneedle [31–34]. Two beads attached to the ends of an actin filament are trapped by a laser. When the actin filament interacts with myosin, the bead is pulled against the trap force. The displacement of a microneedle or bead is determined with nanometer accuracy by a quadrant photodiode detector.

Typical data for the displacement measurement using a laser trap are shown in Fig. 8.8. Fast-rinsing displacements were distinguished from the thermal vibration of the actin filament. During displacement, the thermal motion of the actin filament was less and the duration of the displacement increased as the ATP concentration decreased. This agrees that myosin and actin are strongly bound until another ATP molecule comes after one nucleotide is released. The thermal fluctuation or the stiffness of actin filaments is a good measure of the binding of myosin to actin. The actual displacement of myosin may be greater than the observed displacement of the trapped bead, because of any compliance or loose connections between the myosin molecule and the bead. Therefore, the displacement measured must be corrected for such compliance. Our results show that the displacement produced by a single myosin molecule is 15–20 nm, which is greater than the size of a myosin molecule.

We have consistently obtained values of 10–20 nm for the displacement. However, there are many reports that the displacement is much less, 4–10 nm. The difference may come from a different method of preparing myosin.

It is essential to keep proteins intact. Special care was taken to keep the myosin as intact as possible during measurements. We used myosin filament

a

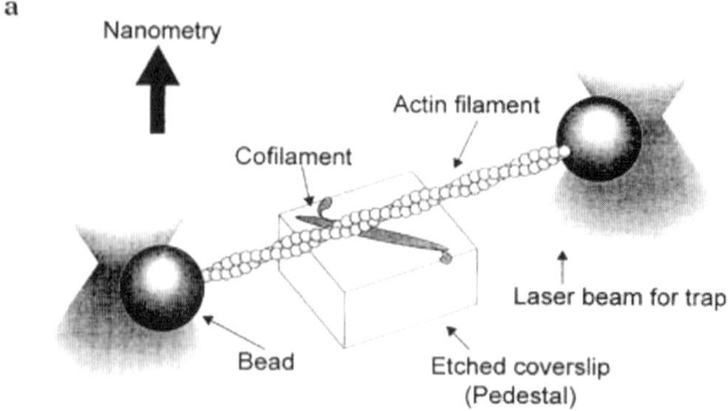

b

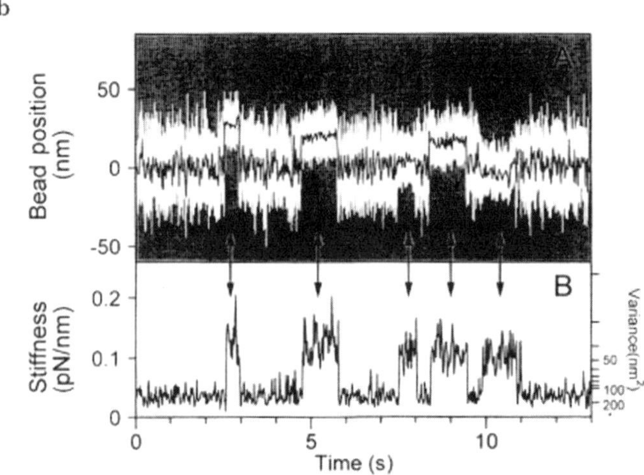

Fig. 8.8. Displacement of a single actin filament produced by a single myosin molecule. A single actin filament is attached to beads trapped by a laser at both ends and allowed to interact with a single one-headed myosin molecule. The displacement of actin is recorded by measuring the position of a bead using a nanometry sensor (**a**). **b** Time record of bead displacement and stiffness that is obtained from the variance in the fluctuation of a bead. White and black lines are raw data and data that have been passed through a low-pass filter of 20 Hz bandwidth, respectively. The bead displacements were distributed due to the randomizing effect of the thermal fluctuation of the beads

rather than single myosin molecules or myosin fragments directly attached to an artificial surface. The use of myosin filaments eliminates possible problems, including (1) effects of random orientation of myosin heads relative to an actin filament and (2) effects caused by the direct attachment of myosin to the artificial surface. We demonstrated that the interaction of myosin with actin depends strongly on the angle between the two filaments [32]. The bead displacement is maximum when two filaments are aligned in parallel, as in the muscle. Bead displacement decreases with an increase in the angle between two filaments and approaches zero when two filaments are perpendicular to one another. When myosin molecules are randomly oriented, the displacement must distribute, depending on molecules and the average value would include an average across various orientations. The true value must be greater than the observed values. The use of filaments has the advantage that the orientation of the myosin head is known. Another possible artifact is the effect of direct attachment of the myosin heads to the artificial surface. The sliding velocity of actin filaments produced by the myosin head directly attached to the glass surface is slower than that by myosin that is attached to the glass surface probably through its rod portion. When the myosin head is attached via the biotin–streptavidin bond and the light chain, which minimize the effects of the artificial surface, the velocity of actin filaments is the same as that of myosin [35]. We reduced the number of myosin heads on the myosin filament by mixing myosin with a large excess of myosin rods [34,36]. In addition, one-headed myosin was used instead of two-headed myosin, because two heads of myosin may make the interpretation more complicated. Mixing of one-headed myosin with myosin rods at a molar ratio of 1:1000 gave a filament of 5–8 μm containing only one to two heads.

Movement of Single Myosin Heads. In addition to the method of manipulating single actin filaments, we were recently successful in developing a new technique to manipulate single myosin S1 (subfragment-1, head part of myosin that contains the binding site for ATP and actin) molecules with a scanning probe [37]. S1 was biotinylated and fluorescent-labeled. A single S1 molecule on a glass surface was captured onto the tip of a scanning probe via the streptavidin–biotin system and a light chain [35]. The single molecule imaging technique was used to confirm that only a single S1 is captured.

The technique for single molecule imaging was developed to detect single fluorophores attached to bio-molecules in aqueous solution [38]. Total internal reflection fluorescence microscopy was used to reduce the background by illuminating only a region within a depth of 100 nm from the interface between the glass slide and the sample solution. Individual protein molecules could be observed as fluorescent spots (Fig. 8.9a). The evidence that a single spot arises from a single dye molecule can be obtained from photobleaching experiments. After irradiation by a laser, the fluorescent intensity drops to a baseline level in a single step. This is in contrast to bulk measurements in which many molecules are included and photobleaching occurs gradually. Additional evidence has been obtained from a comparison of fluorescent images with electron microscopy images [38].

A single S1 molecule attached to the tip of a probe is brought into contact with an actin bundle to interact in the presence of ATP (Fig. 8.10). The

a

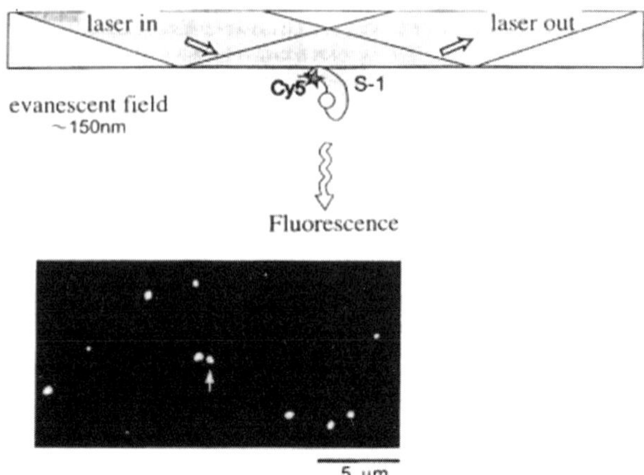

b

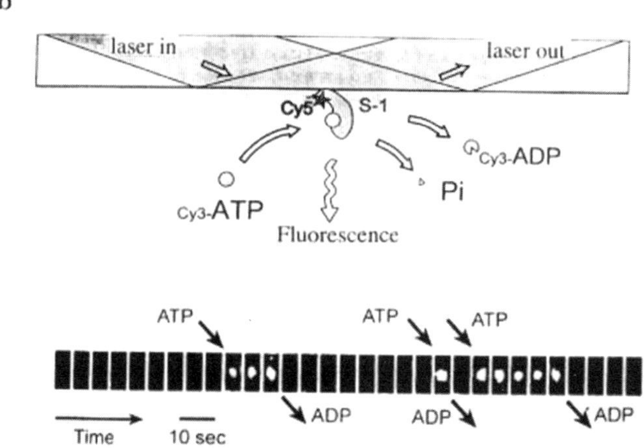

Fig. 8.9. Fluorescent images of single myosin subfragment 1 (S1) molecules and individual ATP turnovers by single S1 molecules. S1 labeled with Cy5 is adsorbed onto the glass surface and the area near the glass surface is illuminated using an evanescent field. S1 molecules are seen as fluorescent spots when Cy5 is excited using a helium–neon laser (**a**). **b** When Cy3-labeled ATP is further bound to S1, the fluorescent spot can be observed, and the fluorescent spots disappear when the nucleotide (mostly ADP) molecule dissociates from the S1. The time record of the Cy3-labeled ATP image at the position of the S1 is indicated by an arrow in **a**. Cy3 was illuminated with an argon laser

a

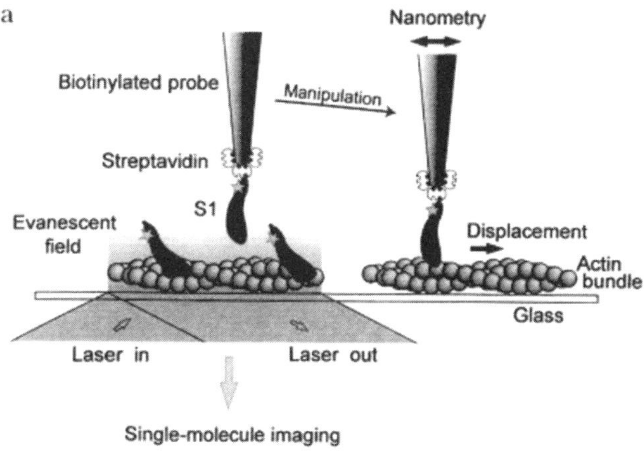

b

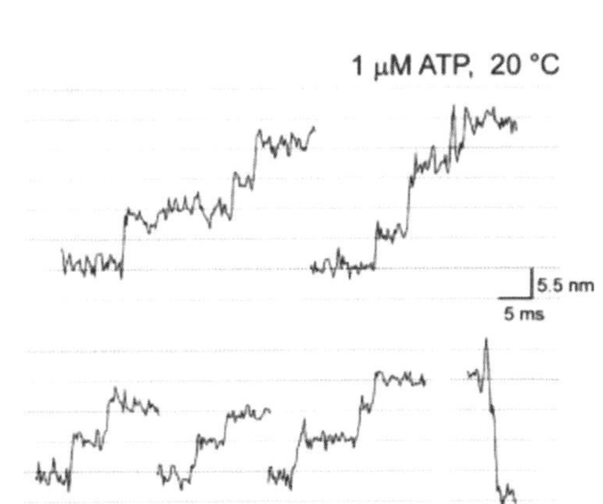

Fig. 8.10. Movement of single myosin heads. **a** A single S1 molecule, which was biotinylated and fluorescent-labeled, was attached at the tail end through biotin–streptavidin to the tip of a scanning probe and allowed to interact with an actin bundle. The displacement produced in the presence of ATP was determined by measuring the position of the probe with nanometer accuracy. **b** Displacement of a single S1 molecule generated by a single ATP molecule is plotted as a function of time. The dashed grid lines indicate a spacing of 5.5 nm

advantage of this method is that this linkage is more rigid than others such as the bead–actin–myosin–head linkage used in optical trap studies. Tight linkage allows greater accuracy in time measurements and spatial resolution. The displacement of the probe was recorded and the data are shown on an expanded timescale (Fig. 8.10b). The displacements that were observed as single large steps with other measurement systems take place stepwise rather than in a single step. The size of the substep is 5.5 nm. The substeps are stochastic, and the number of the steps per displacement varies between one and five, resulting in a total displacement of 5–30 nm. When the maximum load is increased from 0.5 to 5 pN, the step size remains constant, whereas the number of the steps decreases and the dwell time for each step increases. A single step size may reflect an interval between the nearest neighbor actin monomers.

Visualization of Turnover of Single ATP Molecules. The question arises how these mechanical processes are coupled to the chemical reaction of ATP. Is the energy of single ATP molecule used for an individual substep or a group of substeps? We have developed a technique for monitoring the chemical reaction of single ATP molecules by making fluorescent ATP and visualizing it by the single molecule imaging technique [38,39]. Fluorescent ATP was prepared by chemical modification of ATP with the fluorescent dye, Cy3, at its ribose. The fluorescent dye attached to the ATP does not interfere with either the binding to myosin or its ability to be hydrolyzed by myosin. With this fluorescent ATP, the turnover of single ATP molecules can be visualized (Fig. 8.9b): the fluorescent ATP is observed as a spot when it is immobile. However, we cannot detect a spot of fluorescent ATP when it undergoes Brownian motion in solution i.e., when it is not associated with myosin. For the experiments, myosin or S1 is labeled with Cy5. The positions of the myosin molecules are marked when Cy5 is excited, and then the laser is switched to green to excite Cy3-ATP; the fluorescent spots appear when ATP binds to myosin and disappear when they dissociate from myosin. Thus, we can monitor the association of ATP to myosin and its dissociation, most likely, after it is hydrolyzed to ADP. To confirm that this is due to ATP turnover, the time interval during which the nucleotide stayed bound to myosin is measured for individual ATP molecules and compared with the ATPase rate of myosin S1 in solution. This process is stochastic, and distribution of the time intervals can be fitted to an exponential curve. The decay rate constant was consistent with the ATPase rate of myosin S1. The possibility that ATP attaches and detaches without hydrolysis and that the fluorescent ATP molecule is photobleached before it is dissociated from myosin is negligible.

Coupling between Energy Input and Mechanical Work. Now that we have an imaging technique to monitor single ATP turnover as well as a manipulation technique for measuring the force produced by single myosin motors, it is possible to directly determine the coupling between the two by measuring both simultaneously [36].

Figure 8.11 shows traces of the mechanical and biochemical events recorded simultaneously. The top panel shows the displacement, the middle panel the stiffness, and the bottom panel the fluorescence of Cy3-ATP at the position where a single myosin head is located. As mentioned earlier, the stiffness

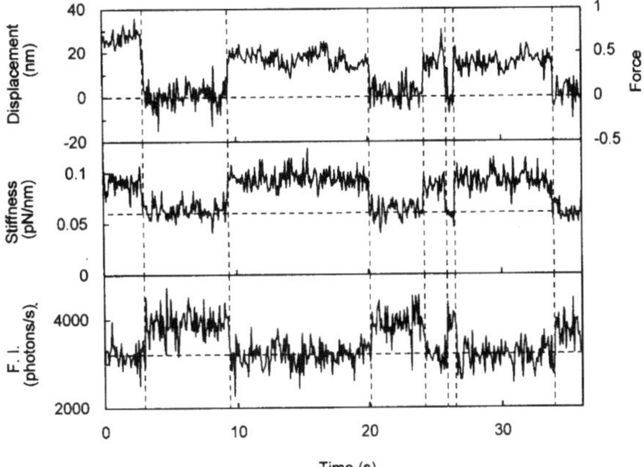

Fig. 8.11. Simultaneous measurements of chemical and mechanical events in forming actomyosin. The time record of displacement and stiffness of an actin filament and the fluorescent change of Cy3-ATP is shown. The mechanical measurements were done by the laser-trap method, as shown in Fig. 8.8. The fluorescent measurement of Cy3-ATP shows association and dissociation of Cy3-ATP to a myosin head in the filament, as in Fig. 8.9b. The measurements were performed by combining the two experiments

or the thermal variation of the actin filament is a good measure of the binding of myosin to actin. A myosin head detaches from an actin filament when a single Cy3-ATP molecule binds to myosin, and a myosin head attaches when Cy3-ATP dissociates, most probably after Cy3-ATP is hydrolyzed to Cy3-ADP and Pi. The possibility that the single spike of Cy3-ATP contains several molecules of Cy3-ATP was ruled out by the fluorescent intensity. Since a group of substeps is observed as a single large step with this resolution of laser trap measurements, a single ATP molecule generates a single large step or a series of substeps. Thus, a single substep does not correspond to a specific step in the chemical reaction.

The timing between the chemical and mechanical events was more carefully examined. The displacement driven by the hydrolysis of Cy3-ATP is maintained until a new Cy3-ATP molecule binds to the myosin head. The detachment of the myosin head from the actin filament monitored by the decrease in the displacement and the stiffness is induced immediately after binding of the Cy3-ATP, which is monitored by an increase in fluorescence (data not shown). ATP is hydrolyzed to ADP and Pi followed by their release from myosin. The increases in the displacement and stiffness or the forces generated are triggered by the decrease in fluorescence or the release of the bound Cy3-nucleotide, most likely Cy3-ADP (Fig. 8.12). In approximately 50% of cases, the force generation occurred at the same time as the release of Cy3-ADP, determined within the accuracy of the experiment (Fig. 8.12a).

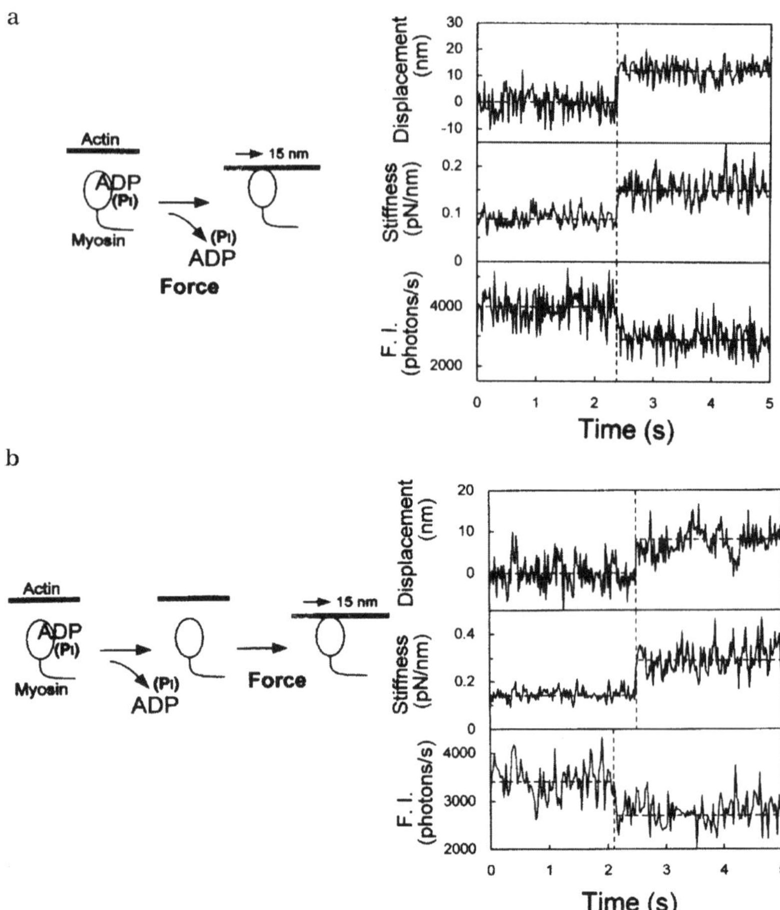

Fig. 8.12. Simultaneous measurements of ADP dissociation and displacement of actomyosin. The time record of the nucleotide dissociation and the force generation events is shown. Note that the timescale is expanded compared with Fig. 8.11. In about half of the cases, the chemical and mechanical events occurred at the same time (**a**). However, in the remaining cases, force generation was delayed after dissociation of the ADP (**b**)

In the other cases, the force generation was delayed as much as 1 s after dissociation of Cy3-ADP (Fig. 8.12b). Thus, the mechanical events are not always coupled with the chemical reaction.

The original "swinging cross-bridge model" or renovated "lever-arm model" has been proposed on the basis of the results of both mechanical and structural studies [40,41]. The elementary steps of the actomyosin ATPase have been characterized using isolated proteins in solution, and the kinetic processes have been related to the structural changes in the myosin head [41,42]. In this model, it is assumed that the chemical, structural, and mechanical events are tightly coupled, and it is obviously expected that a single step causes the displacement of the size of myosin molecules at most, or several nanometers. However, our experiments consistently showed that the displacement generated by the hydrolysis of a single ATP molecule is larger than the size of myosin. The force generation and association of myosin to actin is delayed after the release of ADP. It is also shown that the mechanical events occur repeatedly during the hydrolysis of single ATP molecules. These results cannot be explained by the lever-arm model. Therefore, the energy must have been stored in the myosin and used for generating of the force afterward.

Thus, new technologies of single molecule imaging and manipulation reveal a new insight into the molecular motors, myosin and actin, which cannot be explained by the current model. Brownian movement most probably plays a central role in the new model of molecular motors. For example, the generation of the substeps of myosin is stochastic. To develop a new model, more experiments and further development of single molecule detection techniques will be required.

8.2.2 Membrane Receptors

Overview. Biological cells respond to changes in their environments by analyzing many different kinds of information obtained through membrane receptors on the cell surface. Usually, the number of receptor molecules required to induce a specific cellular response is small ($< 1\ 000$). In addition, the reaction of membrane receptors may be altered in the different space and time within cells. Therefore, it is important to analyze the reaction of membrane receptors, one by one, with spatiotemporal resolution in the living cells.

The reduced excitation volume of the evanescent field raised by total internal reflection (TIR) greatly improved the contrast in fluorescence microscopy enabling the visualization of single fluorescent molecules in aqueous solution [38]. Recently, we extended this technique to visualize single membrane receptor molecules in living cells [43]. In this chapter, we describe the instrumentation and usage of TIR fluorescence microscopy to study the behavior and function of membrane receptors in the cell membrane.

Instrumentation. An objective-type TIR fluorescence microscope based on an inverted microscope (IX-70; Olympus, Tokyo, Japan) [43] was used to visualize the membrane receptors on the cell surface. In this microscope, a thin laser beam illuminated the specimen obliquely through the objective lens (Fig. 8.13a). To retain total internal reflection, the laser beam parallel to the optical axis was focused to a point at the edge of the back focal plane of an oil-immersion objective with a high numerical aperture (PlanApo 60x, NA 1.4; Olympus, Tokyo).

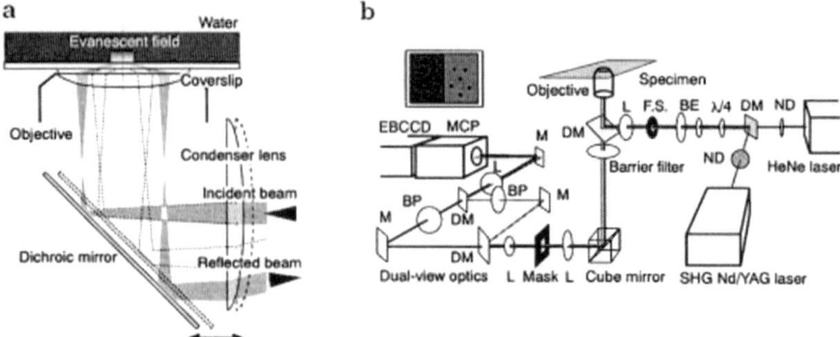

Fig. 8.13. Instrumentation of the objective-type total internal reflection fluorescence microscope used in our study. **a** The optical path around the objective is shown. The incident angle of the excitatory laser beam can be altered by positioning a dichroic mirror and a condenser lens before the objective. **b** A schematic diagram of our TIR microscope system. BE: beam expander, BP: band-pass filter, DM: dichroic mirror, F.S.: field stop, L: lens, $\lambda/4$: quarter-wave plate, M: mirror, MCP: microchannel plate, ND: neutral density filter. See text for details

For dual-color imaging, a green laser beam (532 nm) from a SHG-Nd/YAG laser (model 142-0532-100; Lightwave, Mountain View, USA) and a red laser beam (632.8 nm) from a He–Ne laser (GLG5900; NEC, Tokyo) illuminated the specimen either simultaneously or interchangeably. Typical values for excitatory power density for single molecule imaging were 10–100 $\mu W/\mu m^2$ at the specimen.

Images of the specimen were intensified by a microchannel plate (VS4-1845; Video Scope, Sterling, USA) and acquired by an EB-CCD camera (C7190-20; Hamamatsu, Hamamatsu) at a video rate. A relay lens system with a magnification of 1–7.5× was inserted between the objective and the image intensifier. For dual-color observation and fluorescent resonant energy transfer (FRET) imaging, a dual-view optic [44] was inserted between the microscope and the image intensifier (Fig. 8.13b). Images were recorded in a digital video format (DVC-PRO, Panasonic).

Observation of Membrane Receptors on the Apical and Basal Surfaces of Cells. Many types of cells have their characteristic polarity in the cell membrane. For example, the cell membrane of epithelial cells is differentiated into the apical (faced toward the culture medium in our experimental conditions) and basolateral sides (faced toward the coverslip). Components of the membrane are different on each side, and sometimes biologically important membrane receptors exist only on a specific side, depending on the molecular species. Therefore, it is very important to observe both the apical and basolateral sides.

Traditionally, a TIR microscope has been used to observe the basal cell membrane attached to a substrate, since the typical penetration depth of the evanescent field is less than several hundreds of a nanometer. However, we found that by changing the incident angle of the illuminating lasers, it was possible to image both apical and basal sides of the cell [43]. When the apical surface is illuminated, total internal reflection takes place at the meniscus between the cytoplasm and the extracellular medium (Fig. 8.14a).

The refractive indexes of the coverslips, the cytoplasm [45], and the extracellular medium are 1.52, 1.36–1.37, and 1.33, respectively. When an objective with a numerical aperture (NA) of 1.4 was used, the maximal possible incident angle is 67.1°, which is smaller than the critical angle for TIR between the coverslip and the cytoplasm (63.5–64.3°). The critical angle for TIR between the cytoplasm and the medium is large (76.1–77.9°). However, apical illumination can be achieved using a smaller laser incident angle to the coverslip due to the refraction between the coverslip and the cytoplasm. The critical angle for apical illumination is 61°, which is the same as the critical angle between the coverslip and the medium (Fig. 8.14a).

To demonstrate the effect of the laser incident angle, cells were observed in the presence of a peptide hormone, epidermal growth factor (EGF) conjugated with a fluorophore Cy3 (Cy3-EGF). Images were acquired by changing the incident angle of the laser and the focus of the objective (Fig. 8.14b). In the apical TIR mode, fluorescent spots were shown on both the apical and basal surfaces. In this case, Cy3-EGF on the basal surface was excited by epi-illumination. The background from the cytoplasm was sufficiently small to image the apical surface. When the focus was adjusted to the lateral surface, the cytoplasm of the cells was dark, and spots on the apical and basal surfaces were out of focus. In the basal TIR mode, fluorescent spots were detected only on the basal surface due to the limited excitatory depth of TIR. No fluorescent spots could be observed in the epi-illumination mode due to the high background fluorescence from the medium.

Single-Molecule Visualization of Membrane Receptors. EGF is a small peptide hormone, which induces proliferation of epithelial cells. Since mouse EGF has only one reactive amino residue (at the amino terminus), it can be labeled with amino-reactive fluorophores in a ratio of exactly 1:1.

We observed binding of Cy3-labeled EGF (Cy3-EGF) to the A431 human carcinoma cell line using TIR microscopy (Fig. 8.15a). Cy3-EGF in the culture medium could not be detected as fluorescent spots because of Brownian movement in solution. In addition, the fluorescent background caused by Cy3-EGF in the medium was low due to the low excitatory depth of TIR microscopy. The binding of Cy3-EGF was completely inhibited when excess amounts of nonlabeled EGF were added at the same time as the Cy3-EGF (Fig. 8.15a, inset). This suggests that Cy3-EGF binds specifically to the EGF receptor (EGFR) at the cell surface.

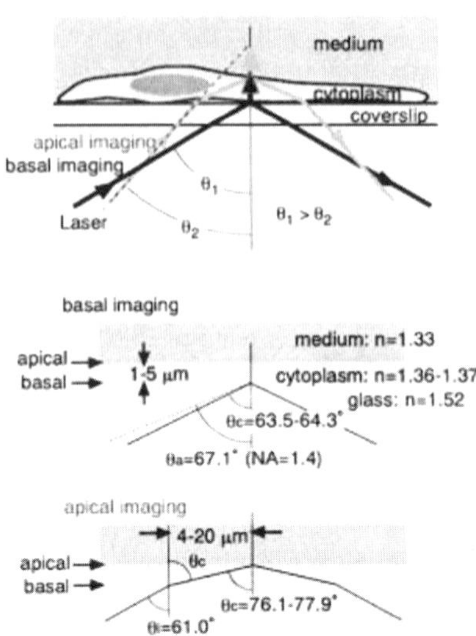

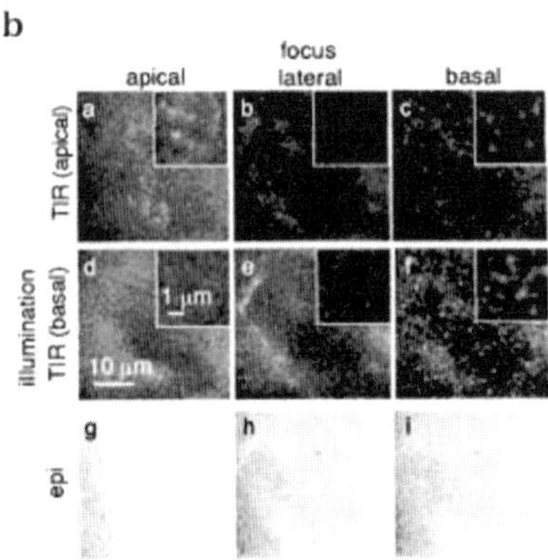

Fig. 8.14. Effects of the laser incident angle on the visualization of membrane receptors in living cells. **a** Optical path for the apical and the basal imaging. In basal imaging, the critical angle ($\theta_c = 67.1°$) equals the incident angle (θ_{c1}) of TIR (*middle panel*). In apical imaging, θ_i ($= 61.0°$) is smaller than θ_c ($= 76.1 - 77.9°$) (*bottom panel*). **b** Comparison of images acquired in apical and basal TIR modes. Cy3-EGF (100 ng/mL) was added to a culture medium of A431 cells. Images of the same field were taken in the apical (a, b, c) and basal (d, e, f) TIR mode, and in the epi-illumination mode (g, h, i). The objective was focused on the apical surface of the cells (a, d, g), on the medial part of the cytoplasm (b, e, h), and on the basal surface of the cells (c, f, i). The background in apical imaging is high due to high concentrations of Cy3-EGF (100 ng/mL), which was used to show clearly the differences in the three illumination modes

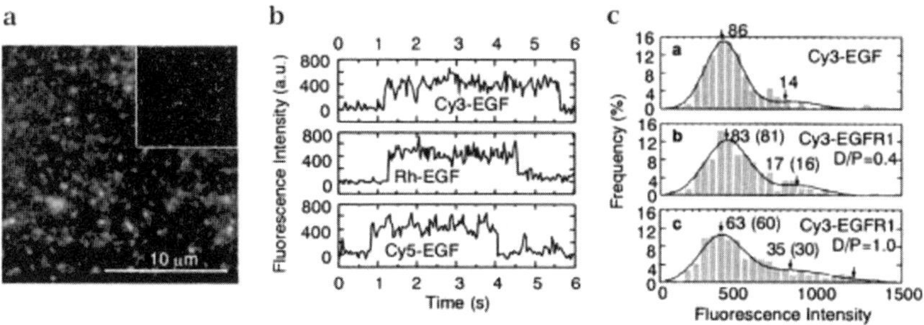

Fig. 8.15. Single-molecule visualization of fluorophore-labeled EGF on the surfaces of living cells. **a** Cy3-EGF (0.2 ng/mL) was added to the culture medium of living A431 cells under a TIR fluorescence microscope. Images were acquired 1 min after the addition of Cy3-EGF. Binding of Cy3-EGF was completely inhibited in the presence of 200 ng/mL of nonlabeled EGF (inset). **b** The fluorescent intensity change in single Cy3, tetramethylrhodamine (Rh), and Cy5-EGF spots on living cells was plotted against time. The spots were photobleached in a single step, indicating that they were single molecules. **c** Distribution of the fluorescent intensity of Cy3-EGF (a) and Cy3-anti-EGFR (EGFR1) (b, c) on the surface of a living cell. Histograms of the fluorescent intensity are fitted to a sum of two (a, b) or three (c) Gaussian distributions (*lines*). The position of the mean (*arrow*) and percentage fraction (*number*) of each Gaussian component are indicated. The numbers in parentheses are the percentage fractions expected from Poisson distributions with averages of 0.4 (b) and 1.0 (c)

The fluorescent intensity change for each Cy3-EGF spot was monitored on living cells (Fig. 8.15). In many cases, a spot appeared suddenly on the cell surface, emitted an almost constant fluorescence and then disappeared suddenly. This disappearance was caused by photobleaching because the duration between appearance and disappearance was proportional to the inverse of the excitatory laser power [43]. Single-step photobleaching strongly suggests that the fluorescent spots represented single Cy3-EGF molecules [38].

A monoclonal anti-EGFR antibody (EGFR1) was conjugated with Cy3 (Cy3-EGFR1), and this antibody was bound to the A431 cells in the same manner as Cy3-EGF. The concentration of Cy3-EGFR1 was reduced (0.1 µg/mL) to avoid aggregation with EGFR induced by EGFR1, which has two binding sites to EGFR. Since EGFR1 has many amino residues, it can be labeled with multiple Cy3 molecules. When the molar ratio of conjugated Cy3 to EGFR1 is small on average, the number of Cy3 molecules bound to one EGFR1 molecule will obey the Poisson distribution. The intensity distribution of Cy3-EGFR1 spots on the cell surface was fitted to the sum of two or three Gaussian distribution functions (Fig. 8.15c, *middle* and *bottom*). The results of fitting are consistent with Poisson distributions of the number of Cy3 to EGFR1 molecules, indicating that the first, the second, and the third components represent EGFR1 conjugated with one, two and three Cy3 molecules, respectively.

The intensity distribution of Cy3-EGF (Fig. 8.15c, top) shows that the mean of the first component is the same as those for Cy3-EGFR1, suggesting that this component contains a single Cy3-EGF molecule. The second component, which emits a signal twice as intense as the first component, thus may represent two Cy3-EGF molecules bound to a dimer of EGFR [43].

From these results, we concluded that single Cy3-EGF molecules could be visualized on the surface of living cells. Using the same technique, it was also possible to visualize single molecules of EGF labeled with other fluorescent dyes such as tetramethylrhodamine (Rh) and Cy5 (Fig. 8.15b).

Single-Molecule Visualization of Membrane Receptors

Applications of Single-Molecule Visualization to Studies of Membrane Receptors. Using the single-molecule visualization technique, we observed localization, movement, oligomerization, and turnover of individual signaling molecules in the cell membranes of living cells [46] and analyzed the process of dimerization of EGFR [43], which is a key step in the signal transduction of the EGF system [47,48]. In addition to these measurements, we can now observe phosphorylation and conformational fluctuation of EGFR by using dual-color and single-molecule FRET TIR microscopy [43].

(a) Detection of phosphorylation of membrane receptors. Dimerization of EGFR induces autophosphorylation on its cytoplasmic tyrosine residues [48]. A monoclonal antibody mAb74 recognizes a conformational change in the cytoplasmic domain of EGFR induced by autophosphorylation [49].

We developed a method for detecting this chemical reaction in semi-intact cells, which was perforated with streptolysin O to introduce Cy3-labeled mAb74 [50]. After stimulation of the semi-intact cells with Cy5-EGF, Cy3-mAb74 was added and binding of Cy3-mAb74 to the EGFR in the basal cell membrane was visualized in single molecules. In the Cy5-channel, some fluorescent spots had approximately double the signal intensity compared to other spots. These bright spots represented Cy5-EGF molecules bound to dimers of EGFR. Fluorescent spots of Cy3-mAb74 co-localized selectively with the spots of dimers of Cy5-EGF, indicating autophosphorylation of EGFR after dimerization (Fig. 8.16).

(b) Detection of conformational fluctuation of membrane receptor complexes. Conformational fluctuations of dimers of the EGF-EGFR complex were measured by single-molecule FRET [51,52]. In this experiment, a mixture of Cy3-EGF and Cy5-EGF was added to the cells, and only Cy3-EGF molecule was excited using the 532-nm laser (Fig. 8.17a). Fluorescent intensities of a single spot at 580 nm (Cy3) and 670 nm (Cy5) were measured simultaneously using dual-view optics (Fig. 8.17b). Anticorrelation of the fluorescent intensity changes of the same spot at the Cy3 channel and the Cy5

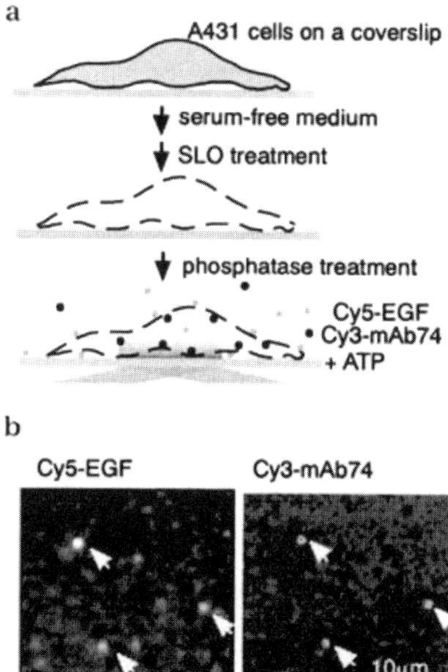

Fig. 8.16. Detection of autophosphorylation of EGF-EGFR complexes in single molecules. A431 cells were perforated by streptolysin O and stimulated by 10 ng/mL Cy5-EGF for 1 min. Phosphorylation of EGFR was detected by Cy3-mAb74 (anti-activated EGFR). Images were acquired 15 min after the addition of Cy3-mAb74 by using dual-view optics. Cy3-mAb74 bound at the position of oligomers of EGF-EGFR complexes (indicated by bright Cy5-EGF spots; *arrows* in b)

channel indicates FRET from Cy3-EGF to Cy5-EGF. The Förster distance between the Cy3/Cy5 pair is 5.8 nm [52]. The fluctuation of the FRET efficiency suggests fluctuations of several nanometers in the distance between two EGF–EGFR complexes.

8.2.3 ATP Synthase

In this section, the single ATPase reaction by single F_1-ATPase is presented. F_1-ATPase is a part of ATP synthase that is a membrane protein and that synthesizes ATP with an electrochemical gradient potential. And F_1-ATPase is a rotatory motor using the energy of ATP hydrolysis. The *Escherichia coli* F_1-ATPase, which was engineered for single molecular analysis, was immobilized on a glass surface, and single ATPase reactions by F_1-ATPase with 100 nM Cy3-ATP were imaged.

Single Molecular Analysis of F_1-ATPase. Recently, studies of linear motor protein molecules such as myosin, kinesin, and dynein at a single mol-

a

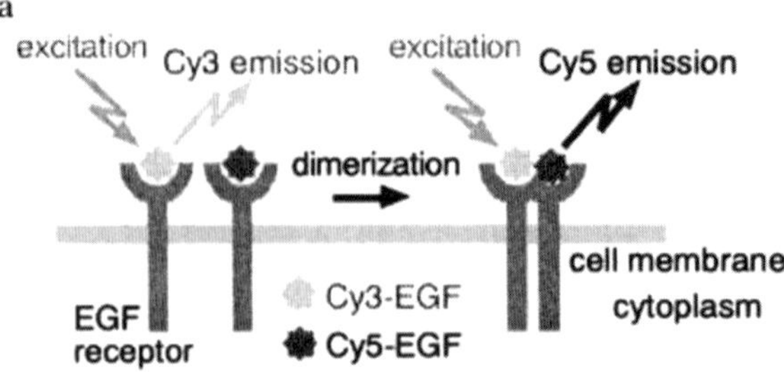

b

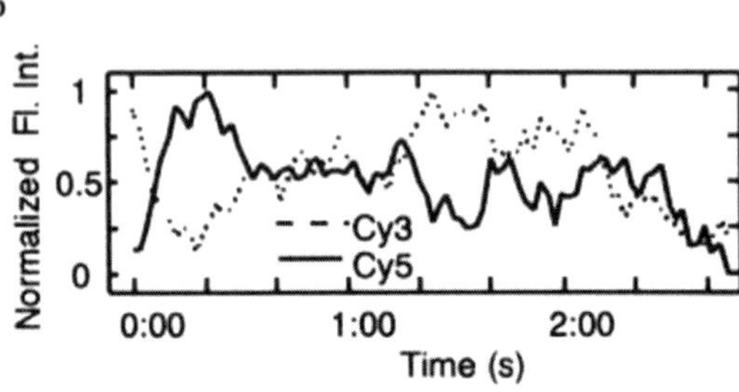

Fig. 8.17. Single-molecule FRETsingle-molecule FRET between Cy3-EGF and Cy5-EGF in a receptor dimer. A mixture of Cy3-EGF (0.2 ng/mL) and Cy5-EGF (0.8 ng/mL) was added to the culture medium of A431 cells. The specimen was excited with a green (532 nm) laser, and fluorescent images of the same field in the Cy3-channel (565–595 nm) and Cy5-channel (650–690 nm) were acquired simultaneously by using dual-view optics. The course of the fluorescent intensity change of a single spot was measured in both the Cy3 and Cy5 channels. To avoid photobleaching, the excitatory laser power was reduced to 10 $\mu W/\mu m^2$, and images were acquired using a four-frame rolling average. The fluorescent intensities in the Cy3 and Cy5 channels were plotted against time after normalizing the peak intensity of each channel

ecule level have made rapid progress (see Sect. 8.2.1 Myosin Actin Motor). In 1995, single motor protein molecules labeled with single fluorophore could be visualized in aqueous solution by a conventional highly sensitive TV camera using total internal reflection fluorescence microscopy (TIRFM) [38]. This made it possible to develop a new approach to investigating motor proteins.

F_1-ATPase is the component of ATP synthases (F_OF_1) [53], which is located in the membranes of mitochondria, chloroplasts, and *E. coli* and which synthesize ATP by an electrochemical gradient of protons. The F_O portion exists across the lipid bilayer and functions as a proton pathway. F_1-ATPase, which is a catalytic sector, contains three α subunits and three β subunits. Its symmetrical structure and chemical experimental data had predicted that the γ subunit, which is at the center of an $\alpha_3\beta_3$ structure, rotates synchronously with the ATP synthesis or hydrolysis. In 1997, Noji et al. confirmed at a single molecular level that F_1-ATPase is the rotatory motor [54]. The rotational motion of fluorescent-labeled marker molecules, which were connected at the γ subunit, were imaged, and the mechanical properties of F_1-ATPase were investigated. Since the chemical reaction and mechanical reaction are related, analysis of the single molecule ATPase reaction will be helpful for further investigation.

Imaging of Single Chemical Reaction. The events of single molecules in the ATPase reaction can be visualized with an evanescent field and the fluorescent ATP analog, 3'(2')-*O*-[*N*-[2-[(Cy3)amino]ethyl]-carbamoyl]-ATP (Cy3-ATP: Fig. 8.18a [39]). Figure 8.18b shows the principle of imaging. When several tens of nM Cy3-ATP is applied to the ATPase on the glass surface, the background fluorescence due to free Cy3-ATP is low because the

a

b

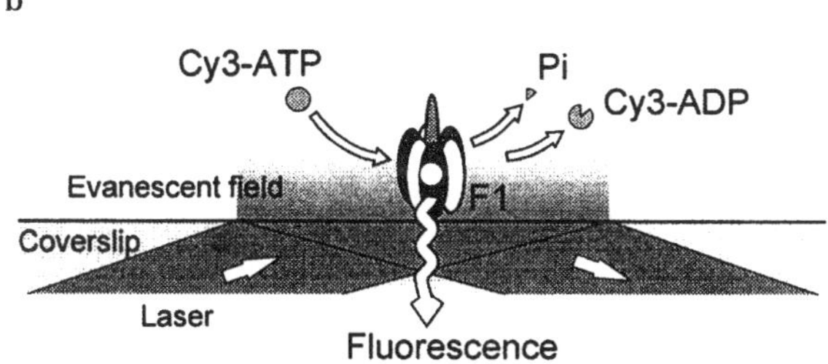

Fig. 8.18. **a** Structure of 3'(2')-*O*-[*N*-[2-[(Cy3)amino]ethyl]-carbamoyl]-ATP (Cy3-ATP). **b** Schematic diagram illustrating how the ATPase reaction can be visualized with Cy3-ATP using objective-type TIRFM. Pi: inorganic phosphate

illumination region is localized near the glass surface. Free Cy3-ATP undergoing rapid Brownian motion is not seen as a discrete spot. When Cy3-ATP is associated with surface-bound ATPase, it can be seen as a clear focused fluorescent spot. If the fluorescent lifetime is longer than the duration of Cy3-ATP/Cy3-ADP association, disappearance of the spot means the dissociation of Cy3-ATP/Cy3-ADP. Hence by observing the presence of focused Cy3 molecules on the surface, individual association–(hydrolysis)–dissociation of Cy3-ATP with a single ATPase molecule can be detected. This imaging technique for a single chemical reaction can only be achieved by using TIRFM. Epifluorescent microscopy excites Cy3-ATP that disperses around the optical axis, and the background fluorescence from these Cy3-ATP hides discrete spots on the glass surface.

Single-Molecule Imaging of Individual ATP Turnovers. Objective-type TIRFM was used to visualize a single Cy3-ATP [39,55]. An objective (PlanApo 100, Olympus), whose numerical aperture (NA) was 1.4, was mounted on the nosepiece of an inverted microscope (IX70, Olympus). The numerical angle θ_a, which is defined as NA $= n \sin\theta_a$ (n: the refractive index of glass is 1.52) is 67.1°. A laser beam from a diode-pumped Nd:YAG laser (wavelength: 532 nm) was reflected by a dichroic mirror and focused on the back focal plane of the objective. The incident beam path was at the edge of the objective, so the incident angle at the glass–water interface was greater than 61.0°, which is the critical angle of the glass–water interface. Fluorescence from Cy3 was imaged by the same objective. The background of scattered light was rejected with a holographic notch filter and a band-pass filter. Cy3-ATP was imaged with a SIT camera (C2400-08, Hamamatsu Photonics) and recorded at a video rate (1/30 s).

Escherichia coli F_1-ATPase was engineered for single molecular analysis [56]. To immobilize F_1-ATPase on the glass surface, His-Tag, which binds to nickel-nitrilotriacetic acid (Ni-NTA), was introduced into the α subunit. The glass surface was coated with Ni-NTA bound bovine serum albumin (BSA), and engineered F_1-ATPase was immobilized on it. The γ subunit Ser-193 is replaced with a Cys residue to make a chemical modification. The ATPase activity of this mutant was almost same as that of the wild type [56].

The events of the ATPase reactions were visualized with Cy3-ATP. To reduce the fluorescent background from the dust attached to the coverslip, coverslips were cleaned with 0.1 M KOH and washed thoroughly with distilled water. Flow cells were constructed from a cleaned coverslip and a smaller coverslip separated by thin strips. Ni-NTA bound BSA was prepared by coupling *N*-(5-amino-1-carboxypentyl) iminodiacetic acid (Dojindo) and BSA with ethyleneglycol-*O,O*'-bis(succinimidylsuccinate). The surface of a cleaned coverslip was coated with Ni-NTA-BSA by introducing 5 mg/mL Ni-NTA-BSA in an assay buffer (25 mM CH_3CH_2COOK, 5 mM $(CH_3COO)_2Mg$, 20 mM HEPES (2-[4-(2-hydroxyethyl)-1-piperazinyl]ethanesulfonic acid) pH = 7.8).

a

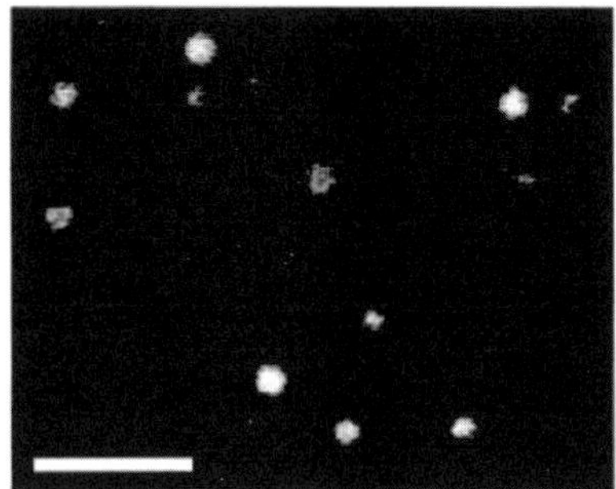

b

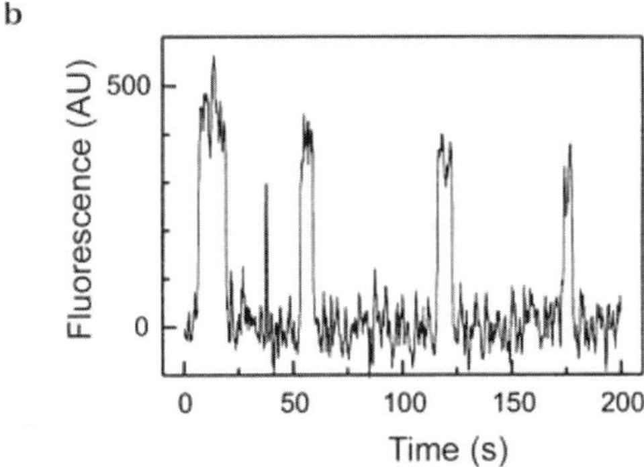

Fig. 8.19. Imaging of a single Cy3-ATP on a single F_1-ATPase molecule. **a** Fluorescence image of single Cy3-ATP molecules bound to F_1-ATPase immobilized on the coverslip. Sixteen video frames were averaged. Scale bar, 2 µm. **b** Fluorescent intensity of single Cy3-ATP molecules associated with a single F_1-ATPase molecule. The raw data of the intensity plot (33 ms time resolution) were passed through a nonlinear median filter of rank 30 (1 s wide). AU: Arbitrary Unit

After rinsing with the assay buffer, 20 pM F_1-ATPase in the assay buffer was introduced and allowed to bind Ni-NTA. After washing out unbound F_1-ATPase with the assay buffer, the bathing solution was replaced with the assay buffer containing 100 nM Cy3-ATP.

Figure 8.19a shows the fluorescent image of Cy3-ATP bound to F_1-ATPase. These fluorescent spots were not observed without introducing F_1-ATPase to the assay chamber. Figure 8.19b shows the fluorescent intensity of single Cy3-ATP molecules (or hydrolyzed Cy3-ADP) on a single F_1-ATPase. The increase of fluorescence corresponds to the association of a single Cy3-ATP molecule with the F_1-ATPase. The power of the incident laser was 0.8 mW at the nosepiece, and an area of approximately 20×50 µm at the specimen plane was illuminated. The fluorescent lifetime of the Cy3 fluorophore

(1/photobleaching rate) was 15 s. The decrease in fluorescence corresponds to the dissociation of Cy3-ADP or the photobleaching of Cy3.

Measurement of Mechanochemical Coupling. Using TIRFM, an individual single ATPase reaction by F_1-ATPase, which was immobilized on a glass surface, was visualized. The γS193C/αHis mutant F_1-ATPase retains rotational motion [56]. In this rotational assay, a fluorescent-labeled actin filament is attached to a γ subunit as an indicator. Single fluorophores and an actin filament, which was labeled with different color fluorophores, were simultaneously imaged [57]. Using this method, the single ATPase reaction and the rotational motion of F_1-ATPase can be visualized, and the relation between the chemical and mechanical reaction of F_1-ATPase can be investigated. Furthermore, an artificial lipid bilayer can be formed on an agarose-coated glass, and single fluorophores in this bilayer can be imaged using TIRFM [58]. Using this technique, F_1-ATPase connected to the F_O sector can be placed on the lipid bilayer, and the ATP synthase can be investigated in the more native environment rather than on a glass surface. Visualization of single molecules and single chemical reactions using evanescent fields has potential in researching ATP synthase and other membrane proteins.

8.3 Cell and Cellular Functions

8.3.1 Near-Field Fluorescent Microscopy of Living Cells.

Expectations for NSOM in Cell Biology. Cell biologists dream about observing single molecules or molecular complexes working in living cells, which would lead them to an understanding of the molecular mechanisms of various cellular processes. To realize such a dream, one needs a microscopic technology that allows one to observe living cells in an aqueous solution with a spatial resolution from one to several tens of nanometers. The NSOM is one of the promising technologies that might satisfy this requirement in the near future.

Conventional fluorescence microscopy allows observing single fluorescent molecules, but its spatial resolution is limited by the diffraction of light (200 nm). Electron microscopy gives resolutions better than 1 nm, but it cannot be used to observe living specimens. Atomic force microscopy (AFM) is another important candidate, but small molecular markers, such as fluorescent tags, cannot be used. NSOM is an optical microscopy that can be applied to living cells in an aqueous environment, and it has no diffractive limit. By using fluorescent probes, it can be used to identify molecules in cellular structures at high spatial resolution.

Recent Observations of the Cytoskeleton by Near-Field Scanning Fluorescence Microscopy. The highest lateral resolution achieved in fluorescence NSOM observations of cellular specimens is 30 nm, which was accomplished using dehydrated cultured fibroblasts [60]. This work compared the NSOM fluorescent image with the confocal fluorescent image of the same actin filaments in lamellipodia and showed that much finer structures can be observed using NSOM. The NSOM was used to detect single molecule fluorescence and the directions of fluorescent emission dipoles of single molecules [59,70]. Hwang et al. [64] detected 100-nm domain structures in the plasma membrane of cultured cells using NSOM. They chemically fixed and stained cells fluorescently, using an anti-HLA class I (a membrane protein) antibody and a fluorescent lipid analog, Bodipy-PC, and observed the cells after dehydration or in aqueous solutions. Enderle et al. [62] observed human red blood cells invaded by the malaria parasite after two-color immunofluorescence staining of the host membrane skeleton and the parasite proteins. After dehydration, the labeled proteins were observed using two excitatory laser beams to detect co-localization of the proteins in the membrane skeletal structure. The spatial resolution was 100 nm. In the above observations, straight-type optical fiber probes coated with aluminum were used, and the tip–sample distances were feedback regulated by the shear-force mode, in which decreases in the lateral vibratory amplitude of the probe were detected when the probe interacted with the sample surface [61].

The first report of using NSOM in liquid (the sample was chemically fixed cells in water) is the work by Muramatsu et al. [65], in which they used a bent-type probe that is fully functional as an AFM probe. The cytoskeletal structures of chemically fixed cells in an aqueous buffer were observed by transmission-mode NSOM at a spatial resolution of 70 nm.

We have been developing NSOM to observe live cells (rather than fixed cells) in aqueous solutions. The specific goal set for this project was to obtain fluorescent images of actin filaments (a cytoskeletal structure) in living cells at a spatial resolution of $\sim$ 100 nm, better than the diffractive limit by a factor of 3–4 [67–69]. There were three major difficulties in such observations. (1) The living cell surface is soft and difficult to scan. (2) The total measurment time after the sample is set on the instrument has to be minimized to keep the cells in good condition during observation. (3) The fluorescent signal intensity is weak. To overcome these points, the following design ideas were employed, respectively: (1) tapping-mode scanning using a bent tip working fully as an AFM probe, (2) ready switching between epifluorescence and NSOM for quickly searching and approaching the cell structure of interest with the NSOM probe, and (3) a trade-off with the spatial resolution by increasing the aperture size.

Instrumentation The NSOM was constructed on the stage of an inverted microscope (IX70, Olympus, Tokyo, Japan). The sample was irradiated with a 488-nm laser through a bent fiber tip coated with gold with a $\sim$ 120 nm-Φ aperture (Seiko Instruments, Muramatsu et al. [66]). The tip–sample distance was feedback regulated by the tapping mode, where the probe was vertically oscillated near the resonant frequency. The fluorescent signal emitted from the sample was collected by an objective lens and was detected by a photomultiplier tube (Hamamatsu-Photonics, R1878, Hamamatsu, Japan). A flat scanner with 50-μm strokes in the X, Y, and Z directions (Nanonics, NIS-50, Lieberman et al. [63]) was employed. The long stroke in the Z direction is important for scanning over living cells, which may be $\sim$ 10 μm tall, and for overcoming problems such as the thermal drift of the probe. This instrument enabled one to switch between epifluorescence and NSOM quickly, so that searching and approaching the cell structures of interest with the NSOM probe could be done easily. An SPI3700 controller (Seiko Instruments, Chiba, Japan) was used as a feedback controller for the piezo stage.

NSOM Fluorescence Observations of Living Cells. We observed cultured living epithelial cells that had been transfected with GFP-actin DNA by NSOM. Cells were cultured on a coverslip and were observed 3 days after passage. To obtain the images shown in Fig. 8.20, the cells were first imaged by NSOM and then were observed by conventional epifluorescence microscopy. The images were recorded with a cooled CCD camera (Roper, MicroMAX-512EBFT). Using NSOM, we observed fine and dark actin filaments, which could not be detected by a conventional fluorescence microscope when the same area was observed (Fig. 8.20b, inset of 8.20a). This was possible, apparently, because the NSOM image provided higher contrast than that of the conventional fluorescence microscope due to localized excitation by the NSOM. The spatial resolution was around 120 nm, as estimated from the half width of the cross section of the smallest structure in Fig. 8.20b. The movement of the actin filaments was observed by repeating the scans. These results indicate that the fluorescent mode of NSOM has become a useful tool for observing living cells in aqueous solutions, and thus NSOM has great potentials in cell biology.

8.3.2 Dynamics of Cell Membranes

Neuronal growth cones of a rat hippocampal neuron were stained with membrane-labeling fluorescent dyes (DiI or octadecyl rhodamine B). The dynamic changes of cell/substrate contacts of the neuronal growth cone were observed using the multi-imaging capability of the TIRFM system.

To understand integrin dynamics, human umbilical vein endothelial cells (HUVECs) were stained with FITC-labeled anti-CD29 (β1 subunit of integrin) and plated on a glass coverslip coated with fibronectin to visualize the clustering processes of integrins in living HUVECs. The dynamics of integrins in HUVECs were also observed by total internal reflection fluorescence microscopy (TIRFM).

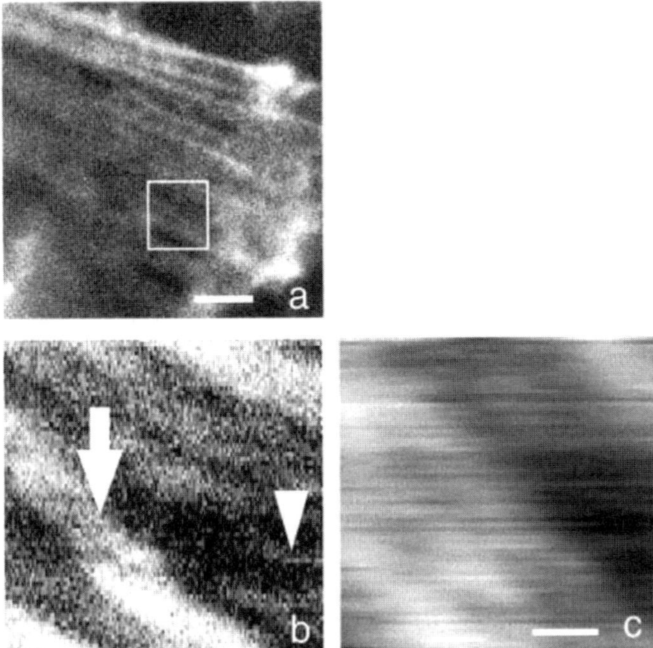

Fig. 8.20. **a** An image obtained by epifluorescence and a cooled CCD camera. GFP-actin filaments can be observed, but most of them are stress fibers, which are bundles of many actin filaments. Bar: 5 μm. **b** An NSOM fluorescent image of GFP-actin. The area of the inset in **a** was scanned. In addition to stress fibers (arrows), thin actin fibers (arrow heads) can be observed. **c** An AFM image that was simultaneously obtained with the NSOM image is shown. The ridges correspond to the stress fibers observed in **b**. Bar (**b, c**): 1 µm

Dynamics of the Membrane of Neuronal Growth Cones. Neurons protrude growth cones during development and also during nerve sprouting following neuronal damage [71]. Previous studies suggest that growth cones guide developing neurites along a precisely defined pathway and make synapses with their target cells. Thus, the growth cone plays an important role in pathway guidance. A number of possible mechanisms underlying the guidance of growth cones have been suggested. These include the ideas of selective adhesion [78], chemotaxis, electric fields [72], and the adhesive contact between the growth cone and neuron [80]. Taken together, these studies strongly suggest that adhesive contact between the growth cone membrane and substrate is necessary for the dynamic movement of growth cones. To fully understand the dynamic contacts between the growth cone and substrate, it is crucial to estimate the distance between the substrate and growth cone membrane of living neurons. A detailed analysis of ripping of the cell from the substrate and the formation of new contacts at growth cones would be necessary to understand the mechanisms of growth cone movement.

Interference reflection microscopy (IRM) [75] and total internal reflection fluorescence microscopy (TIRFM) have been found useful in visualizing the cell membrane–substrate contact region. In these techniques, illumination of a water-covered glass at angles greater than the critical angle results in an "evanescent field" extending several hundred nanometers into the aqueous phase. Only specimens in the vicinity of the glass/water interface are illuminated by the evanescent field and thus can be observed. Because the evanescent field declines sharply with respect to the z axis, TIRFM is very sensitive to the distance between the neuron and the substrate, more so than even IRM [74]; the cell–substrate contact region is selectively illuminated, thus minimizing the exposure of the cell interior to the excitatory light. As a result, fluorescence from fluorophores both in the bulk solution and inside the cells is suppressed, reducing background fluorescence significantly.

Since TIRFM visualizes only cell–substrate contact regions, observations of the same cell with epifluorescence and DIC are crucial for the correct interpretation of TIRFM images; DIC images give information on cellular structures, and epifluorescence images give the dye distribution in the entire cell.

In this report, TIRFM, video-enhanced DIC, epifluorescence, and IR contrast optics, were combined in a single microscope. The addition of a nanometer precision positioning system allowed examining the evanescent field depth in situ and manipulating subcellular structures. Rat hippocampal neurons were cultured on a coverslip, and their neuronal process and growth cones were observed with this system. Time-lapse TIRFM was used to image the region where neurons made close contact with the substrate and to analyze dynamic changes under high potassium stimulation. The materials and methods of this study are mentioned elsewhere [81].

Cell Membrane Labeling and Observation with a TIRFM and an IR Contrast Microscope. The cell membrane was fluorescent-labeled by microtube (0.1 mm diameter) pressure application of 1,1'-dioctadecyl-3,3,3',3'-tetramethylindocarbocyanine perchlorate [DiI_{18}] or $\mathrm{FM}^{\mathrm{TM}}$-DiI (water soluble derivative of DiI) in dimethyl sulfoxide (DMSO, 10 µg/50 µL). For time-lapse video microscopy, cells were labeled with octadecyl rhodamine B chloride (O246, Molecular Probes, Inc., Eugene Oregon USA). O246 (5 µL of stock solution 1 mg/mL N,N-dimethylformamide) was dissolved in the culture medium (5 mL) and applied to the specimen for 5 min. All of these dyes localize in the lipid bilayer of the cell membrane.

A TIRFM with a nanometer precision positioning system was constructed. TIRFM optics were incorporated into an inverted microscope (Nikon Diaphot TMD-300, Japan) along with DIC optics and epifluorescence illumination. The basic microscope was equipped with a water-immersion condenser

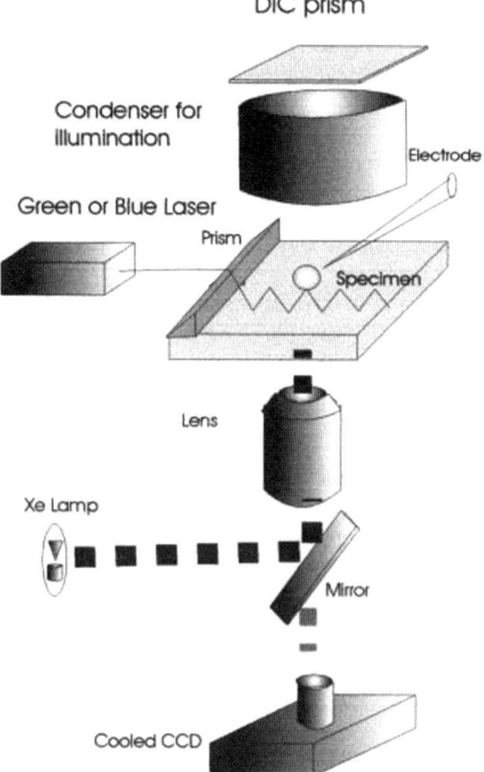

Fig. 8.21. Schematic of the multifunctional microscope used in this study. A laser beam (frequency-doubled Nd:YAG, 532 nm) was introduced into the glass substrate plate with an entrance prism. Neurons or human umbilical vein endothelial cells (HUVECs) were plated on a coverslip. The TIRF images of cells were observed with a water-immersion objective lens placed below the substrate. A glass capillary attached to piezo positioning equipment is to the right of the cell for stimulations

(NA = 0.9) for DIC illumination and a 60× water-immersion objective lens (NA = 1.2; Nikon, Japan) for DIC observation. This observation chamber consists of a 0.22-mm thick soda-lime glass coverslip and a rectangular prism (Fig. 8.21). Neurons were cultured in this chamber for 3–12 h before experiments. During observation, the temperature of the chamber was kept between 28 and 34°C. For TIRFM work, 532-nm light (3–50 mW) from an YAG laser (Coherent, DPSS 532, USA) was collimated by a lens (7 times beam expander; Meles Griot, England) to about a 5-mm diameter beam and directed into the rectangular prism (2 mm high BK 7 glass prism, $n = 1.522$, Nihon Ryokyo Inc., Japan) by an achromatic lens ($f = 200$ mm) and a mirror (Fig. 8.21). The fluorescent image was captured with either a cooled CCD camera (Micromax, Princeton Instruments, USA) or a SIT camera (C2400, Hamamatsu Photonics, Japan). Images of neurons were processed with a MetaMorph imaging system (Universal Imaging Co., USA). This setup is particularly well adapted for viewing cells growing on coverslips. This microscope arrangement is also extremely convenient, since samples can be manipulated from the upper free space.

The incident angle (θ) in our experimental condition was 66°. The depth (D_p) of the evanescent light penetration (1/e of surface light intensity) is estimated at 106 nm for 532-nm light using

$$D_\mathrm{p} = \frac{\lambda}{4\pi n_1}\sqrt{\sin^2\theta - \left(\frac{n_2}{n_1}\right)^2}\,, \tag{8.1}$$

where λ is the wavelength of the laser light and n_1 and n_2 are the refractive indexes of glass (1.52) and water (1.33).

Observations with TIRFM, Epifluorescence, and DIC Microscopes. Typical TIRFM, epifluorescent, and DIC images of the same neuron are shown in Fig. 8.22. The DIC image shows the soma, neurite, flat growth cone, and extended filopodia (Fig. 8.22c). The epifluorescent image shows the bright soma region and uniform staining of processes, growth cone, and filopodia (Fig. 8.22b). The TIRFM image of the same neuron was quite distinct from the DIC and epifluorescence images (Fig. 8.22a). In the TIRFM images, the center of the growth cone and the tip of the filopodium were bright in the majority of cells. The tip of the filopodium exhibited a bright

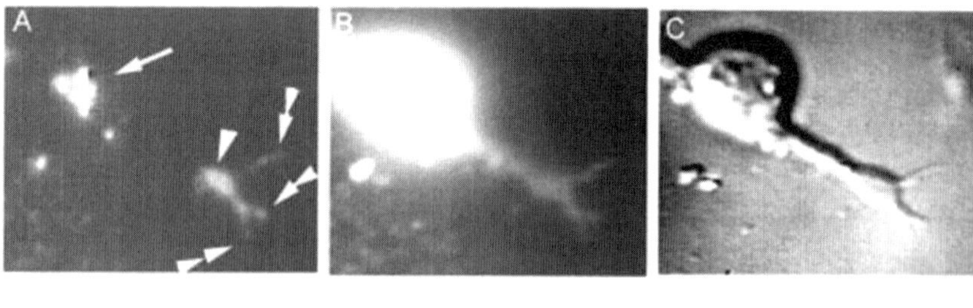

Fig. 8.22. Images of a single neuron stained with fluorescent dye (FM-DiI) and observed under total internal reflection fluorescence (**a**), epifluorescence (**b**), and video-enhanced DIC (**c**) microscopes. The soma, neurite, growth cone, and filopodia (**c**) are clearly visible in the DIC image. The epifluorescent image shows staining of soma, neurite, growth cone, and filopodia (**b**). In the TIRFM image, areas corresponding to the soma (arrow), growth cone (arrow head), and tip of the filopodium (double arrow heads) are bright, indicating substantial cell/substrate contact in these regions (**a**)

fluorescent spot with the bright area sometimes extending the entire length of the filopodium. Often, neural processes were not illuminated with evanescent light. Varicose cell structures observed along the neurite were illuminated with evanescent light and observed with TIRFM. Soma regions were also bright, and several bright spots or large soma area were observed. These TIRF images showed the regions that had the closest contact to the substrate. Similar images of growth cones were observed when cells were stained with rhodamine-conjugated con-A, which labels cell surface glycoproteins [73], or with O246.

Membrane Approach to the Substrate Following Excitatory Stimulation of Growth Cone. While high potassium solution was applied locally to the growth cone for 10–20 s to excite the neuron, TIRFM images were recorded at 5 s intervals. The high K^+ stimulation induced a transient increase in the fluorescent intensity at the growth cone. The maximum increase in fluorescent intensity was 64% of control after background subtraction 14%, $n = 12$). This value (64%) corresponds to a 46-nm movement of the cell membrane in the direction of the substrate. The fluorescent intensity returned to the basal level 3–5 min after the end of the high potassium stimulation. No significant increase of dynamic contact was observed in the growth cones when control-modified Kreb's solution was applied locally to the growth cone, excluding the possibility that the flow of the medium pressed the neurite down to the substrate during the ejection.

The TIRF and IR contrast images in this study showed that the ventral membrane of the growth cone and filopodia are attached to the substrate. The contact of the growth cone to the substrate in the dorsal root ganglion neurons (first reported using an IR contrast microscope [77]) it is thought, is due to an adhesive molecule, integrin, located on the tips of the growth cone filopodia [76]. This adhesion and the subsequent production of tension by filopodia results in forward motion of the growth cone [79]. These observations, including the present results, support the idea that tips of filopodia contact a target and in some case initiate wider contact of the growth cone with the target to produce dynamic movements.

Two types of contact, focal contact and matrix contact, have been reported in various types of cells. In focal contact, the membrane is separated from the substrate by 15–50 nm. In matrix contact, the cell membrane is separated from the substrate by 100 nm or more. In this study, TIRFM was used to estimate the separation between the ventral membrane of the growth cones and the substrate. TIRFM images were obtained only when the membrane was located less than 200 nm from the substrate. The mean separation during excitatory stimulation was estimated at around 50 nm in this study. This means that the growth cone membrane before stimulation was located between 50 and 200 nm from the substrate. Thus, the contact could be classified as matrix contact. This study has shown that TIRFM is adequate for

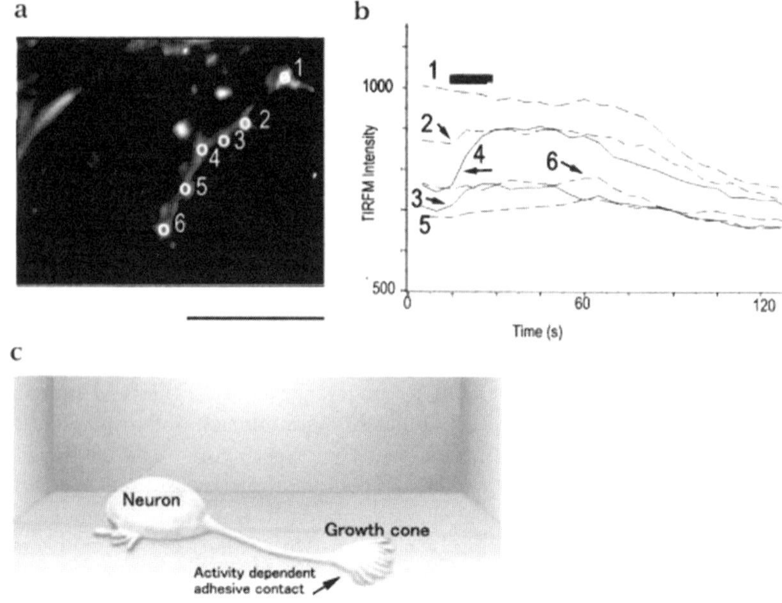

Fig. 8.23. The fluorescent intensity under TIR excitation increased in the ventral membrane of the growth cone following excitatory stimulation. Excitatory simulation of the growth cone was done by pressure ejection of a high potassium solution. **a** A TIRFM image of a growth cone. Fluorescent intensities were measured at regions denoted by 1 to 6. The changes in the fluorescent intensity are shown in **b**. High potassium stimulation was applied during the period shown by the horizontal bar. The bar in **a** denotes 10 µm. The schematic drawing shows that the growth cone approaches the substrate when excited

observing dynamic changes in membrane/substrate contact regions in living neurons. One important future question is the cellular and molecular mechanisms of the dynamic changes in the separation between the membrane and substrate. The following section briefly reports recent advances in observing of adhesive contact by TIRFM.

Clustering of Integrins upon Adhesion Contact in Endothelial Cells Studied by Multimode Imaging Microscopy. Clustering of adhesive molecules is necessary for strengthening adhesive contacts since numbers of molecules would share the detaching force on the cells. This section describes the dynamics of integrins in HUVECs observed by total internal reflection fluorescence microscopy (TIRFM).

To study integrin dynamics, human umbilical vein endothelial cells (HUVECs) were stained with FITC-labeled anti-CD29 (β1 subunit of integrin) and plated on a glass coverslip coated with fibronectin to visualize integrins in living HUVECs. We especially used TIRFM to selectively visualize labeled integrins at the basal surface of the cells. Movements of labeled integrins were recorded with a time lapse video system. The average size of the clusters was several micrometers long and approximately 1 μm wide. The number of the clusters increased, and each cluster elongated during 6 hours of recordings (Fig. 8.24). These cultured cells were fixed and stained with rhodamine-labeled phalloidin to visualize actin filaments. Actin filaments were located near integrin clusters, and fluorescent spots of integrins were associated with these actin fibers. The elongation of adhesion plaque was inhibited by cytochalasin D and ML-9 (a myosin phosphorylation inhibitor). These recent results suggest that clustering of integrins is mediated by an actin fiber-based transporting system in HUVECs [83]. The cellular mechanism for integrin cluster formation is left for future studies using TIRFM and advanced optical methods, e.g., SNOM.

8.3.3 Near-Field Imaging of Neuronal Cell and Transmitter

Imaging of Neuronal Cell. There is great interest in increasing the understanding of the mechanisms involved in information processing in neurons. A good example is the area of long term potentiation (LTP). LTP is an increase in signal throughput efficiency that the synapse undergoes after receiving a large number of signals. This increase in throughput efficiency is largely believed to be the basis for memory in the brain. The exact mechanism of LTP is, however, still unclear. It is believed that structural changes such as the concentration of NMDA (*N*-methyl-*D*-aspartate) glutamate receptors at synaptic interfaces play an important role. There has been some excellent research using electron microscopy, confocal microscopy, and AFM to this end. However, electron microscopy cannot look at live cells, and hence it is difficult to analyze live cell function by this method. Confocal microscopy has provided beautiful pictures of neuronal cells but suffers from the diffractive limit of light and it is not possible to make simultaneous topographic images. AFM can provide extremely high-resolution images, but topographic images alone can be hard to interpret. Hence, SNOAM, with its high optical resolution and topographic imaging ability is extremely useful.

Neuron imaging is different from imaging chromosomes or *E. coli*, because they are larger. The overall size of a neuron with its dendrites and axon can be tens of microns. However interesting features such as synaptic junctions and receptors are on the nanometer scale. Therefore, it is useful to be able to do initial scans over a large area and then successively zoom up to more interesting areas. In Fig. 8.25, there is a picture of two intertwined neurons. The initial scan is 72 microns for the entire area (a). The second scan zooms up on an area where the dendrites overlap (b). Then an area just below the overlap is both topographically and optically imaged (c,d). The optical image gives a mix of an autofluorescent signal combined with the transmission properties of light through the membrane. The flat areas

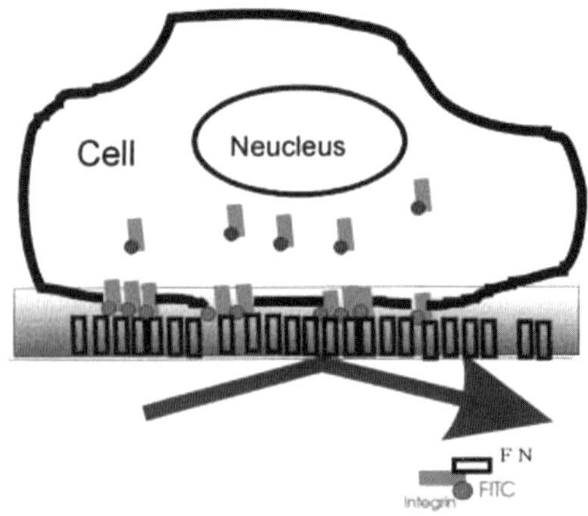

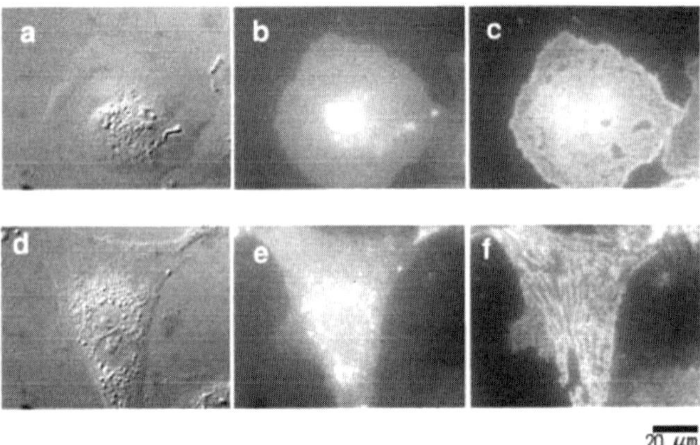

Fig. 8.24. Images of human umbilical vein endothelial cells (HUVECs) plated on a coverslip. **a** Image of TIRFM observation of HUVECs with an inverted microscope. **a** and **d** show the DIC images of the plated cells after 1 hour and 6 hours. **b** and **e** shows the epifluorecent images of the same cells. **c** and **f** show TIRFM images of fluorescent-labeled integrin. Note that the integrin clusters are imaged clearly by TIRFM compared with epifluorescenct images

of the cell give the highest throughput, whereas the curved areas and the uncovered glass give a lower throughput. Thus, it is possible to correlate optical and topographic images. Both images alone would be harder to interpret.

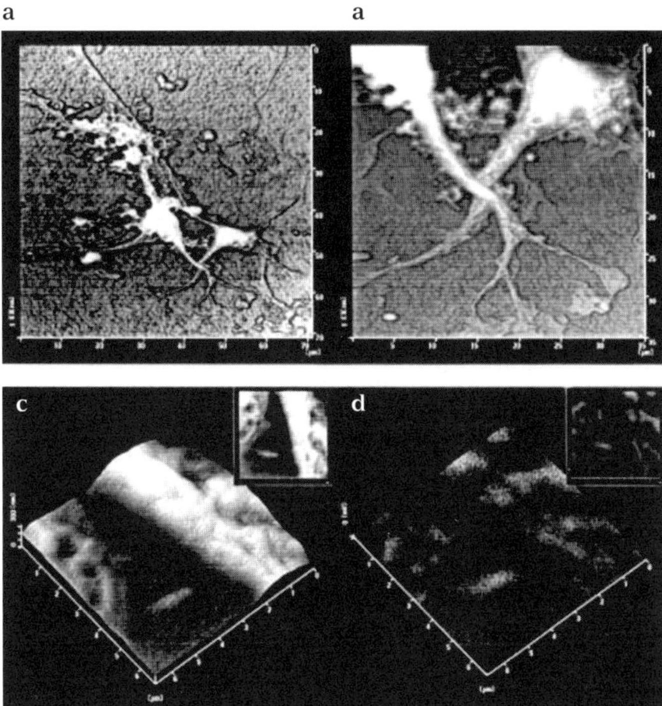

Fig. 8.25. A picture of two intertwined neurons. **a** The initial topographic scan. **b** The second scan zooms up on an area where the dendrites overlap. Note that the growth cones are clearly visible at the bottom right. **c** The topographic image of the membrane just below the overlap. **d** The optical image of **c**. Note that the darker regions indicate either no cell or a steep angle of the cell

For this, neurons were extracted from the hippocampal region of chick embryos. Embryos were extracted from chicken eggs grown typically for 8–12 days in an incubator. Cells were incubated in a DMEM medium with 10% BFS at 37°C and 5% CO_2 for 3–7 days to form neural networks. The cells in the above experiment were then fixed with 10% paraformaldehyde and dried. The SNOAM system for the above experiment was used in the tapping mode with nonfluorescent optics.

Imaging of Synaptic Glutamate Release. Glutamate plays the role of a neurotransmitter in the hippocampal region of chick embryos [83]. In this preliminary research [84], glutamate release from neurons has been imaged with Amplex Red by SNOAM. Amplex^{tm} Red was used to image the glutamate release. The reaction scheme is explained in Fig. 8.26 [85,86]. This is a one off reaction and hence is not applicable to real-time monitoring.

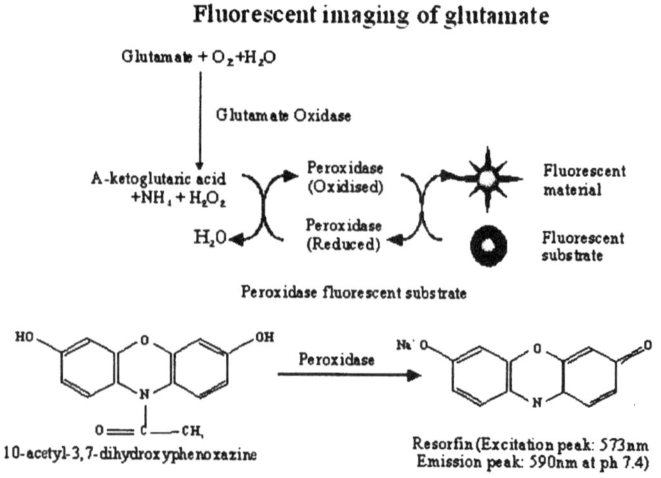

Fluorescent imaging of glutamate

Glutamate + O_2+H_2O

Glutamate Oxidase

A-ketoglutaric acid +NH_4 + H_2O_2

H_2O

Peroxidase (Oxidised)

Peroxidase (Reduced)

Fluorescent material

Fluorescent substrate

Peroxidase fluorescent substrate

Peroxidase

10-acetyl-3,7-dihydroxyphenoxazine

Resorfin (Excitation peak: 573nm Emission peak: 590nm at ph 7.4)

Fig. 8.26. Fluorescent imaging of glutamate. For this, neurons were extracted from the hippocampal region of chick embryos. Embryos were extracted from chicken eggs grown typically for 8–12 days in an incubator. Cells were incubated in a DMEM medium with 10% BFS at 37°C and 5% CO_2 for 3–7 days to form neural networks. The cells in the above experiment were then fixed with 10% paraformaldehyde and dried. The SNOAM system for the above experiment was used in a tapping mode with nonfluorescent optics

However, it is useful to demonstrate the validity of the technique. Cultured neurons have relatively high autofluorescence in the 488-nm region; hence, a fluorescent compound with a stimulation peak in the green region is desired. Amplex^{tm} Red has a stimulation peaks of 563 and 587 nm, respectively, and hence is ideal in this case.

Figure 8.26 shows the reaction scheme for detecting glutamate using Amplex^{tm} Red, glutamate oxidase, and peroxidase.

Neurons were extracted from the hippocampal region of chick embryos. Embryos were extracted from chicken eggs grown typically for 8–12 days in an incubator. Cells where incubated in a DMEM medium with 10% BFS at 37°C and 5% CO_2 for 3–7 days to form neural networks. The medium was changed every 72 hours, and the life span is about 2 weeks, 3 at the most. However, usually after 3–4 days there is a sufficient neural network for analysis.

First, the culture medium was replaced with a reaction buffer consisting of 170 mM NaCl, 3.5 mM KCl, 0.4 mM KH_2PO_4, 20 mM HEPS, 5 mM $NaHCO_3$, 5 mM glucose, 1.2 mM Na_2SO_4, 1.2 mM $MgCl_2$ and 1.2 mM $CaCl_2$ in 1 L of MilliQ water. Then 50 µL of 1 unit/mL peroxidase, 50 µL 0.25 u/mL glutamate oxidase and 10 µL 50 u/mL Amplex Red were added to the neuron culture. KCl was then added to the neurons to stimulate action potentials and hence glutamate release. The release of glutamate has a chain effect which causes the 10-acetyl-3,7-dihydroxyphenoxazine form of Amplex Red to turn into the fluorescent resorufin. After 30 minutes, 10% formalin solution was added to fix the neurons. Resorufine will spread through the solution and raise the general fluorescent intensity. However enough will remain around the neurons to show contrast. For these results, SNOAM was operated in the cyclic-contact mode, with 545–580 nm filters to obtain the fluorescent signal.

The technique is limited by the use of Amplex Red^{tm} rather than the SNOAM system. Calcium dyes such as Fluo-3/AM and Fluor-2/AM can be used to monitor calcium transients in real time, but at present there is no equivalent for glutamate. Of course, it can be argued that this is largely possible with confocal microscopy. However, the advantages of SNOAM over confocal microscopy are the simultaneous topographic image and added functionality. Added functionality can be realized by being able to stimulate neurons, either by pressure or electrically with the SNOAM probe. Thus the SNOAM technique will continue to develop as a useful tool for neurobiology.

The synaptic area is very small, however. Typical distances between pre- and postsynaptic membranes are typically 50 nm. Hence, SNOAM has advantages over its far-field counterparts. Another important consideration is fluorescent tagging of interesting areas, preferentially in real time. It is possible to look at calcium transients in real time, and it is possible to tag receptors such as the NMDA (*N*-methyl-D-aspartate) receptor which is linked to long term potentiation and memory.

Acknowledgments. The authors thank Prof. T. Yanagida and members of his laboratories in Osaka University and ICORP, Dr. M. Murata (National Institute of Physiological Sciences) Dr. M. Tokunaga (National Instutute of Genetics), members of the Futai Laboratory (Institute of Science and Industrial Research, Osaka University), Mr. Kawakami, and Dr. M. Sokabe (Nagoya University) for engineering F_1-ATPase, Streptolysin O, technical advice, and critical discussions.

References

1. S. Mononobe, R. Uma Maheswari, and M. Ohtsu: Fabrication of nanometer-level resolving probe with metal-dielectric-metal coat for near-field imaging of single strand DNA molecules. Near Field Optics-5, Tech. Dig. 5th Int. Conf. Near Field Opt. Rel. Tech., PC6. 1998
2. M. Naya, R. Micheletto, S. Mononobe, R. Uma Maheswari, and M. Ohtsu: Appl. Opt. **36**, 1681 (1997)
3. M. Schwartz, D. Axelrod, E. L. Feldman, and B. W. Agranoff: Brain Res. **194**, 171 (1980)
4. D. Gingell, O. S. Heavens, J. S. and Mellor: J. Cell Sci. **87**, 677 (1987)
5. T. Funatsu, Y. Harada, M. Tokunaga, K. Saito, and T. Yanagida: Nature **374**, 555 (1995)

6. E. Betzig and J. K. Trautman: Science **257**, 189 (1992)
7. E. Betzig and R. J. Chichester: Science **262**, 1422 (1993)
8. A. B. Lawrence, E. C. Joseph, and N. F. Phillip: Anal. Chem. **68**, 185R (1996)
9. U. T. Durig, D. W. Pohl, and F. Rohner: Appl. Phys. Lett. **59**, 3318 (1986)
10. N. F. van Hulst, M. H. P. Moers, O. F. J. Noordman, R. G. Tack, F. B. Segerink, and B. Bolger: Appl. Phys. Lett. **62**, 461 (1993)
11. H. Muramatsu, N. Chiba, T. Ataka, H. Monobe, and M. Fujihira: Ultramicroscopy **57**, 141 (1995)
12. H. Muramatsu, N. Chiba, K. Homma, K. Nakajima, T. Ataka, S. Ohta, A. Kusumi, and M. Fujihira: Appl. Phys. Lett. **66**, 3245 (1995)
13. N. Chiba, H. Muramatsu, T. Ataka, and M. Fujihira: Jpn. J. Appl. Phys. **4**, 321 (1995)
14. P. K. Hansma, J. P. Cleveland, M. Radmacher, D. A. Walters, P. E. Hillner, M. Bezanilla, M. Fritz, D. Vie, and H. G. Hansma: Appl. Phys. Lett. **64**, 1738 (1994)
15. C. A. J. Putman, K. O. Van der Werf, B. G. De Grooth, N. F. Van Hulst, and J. Greve: Appl. Phys. Lett. **64**, 2454 (1994)
16. M. Chalfie, Y. Tu, G. Euskirchen, W. W. Ward, and D. C. Prosher: Science **263**, 802 (1994)
17. G. Hans-Hermann and K. Christoph: FEBS Lett. **389**, 44 (1996)
18. H. Muramatsu, N. Chiba, T. Umemoto, K. Homma, K. Nakajima, T. Ataka, S. Ohta, A. Kusumi, and M. Fujihira: Ultramicroscopy **61**, 265 (1995)
19. S. Iwabuchi, E. Tamiya, et al.: Nucleic Acids Res. **25(8)**, (1997)
20. H. Muramatsu, K. Homma, N. Yamamoto, J, Wang, K. Sakata-Sogawa, and N. Shimamoto: Mater. Sci. (in press)
21. For example, C. R. Bagshaw: *Muscle Contraction*, 2nd ed. (Chapman & Hall, London 1984)
22. A. F. Huxley and R. Niedergerke: Nature **173**, 971 (1954)
23. H. E. Huxley and J. Hansen: Nature **173**, 973 (1954)
24. T. Yanagida, M. Nakase, K. Nishiyama, and F. Oosawa: Nature **307**, 58 (1984)
25. S. J. Kron and J. A. Spudich: Proc. Natl. Acad. Sci. USA **83**, 6272 (1986)
26. Y. Harada, K. Sakurada, T. Aoki, and T. Yanagida: J. Mol. Biol. **216**, 49 (1990)
27. A. Kishino and T. Yanagida: Nature **334**, 74 (1988)
28. A. Ishijima, T. Doi, K. Sakurada, and T. Yanagida: Nature **352**, 301 (1991)
29. A. Ishijima, Y. Harada, H. Kojima, T. Funatsu, H. Higuchi, and T. Yanagida: Biochem. Biophys. Res. Commun. **199**, 1057 (1994)
30. A. Ishijima, H. Kojima, H. Higuchi, Y. Harada, T. Funatsu, and T. Yanagida: Biophys. J. **70**, 383 (1996)
31. K. Svoboda, C. F. Schmidt, B. J. Schnapp, und S. M. Block: Nature **365**, 721 (1993)
32. J. T. Finer, R. M. Simmons, and J. A. Spudich: Nature **368**, 113 (1994)
33. J. E. Molloy, J. E. Burns, J. Kendrick-Jones, R. T. Tregear, and D. C. S. White: Nature **378**, 209 (1995)
34. H. Tanaka, A. Ishijima, M. Honda, K. Saito, T. Yanagida: Biophys. J. **75**, 1886 (1998)
35. A. H. Iwane, K. Kitamura, M. Tokunaga, and T. Yanagida: Biochem. Biophys. Res. Commun. **230**, 76 (1997)
36. A. Ishijima, H. Kojima, T. Funatsu, M. Tokunaga, H. Higuchi, and T. Yanagida: Cell **92**, 161 (1998)

37. K. Kitamura, M. Tokunaga, A. H. Iwane, and T. Yanagida: Nature **397**, 129 (1999)
38. T. Funatsu, Y. Harada, M. Tokunaga, K. Saito, and T. Yanagida: Nature **374**, 555 (1995)
39. M. Tokunaga, K. Kitamura, K. Saito, A.+H. Iwane, and T. Yanagida: Biochem. Biophys. Res. Commun. **235**, 47 (1997)
40. H. E. Huxley: Science **164**, 1356 (1969)
41. R. Dominiquez, Y. Freyzon, K. M. Trybus, und C. Cohen: Cell **94**, 559 (1998)
42. R. W. Lymn and E. W. Taylor: Biochemistry **10**, 4617 (1971)
43. Y. Sako, S. Minoguchi, and T. Yanagida: Nature Cell Biol. **2**, 168 (2000)
44. K. Kinoshita, H. Itoh, S. Ishiwata, T. Nishizaka, and T. Hayakawa: J. Cell Biol. **115**, 67 (1991)
45. A. B. Mathur, G. A. Truskey, and W. M. Reichert: Biophys. J. **78**, 1725 (2000)
46. Y. Sako, K. Hibino, T. Miyauchi, M. Miyamoto, M. Ueda, and T. Yanagida: Single Mol. **1**, 151 (2000)
47. Y. Yarden and J. Schlessinger: Biochemistry **26**, 1443 (1987)
48. Y. Yarden and J. Schlessinger: Biochemistry **26**, 1434 (1987)
49. R. Campos-Gonzalez and J. R. Glenney, Jr.: Growth Factors **4**, 305 (1991)
50. F. Kano, K. Takenawa, A. Yamamoto, K. Nagayama, E. Nishida, and M. Murata: J. Cell Biol. **149**, 357 (2000)
51. S. Weiss: Science, **283**, 1676 (1999)
52. Y. Ishii, T. Yoshida, T. Funatsu, T. Wazawa, and T. Yanagida: Chem. Phys. **247**, 163 (1999)
53. P. D. Boyer: Biochim. Biophys. Acta **1365**, 3 (1998)
54. H. Noji, R. Yasuda, M. Yoshida, and K. Kinosita Jr: Nature **386**, 299 (1997)
55. D. Axelrod: Methods Cell Biol. **30**, 245 (1989)
56. H. Omote, N. Sambonmatsu, K. Saito, Y. Sambongi, A. Iwamoto-Kihara, T. Yanagida, Y. Wada, and M. Futai: Proc. Natl. Acad. Sci. USA **96**, 7780 (1999)
57. K. Saito, M. Tokunaga, A. H. Iwane, T. Yanagida: J. Microsc. **188**, 255 (1997)
58. T. Ide, T. Yanagida: Biochem. Biophys. Res. Commun. **265**, 595 (1999)
59. E. Betzig and R. J. Chichester: Science **262**, 1422 (1993)
60. E. Betzig, R. J. Chichester, F. Lanni, and D. L. Taylor: Bioimaging **1**, 129 (1993)
61. E. Betzig, J. K. Trautman, T. D. Harris, J. S. Weiner, and R. L. Kostelak: Science **251**, 1468 (1991)
62. T. Enderle, T. Ha, D. F. Ogletree, D. S. Chemila, C. Magowan, and S. Weiss: Proc. Natl. Acad. Sci. USA **94**, 520 (1997)
63. K. Lieberman, N. BenAmi, and A. Lewis: Rev. Sci. Instrum. **67**, 3567 (1996)
64. J. Hwang, L. A. Gheber, L. Margolis, and M. Edidin: Biophys. J. **74**, 2184 (1998)
65. H. Muramatsu, N. Chiba, K. Homma, K. Nakajima, T. Ataka, S. Ohta, A. Kusumi, and M. Fujihira: Appl. Phys. Lett. **66**, 3245 (1995)
66. H. Muramatsu, N. Chiba, K. Homma, T. Ataka, M. Shigeno, H. Monobe, and M. Fujihira: Ultramicroscopy **71**, 73 (1998)
67. S. Ogawa, S. Ohta, G. Marriott, and A. Kusumi: Cell Struct. Function (in press)
68. S. Ohta-Iino, S. Ogawa, G. Marriott, and A. Kusumi: Biophys. J. (Abstract) (2000)
69. S. Ohta-Iino, S. Ogawa, G. Marriott, and A. Kusumi: Near-field fluorescence microscopy of living cells (2000, submitted)

70. X. S. Xie, and R. C. Dunn: Science **265**, 361 (1994)
71. M. C. Brown, R. L. Holand, and W. G. Hopkins: Annu. Rev. Neurosci. **4**, 17 (1981)
72. R. W. Devenport and C. D. McCaig: J. Neurobiol. **24**, 89 (1993)
73. E. L. Feldman, D. Axelrod, M. Schwartz, A. M. Heacock, and B. W. Agranoff: J. Neurobiol. **12**, 591 (1981)
74. D. Gingell, O. S. Heavens, and J. S. Mellor: J. Cell Sci. **87**, 677 (1987)
75. D. Gingell and I. Todd: Biophys. J. **26**, 507 (1979)
76. P. N. Grabham and D. J. Goldberg: J. Neurosci. **17**, 5455 (1997)
77. R. W. Gundersen: J. Neurosci. Res. **21**, 298 (1988)
78. J. A. Hammarback, J. B. McCarthy, S. L. Palm, L. T. Furcht, and P. C. Letourneau: Dev. Biol. **126**, 29 (1988)
79. P. Lamoureux, R. M. Buxbaum, and S. R. Heidemann: Nature, **340**, 159 (1989)
80. H. Tatsumi and Y. Katayama: Neurosci. Res. **99**, 855 (1999)
81. H. Tatsumi, Y. Katayama, and M. Sokabe: Neurosci. Res. **35**, 197 (1999)
82. K. Kawakami, H. Tatsumi, and M. Sokabe: J. Cell Sci. **114**, 3125 (2001)
83. Ryo Tokioka et al.: Developmental Brain Research **74**, 146 (1993)
84. P. Degenaar et al.: *Proc. SPIE – Progress in Biomedical Optics and Imaging II, vol. 1*, **16**, 189 (2000)
85. J. G. Mohanty et al.: J. Immunol. Methods **202**, 133 (1997)
86. M. Zhou et al.: Anal. Biochem. **253**, 162 (1997)

9 Near-Field Imaging of Quantum Devices and Photonic Structures

M. Gonokami, H. Akiyama, and M. Fukui

9.1 Spectroscopy of Quantum Devices and Structures

The requirement of spatial resolution in the nanometer regime has led to the development of electron, X-ray and scanning probe microscopy. However, these techniques do not possess the advantages of optical microscopy, such as nondestructiveness, low cost, and versatility. Therefore, the combination of the interactive mechanism of optical microscopy and the high spatial resolution of scanning probe methods is important, and NSOM was invented. This technique is used in many applications in semiconductor physics, biology, and other fields.

After the invention of room temperature NSOM in optical frequencies, low temperature NSOM was constructed at Bell Labs by Grober et al. [1], combining the inherent high spatial resolution in NSOM and the high spectral resolution and high quantum efficiency available at cryogenic temperatures to investigate localized spectroscopy of quantum wells (QWs), quantum wires (QWRs), and quantum dots (QDs). Using this technique, the authors [1] investigated cleaved edge overgrowth QWs and identified the luminescent peaks from the (001)-oriented multiple QWs, the (110)-oriented single QWs, and the QWRs. They have observed that the line width of the emission from the QWR is related to roughness in the single QW and the quenching of the emission from single QW and multiple QW near the QWRs is due to the diffusion of photoexcited carriers into the QWRs. Hess et al. [2] investigated the low-temperature photoluminescence (PL) of QWs by applying a magnetic field using this technique. They imaged the spectrally distinct emission lines and established that these luminescent centers arise from the excitons localized laterally at local potential minima formed by interface fluctuations of QWs.

NSOM has also been extended to probe single-molecule dynamics. Betzig et al. [3] showed that this technique can locate single chromosphores as well as their orientations. Using NSOM, they imaged individual carbocyanine dye molecules in a submonolayer spread and determined the orientation of each molecular dipole. Trautman et al. [4] measured the time-dependent emission spectrum of a single molecule in air at room temperature using NSOM. Xie et al. [5] investigated the room temperature dynamics of single sulforhodamine 101 molecules on two different timescales with NSOM, which provided insight into their spectroscopic properties. Ambrose et al. [6] measured the fluorescent lifetimes of single Rhodamine 6G molecules with pulsed laser excitation employing NSOM.

NSOM has also been employed to investigate the coherent nonlinear optical response of individual excitons in semiconductor QDs. Bonadeo et al. [7] combined single quantum dot probing with nonlinear optical spectroscopy to demonstrate the similarities between the coherent nonlinear optical response of individual excitons in GaAs QDs and atoms in their coherent optical interactions. Nechay et al. [8] developed the near-field degenerate pump-probe method to investigate the carrier dynamics in GaAs single quantum wells locally patterned by focused-ion-beam (FIB) implantation. They observed that lateral carrier diffusion plays an important role in the decay of the excited carriers.

Overall, room- and low-temperature NSOM are used all over the world to investigate various quantum devices and structures, such as quantum wells, wires, and dots by cw and also by ultrafast spectroscopy.

Other types of near-field spectroscopy such as solid immersion microscopy [9] have been developed to investigate a variety of systems. The method has been developed to strike a trade-off between the need for high spatial resolution and large transmission efficiency. We discuss the recent progress in the following sections.

9.1.1 Near-Field Microscopy with a Solid-Immersion Lens

A solid-immersion lens (SIL) is an aberration-free lens of high-refractive-index material with a truncated-sphere shape placed in contact with a sample. Its basic principle is the same as that of the oil-immersion microscope and hence is well known; all-solid oil-free operation enables such advantages as no contamination of specimen and usefulness in vacuum and at low temperature.

The SIL technique had originally been investigated for confocal microscope and optical data storage applications [9–15] but has recently been developed in various applications in NSOM probes [16], lithography [17], microscopy and spectroscopy of semiconductors [18,19], and, in particular, photoluminescent (PL) microscopy of nanostructures in vacuum and/or at low temperatures [20–23].

Besides high-n glass (n = 1.6–2.0), sapphire (n = 1.76) [23] and GaP (n = 3.1–3.4, depending on the wavelength) [24,25] have been used as the material of SILs for visible and near-infrared light, and an effective NA as high as 2.0 has been achieved [25].

There are two types of SILs, as shown in Fig. 9.1. One is the hemispheric SIL, and the other is the Weierstrass-sphere SIL, which is a superhemispheric SIL whose thickness is equal to $(1 + 1/n)a$, where n and a are the refractive index and the radius of the SIL, respectively.

For these two special cases, no aberration exists in imaging the central region of the bottom surface. In addition, there is no chromatic aberration

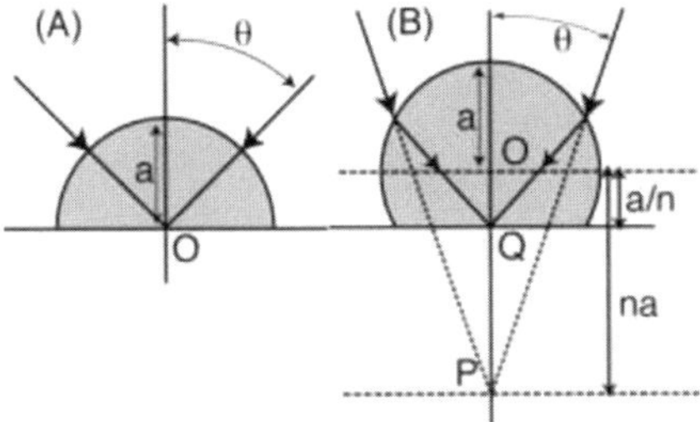

Fig. 9.1. Schematics of two types of SILs: **a** hemisphere and **b** Weierstrass sphere (superhemisphere)

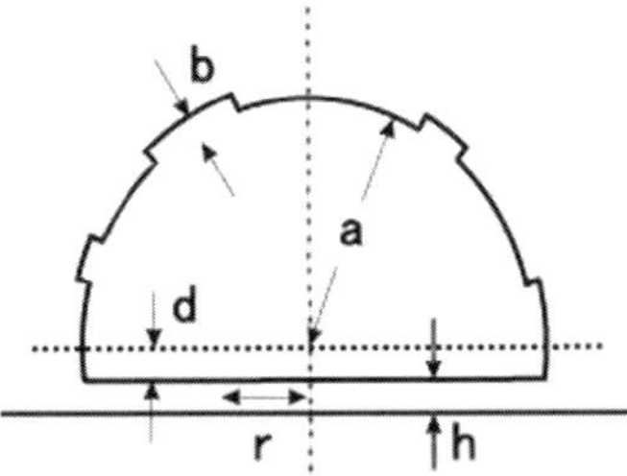

Fig. 9.2. Schematic of typical errors in a hemispheric SIL: an aspheric error b, a thickness error d, and an air gap h

in the hemispheric SIL since all light rays are incident normally to the hemispheric surface and converge at the center. The magnification and resolution improvement factors are both n in hemispheric SILs, whereas they are n^2 in Weierstrass-sphere SILs. When one prepares a SIL by polishing a ball lens or by molding and uses it in practice, however, deviation from the ideal design induces various aberrations and hence degrades microscopic images, which we discuss later for the hemispheric SIL.

Figure 9.2 shows the four typical errors, or factors that degrade the resolution in a hemispheric SIL with radius a: an aspheric error b, a thickness error d, the field-of-view diameter $2r$, and an air gap h. The allowances for the first three parameters can be estimated via the "quarter-wavelength condition" on aberrations $W < \lambda/4$, which are [26]

$$|b| < \frac{\lambda}{4(n-1)}, \quad |d| < [\frac{2a\lambda}{n(n-1)}]^{1/2}, \quad 2r < [\frac{2a\lambda}{n(n-1)}]^{1/2}. \tag{9.1}$$

They become larger for smaller n, larger λ, and larger a. To use a SIL with larger a, attention should also be paid to the bottom-surface flatness of the SIL and the working distance of an objective lens. In a typical SIL with $a = 1$ mm and $n = 2$ for $\lambda = 600$ nm, the accuracy of a spheric surface should be within $\pm\lambda/4$ or 150 nm, which is achieved in some commercial ball lenses. In the same case, the tolerable thickness error $|d|$ and the field-of-view

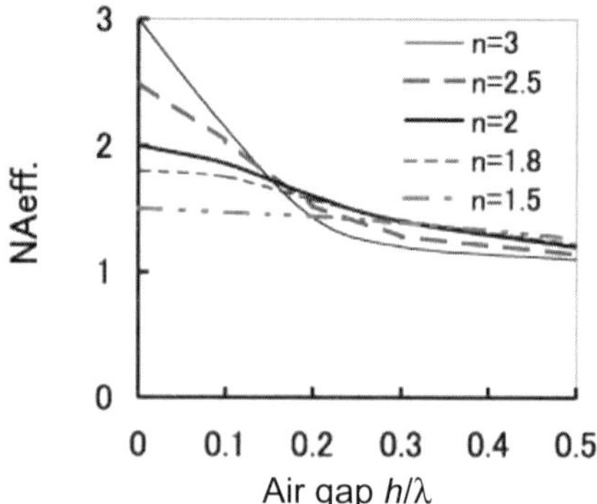

Fig. 9.3. Effective numerical aperture $\mathrm{NA_{eff}} \equiv 0.51\lambda/\mathrm{FWHM}$ as a function of h/λ for several values of n, where FWHM is the full width of the half maximum of the point-spread function

diameter $2r$ are 24 μm. This is large enough for fabrication accuracy of the order of 1 μm and the size of the region of interest.

The allowance for h is strongly limited by the air-gap transmission coefficient rather than by the aberration. The air gap reduces the amplitude of a plane wave with large θ, or the evanescent wave, and hence limits the effective numerical aperture. In Fig. 9.3, we plot the effective numerical aperture $\mathrm{NA_{eff}} \equiv 0.51\lambda/\mathrm{FWHM}$ as a function of h/λ for several values of n. Since $\mathrm{NA_{eff}}$ decays rapidly from the ideal value of n to 1 for increased h/λ, we need tight contact between an SIL and an object below the spacing of about 0.2 λ. Though a SIL with higher n can have better resolution, allowances for air gap h, as well as aspheric error b, thickness error d, and field of view 2 r, become narrower to keep the high resolution.

The important advantages of SILs in fluorescence microscopy are the high spatial resolution and also the great improvement in collection efficiency [18,27,28]. This effect is due to the characteristic radiation by the dipole located near the surface of a dielectric medium, which is emitted predominantly toward the high-refractive-index medium [29–31]. Figure 9.4 shows the calculation of the emission patterns of randomly oriented dipoles at distances $z = 0$ μm (solid line) and 0.11 μm (dotted line) from the surface of a SIL with $n = 1.845$ [27].

Figure 9.5 shows the calculated percentages of fluorescence collected by objective lenses with NA = 1, 0.8, and 0.55 on the SIL ($n = 1.845$) and the air sides for $z = 0$–0.22 μm. For example, at $z = 0.11$ μm, about 62% of the total power is emitted toward the SIL and 38% into the air. Then, a NA = 0.8 (0.55) objective collects 59% (37%) of fluorescence through the SIL and 23% (11%) via air, respectively. The percentage reaches a maximum of 89% at $z = 0$ for NA = 1.

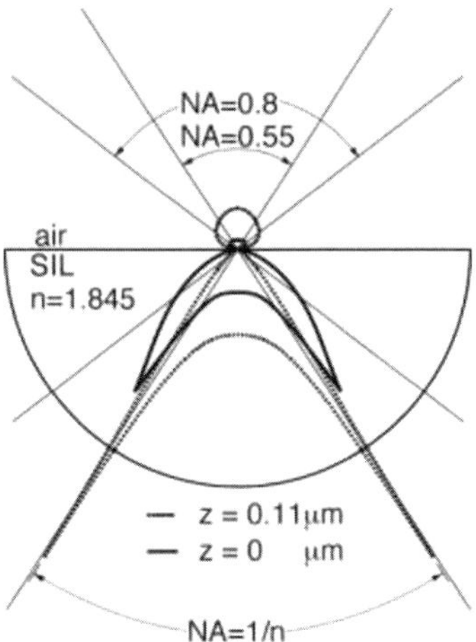

Fig. 9.4. The emission patterns of randomly oriented dye molecules located in the air region at distances $z = 0$ (solid line) and 0.11 µm (dotted line) from the SIL–air surface. The refractive index n of the SIL is 1.845

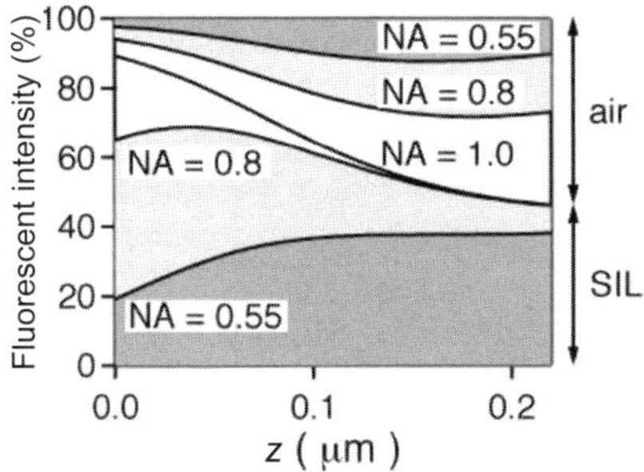

Fig. 9.5. Normalized fluorescent intensities collected by objective lenses with NA = 1 (*the center line*), 0.8 (*second pair of lines from the center*) and 0.55 (*third pair of lines from the center*) via the SIL (*axis from the bottom to the top*) and air (*axis from the top to the bottom*) for $n = 1.845$, plotted between $z = 0$ and $z = 0.22$ µm

Figure 9.6 shows the n-dependence of the percentages of fluorescence for $z = 0$ collected by objective lenses with NA = 1, 0.95, 0.8, and 0.55. Using SILs made of materials with a larger refractive index n and objective lenses with higher NA, we can approach 100% collection of fluorescence.

This calculation was indeed confirmed by Koyama et al. [27,28] by measuring the light-collection efficiency and Fourier images of red-fluorescent-dye-doped ($\lambda \sim 600$ nm) polystyrene beads 0.11–0.22 µm in diameter, which are directly attached to the flat surfaces of hemispheric SILs. For example, the measured efficiency improvement factor was 7.6 ± 0.2 for the beads 0.11 µm

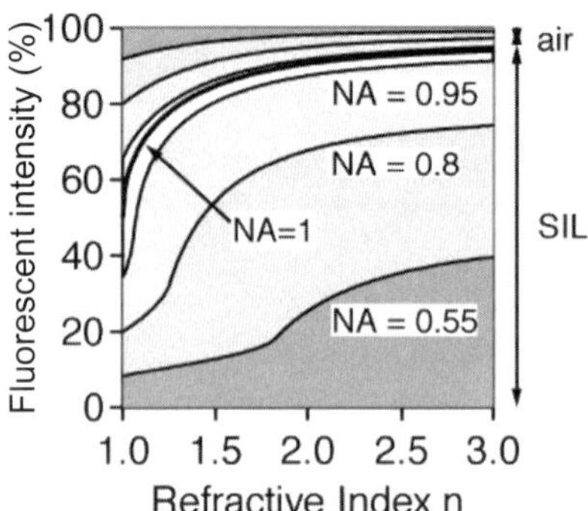

Fig. 9.6. Normalized fluorescent intensities collected by objective lenses with NA=1 (*the center line*), 0.95, 0.8, and 0.55 (*the second to fourth pairs of lines from the center*) via the SIL (*axis from the bottom to the top*) and air (*axis from the top to the bottom*), for $z = 0$ μm plotted against $n = 1$–3

in diameter. The experimental results were well explained by the calculated efficiency by assuming representative z values of 0.02 μm. The estimated collection efficiencies in the SIL and conventional microscopy were 62% and 8%, respectively.

They measured the Fourier images of emission as the fluorescent intensity on the back focal plane of the objective via relay lenses [28], where the original emissive intensity in the direction of the polar angle θ is transformed to the intensity at the radius proportional to $\sin\theta$.

Figure 9.7a shows the calculated emissive pattern for SILs with $n = 1.845$ (*solid curve*) and 1.687 (*dashed curve*) when $z = 0.06$ μm. Figure 9.7b shows a Fourier image of a 0.22-μm bead on a $n = 1.687$ SIL obtained on the air side. Figure 9.7c,d shows Fourier images of a 0.22-μm bead obtained on the SIL side with $n = 1.687$ and 1.845. Each has a bright ring with each radius which corresponds to the emission peak at each critical angle in the calculated patterns in (a).

9.1.2 Solid-Immersion Microscopy of GaAs Nanostructures

A SIL was first applied to fluorescenct, or photoluminescent (PL) microscopy of quantum device structures for GaAs quantum wells (QWs) by Sasaki et al. [18] at room temperature and by Yoshita et al. [20,21] at low temperatures. They demonstrated imaging and spectroscopy with uniform and point excitation with SILs. Wu et al. [23] measured local PL excitatory spectra and spatial correlation between excitation and detection of PL to show localized and delocalized states in narrow QWs. Vollmer et al. [24] first demonstrated ultrafast spectroscopy of QWs with SIL.

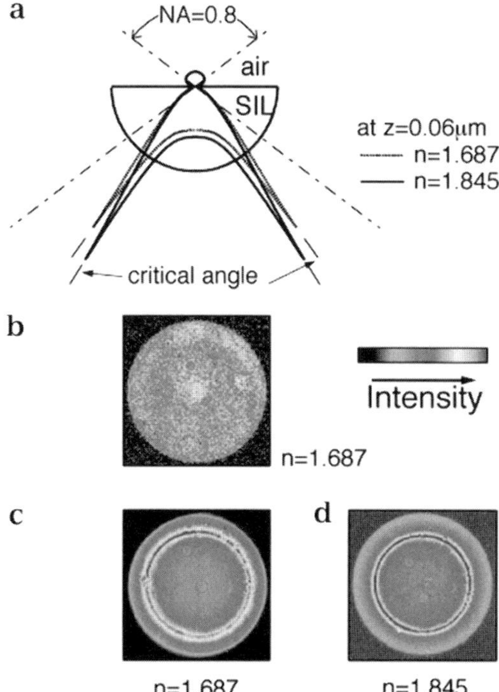

Fig. 9.7. a Calculated emission patterns from randomly oriented dipoles located in the air region at a distance $z = 0.06$ μm from SIL–air surfaces. The dotted and solid lines are calculated for refractive indexes n of the SIL, 1.687 and 1.845, respectively. **b**–**d** Observed Fourier images (on the back focal plane of the NA = 0.8 objective) for the fluorescence of a 0.22-μm diameter microbead, on the air side (**b**) or the SIL side with an $n = 1.687$ (**c**) and an $n = 1.845$ (**d**) SIL

Now we describe the work by Yoshita et al. [20,21] to describe imaging and spectroscopy with uniform and point excitation with SILs at low temperatures.

They used a Weierstrass-sphere SIL made from a TaF-3 glass ball lens (Nippon electric glass) with refractive index $n = 1.8$ and diameter of 750 μm (radius $r = 375$ μm). In their micro-PL setup, the SIL and a sample were placed in a cryostat. An objective lens was placed outside the cryostat, and observations were made through an optical window of the cryostat. They used the objective lens (Union, PLLWDC40×) with a nominal magnification factor of 40, working distance of 10 mm, and numerical aperture (NA) of 0.5, in which an aberration caused by the optical window was compensated for. Two-dimensional images were detected by a −40°C-cooled CCD camera (Prinston Instruments, TE/CCD512TKM/1).

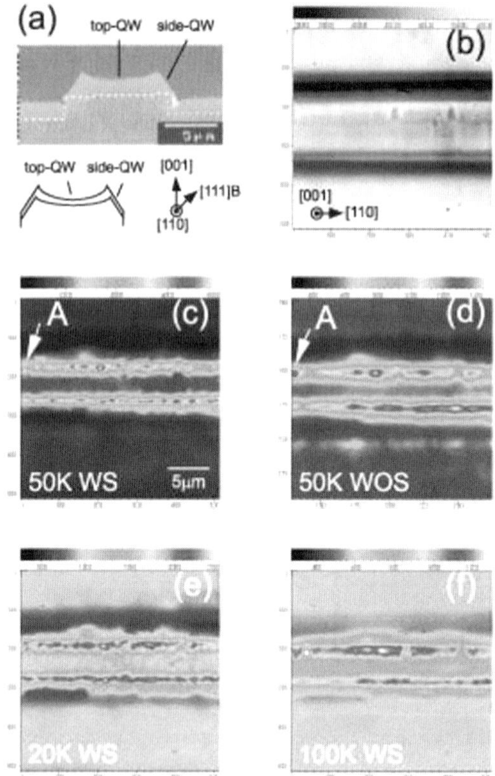

Fig. 9.8. a A cross-sectional SEM image of a GaAs QW sample grown on a patterned GaAs (001) substrate with 10-μm wide mesa stripes in the [110] direction. **b** The reflection image of the sample. The PL images measured **c** with and **d** without the SIL at 50 K, **e** with the SIL at 20 K, and **f** with the SIL at 100 K under uniform excitation

In the reflection measurement, light from a 100-W mercury-arc lamp filtered at 546 nm by an interference filter illuminated the sample surface under coherent illumination. In the uniform excitatory PL measurement, excitatory light from the lamp filtered at a wavelength from 510 to 560 nm by a band-pass filter and reflected by a dichroic mirror illuminated the sample, whereas PL light reached the CCD camera through a dichroic mirror and a short-cut glass filter. For point excitation, light from a He–Ne laser was focused on the sample surface with a spot size of 0.4 μm via an objective lens and the SIL.

The sample used was a facet-growth GaAs QW. A GaAs QW with a nominal vertical thickness of 5 nm sandwiched by AlAs barriers was grown on a patterned GaAs (001) substrate with 10-μm wide mesa stripes in the [110] direction by molecular beam epitaxy (MBE) following the growth of a 1.5-

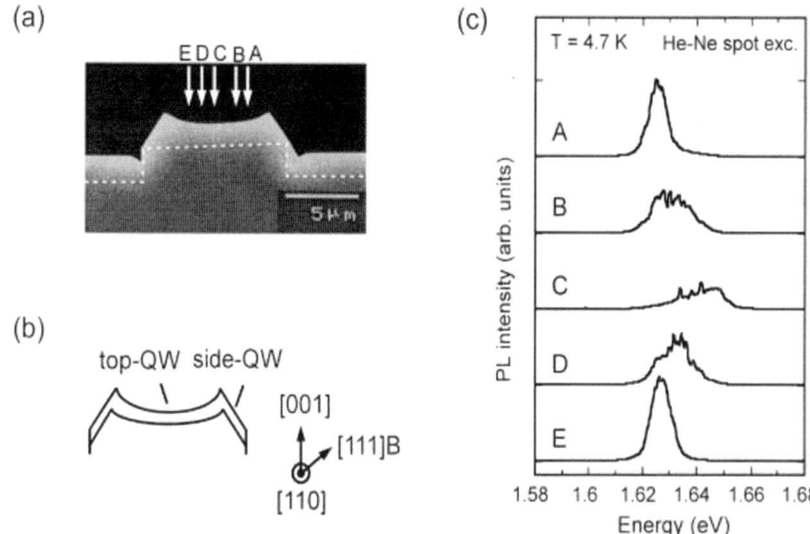

Fig. 9.9. **a** A cross-sectional scanning electron microscope image of a GaAs QW sample grown on a patterned GaAs (001) substrate with 10-μm wide mesa stripes in the [110] direction and **b** a schematic of QW thickness variation. The dashed line indicates the original mesa-patterned substrate. **c** PL spectra measured with point excitation at positions on the top-QW indicated as *A* to *E* in the SEM image in **a**. The spot size of the excitation was 0.4 μm

μm thick GaAs buffer layer. Its cross-sectional scanning electron microscope (SEM) image is shown in Fig. 9.8a.

As a result of the facet-growth process with atom migration between facets, it is expected that the top QW on the top curved surface should get thicker and hence, have lower quantization energy near both edges than at the center on the mesa stripe, as schematically shown in Fig. 9.8a.

The PL images observed with and without the SIL at the same position at 50 K under uniform excitation are shown in Fig. 9.8c,d, respectively, against a common scale for the position. Figure 9.8e,f shows PL images with the SIL at 20 and 100 K at the same position in Fig. 9.8c, respectively. Note that all of these PL images with varied temperatures were stably obtained for the same region as well for the reflection image shown in Fig. 9.8b. The additional magnification factor of the transverse scale by using the SIL was 3.39, which is fairly in good agreement with the theoretical value of 3.24 ($= n^2$).

It is clear from a glance at the PL images in Fig. 9.8c,d that the spatial resolution is enhanced by using the SIL. The enhancement of the spatial resolution was confirmed in all PL images observed at temperatures from 4 to 200 K.

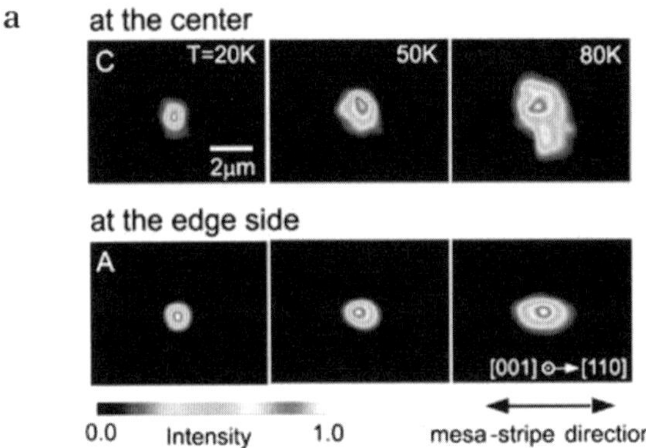

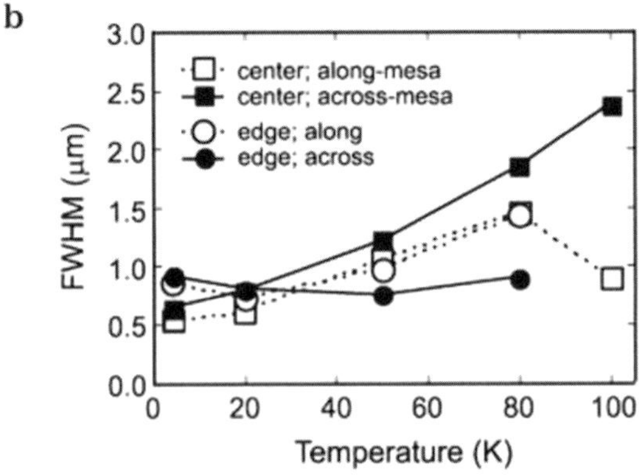

Fig. 9.10. **a** PL images with point excitation at positions C and A at 20, 50, and 80 K, respectively. **b** The full widths at half maximum (FWHMs) of the cross-sectional intensity profiles of the PL images in and across the mesa-stripe direction

Figure 9.9 shows the PL spectra under point excitation at five different positions on the top QW, which are indicated as A to E in the SEM image. The PL peak shifted to the lower energy side as the excitatory position was moved from the center to the edges of the mesa, which indicates the well thickness variation in the top QW across the mesa. The well thicknesses estimated from the PL peak energy at the center (C) and at the edge (A) were 5.2 and 5.8 nm, respectively, and the thickness variation was 12%.

If such well thickness variation exists in the top QW, it is expected that photocarriers generated at the center of the top QW tend to flow into the lower energy region at both sides.

Figure 9.10a shows the PL images observed from the top of the mesa under point excitation at three different temperatures. The upper images were obtained with the point excitation at the center (position C in Fig. 9.9)

of the top QW and the lower images were obtained at the edge (position A in Fig. 9.9).

At 20 K, the PL images are small circular spots. At higher temperatures, however, PL images become broader and anisotropic. At the edge, the PL image spread only in the mesa-stripe direction, whereas at the center, it spread dominantly across the mesa-stripe direction.

The diffusion length is estimated from the full widths at half maximum (FWHMs) of the PL intensity profiles in and across the mesa-stripe direction from the PL images, as shown in Fig. 9.10b. Since the spatial resolution of the PL images under point excitation was estimated at 0.6 µm, the FWHMs beyond 0.6 µm originated from the carrier transport, and the additional portion indicates the length of carrier transport (or transport length of carriers).

At the center of the mesa, carriers preferably transfer from the center to the edge sides, that is, to the lower quantization energy sides due to the drift effect caused by the variation in well thickness in addition to the diffusion effect. At the edge, on the other hand, since the quantization energy is the lowest near the edge sides, carriers can diffuse only within the same energetic states in the mesa-stripe direction and cannot flow across the mesa direction.

9.1.3 Time-Resolved Spectroscopy of Single Quantum Dots Using NSOM

Low-dimensional structures such as quantum dots (QDs) have been attracting a lot of attention because of the three-dimensional confinement of charge carriers which leads to interesting physical effects. The QDs are characterized by a δ-like density of states [32], which gives rise to high photon emission efficiencies and makes them a promising candidate for optoelectronic devices. Self-assembled quantum dots (QDs), which can be prepared in appropriate number with high quality [33], have enabled realization of practical quantum dot devices. Recently QDs whose photoluminescence at full width at half maximum (FWHM) was 21 meV, which is smaller than that of a quantum well, have been obtained [34].

However, before employing the QDs in practical devices, understanding carrier relaxation in QDs is important because the bottleneck effect that is expected to slow the carrier relaxation time drastically in QDs [35,36] is detrimental for device applications. Time-resolved spectroscopy is an efficient tool for elucidating the relaxation mechanism in QDs. Several groups have performed time-resolved optical measurements to investigate the relaxation and recombination mechanism. Experiments have been performed both on a large number of dots [37–40] and on a single dot selected by a technique based on masking, combined with the use of an optical microscope [41].

Recently, near-field photoluminescent spectra have been investigated, and the importance of near-field microscopy and the possibility of intradot carrier relaxation have been also pointed out [42]. The combination of time-resolved

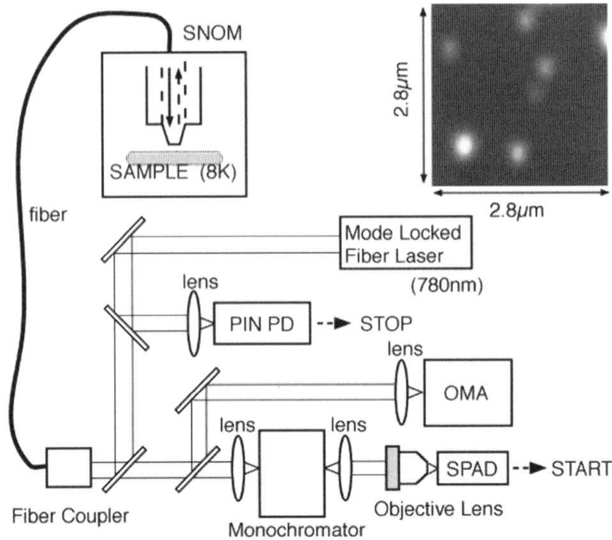

Fig. 9.11. Experimental setup. The inset shows a time-integrated photoluminescent image taken at 8 K without a monochromator

spectroscopy with near-field optics will allow one to probe the dynamic process of carrier relaxation into and within a single quantum dot.

Here, the authors will present the experimental study of InGaAs/GaAs single quantum dots by time-resolved spectroscopy using a highly sensitive near-field probe with an optimized apex shape [43]. The optical response of an individual dot in the time and frequency domains is presented. The authors will also present and discuss a model for relaxation, which accounts for cascade relaxation between neighboring emission levels, state filling, and the existence of fast relaxation processes, and carrier feeding from a wetting layer is developed. The model reproduces the experimental features very well, including excitatory power dependence.

A typical experimental setup for the time-resolved measurements is shown in Fig. 9.11. The sample is $In_{0.5}Ga_{0.5}As$ self-assembled QDs with a density of $\leq 10^{10}$ cm^{-2} and a diameter of about 40 nm, grown on a (100) GaAs substrate by gas-source molecular beam epitaxy. The growth sequence and growth parameters are similar to those described in [44]. The second harmonic of the mode-locked fiber laser is used as an excitatory source to generate carriers in InGaAs QDs, GaAs, and AlGaAs barrier layers. High sensitivity and high spatial resolution can be obtained by using a double-tapered fiber probe (details in [45]). The size of the image of a single dot (see the inset of Fig. 9.11) is limited by the spatial resolution ($\sim$ 300 nm). Since the density of the quantum dots is not uniform, a single dot can be selected, within our spatial resolution, by scanning the right spot on the sample. The illumination and

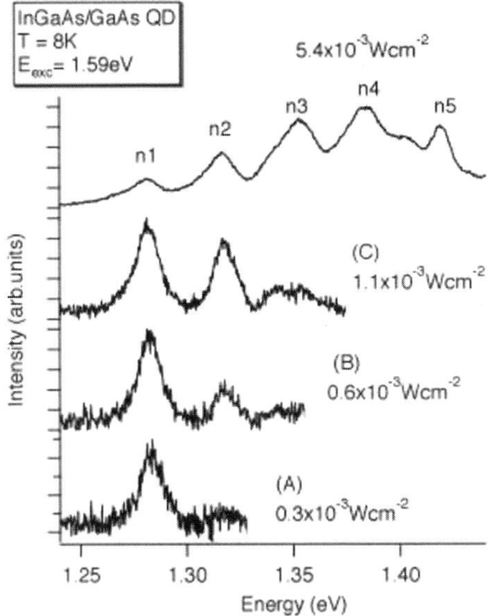

Fig. 9.12. Low-temperature emission spectra of a single InGaAs/GaAs self-assembled QD sample for different excitatory intensities. Peaks appear gradually with the increase in the excitatory intensity

collection hybrid mode can be used to excite and collect the emission by using a single QD by using the fiber probe. This method avoids the deterioration of resolution due to carrier diffusion. The collected emission is transmitted to a fiber coupler and to the detection system. The detection system consists of a monochromator with a spectral resolution of about 10 meV and a Si single-photon avalanche diode (SPAD) [46]. The instrumental response is 270 ps, and the time resolution of the system is about 100 ps, using a deconvolution program. An optical multichannel analyzer (OMA) with a spectral resolution of approximately 4 meV is used for spectral measurements.

The emission spectrum is shown in Fig. 9.12; it clearly shows distinct lines with a period of approximately 35 meV, which are attributed to 0-D electron–hole transitions. Higher energy peaks gradually appear with the increase in the excitatory power, and the energetic center of the spectrum is shifted to these higher energy emission peaks. At an excitatory intensity of 5.4×10^{-3} Wcm^{-2}, five peaks are observed. The time-resolved measurements at energies corresponding to the peaks of the emission spectrum are shown in Fig. 9.12. The solid curves represent the experimental results of the time-correlated single photon counting measurements [47]. The decay of each state slows down and the emission clearly shows plateau-like behavior as the excitatory intensity increases. This indicates that the emission is from a single dot. A comparison of the time response functions of different emission peaks reveals that the lower energy peak exhibits slower decay and less resemblance to a single exponential curve. This indicates that the effect of filling lower energy states in a single dot cannot be ignored.

To examine the relaxation process in an individual dot, a model, based on the rate equation approximation, can be employed. We will limit ourselves to the weak excitation condition where the emission from three lowest levels [(A)–(C) in Fig. 9.12] is observed. We assume that (i) state filling and cascade relaxation between neighboring levels are dominant [37,48]; (ii) at weak excitation, in comparison with [39,40,49] the radiative, nonradiative, and interlevel recombination rates are independent of the excitatory intensity; (iii) the initial occupation of each state is determined by the faster relaxation processes [39,50]; (iv) the wetting layer is responsible for subnanosecond carrier feeding into each state; and (v) relaxation of the carriers in the wetting layer is not affected by the occupation rate of each state because they can relax by other nonradiative processes. These assumptions enable us to represent equations for the temporal evolutions of occupation as follows;

$$\begin{aligned} \dot{n}_j &= -\tau_j^{-1} n_j - \tau_{j-1\ j}^{-1} n_j (1 - n_{j-1}) \\ &\quad + \tau_{j\ j+1}^{-1} n_{j+1} (1 - n_j) + \tau_{j\mathrm{w}}^{-1} n_\mathrm{w} (1 - n_j) \qquad (9.2) \\ j &= 1, 2, 3, \cdots, \ \tau_0^{-1} = 0\,. \qquad (9.3) \end{aligned}$$

Here, n_j and τ_j^{-1} are the occupation and recombination rates of the jth state. τ_{ij}^{-1} is the interlevel relaxation rate between the n_i and n_j states. The subscript w stands for the wetting layer. The occupation of the wetting layer, it is assumed, decays exponentially with a time constant τ_w. Upon weak excitation, the occupation n_j can be set to 0 for $j \geq 4$ in (9.2).

The recombination times τ_1, τ_2, and τ_3 can be obtained by fitting the time-resolved signal to the convolution of a single exponential function and the instrumental response function. The value of τ_{12} can be obtained by examining the spectrum at the excitatory intensity (B), at which n_3 does not contribute much to the carrier feeding to the second level. Then, assuming that τ_{12} is constant, we examine the experimental results of (C) to obtain τ_{23}.

The rate equations can be solved numerically with the initial occupation of each level as fitting parameters. The results of numerical simulation are shown in Fig. 9.13 for the following parameters: $\tau_1 = 590$ ps, $\tau_2 = 360$ ps, $\tau_3 = 360$ ps, $\tau_\mathrm{w} = 290$ ps, $\tau_{12} = 80$ ps, and $\tau_{23} = 200$ ps.

The results of numerical simulation do not change considerably within the range 50 ps $< \tau_{12} <$ 100 ps and 140 ps $< \tau_{23} <$ 260 ps. Note that the fitting parameters satisfy the relationships $\tau_{ij} \ll \tau_i$. This corresponds to the fact that upon low excitation, emissions from upper levels are not observable, which indicates that the interlevel relaxation time must be much shorter than the recombination time.

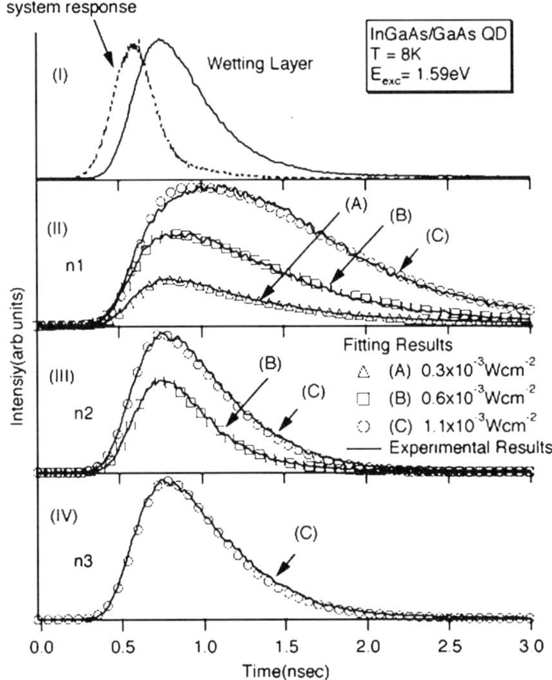

Fig. 9.13. (I) System and wetting layer response functions. (II)–(IV) Time-resolved emission spectra for the three lowest QD transitions, n_1, n_2, and n_3, respectively. (A), (B), and (C) correspond to the excitatory intensities in Fig. 9.12. The height of each spectrum is differentiated so as to identify the excitatory intensity dependence. The system response is the function used in the deconvolution calculation. Note that the rise times of the four lower spectra are identical. This indicates that these levels are filled simultaneously within the time resolution

9.2 Observation of Polysilane by Near-Field Scanning Optical Microscope in the Ultraviolet (UV) Region

The physical and chemical properties of semiconducting polymers with large energy gaps have recently been studied extensively realize linear and nonlinear optical devices controlled by blue light. Among the various studies of this class of polymers, direct observation of the actual conformation of these molecules is most important because structural changes in the global and local molecular conformations greatly affect th optoelectronic properties of π- and σ-conjugated polymers [51]. Development of the near-field scanning optical microscope (NSOM) for the blue and ultraviolet (UV) regions is essential for direct observation of the conformation of semi-conducting polymers

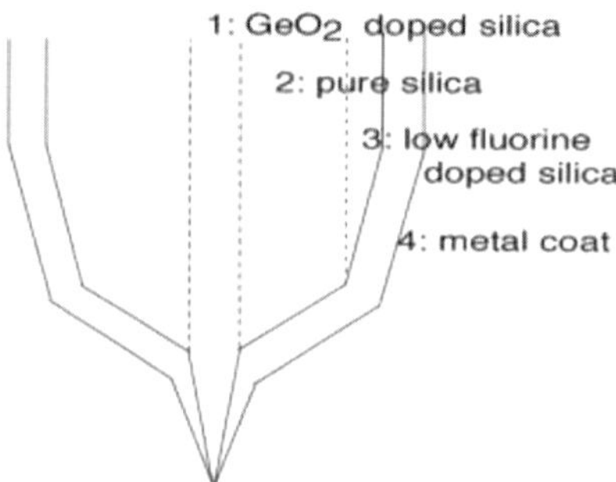

Fig. 9.14. Schematic structure of a fiber probe used for a UV-NSOM system

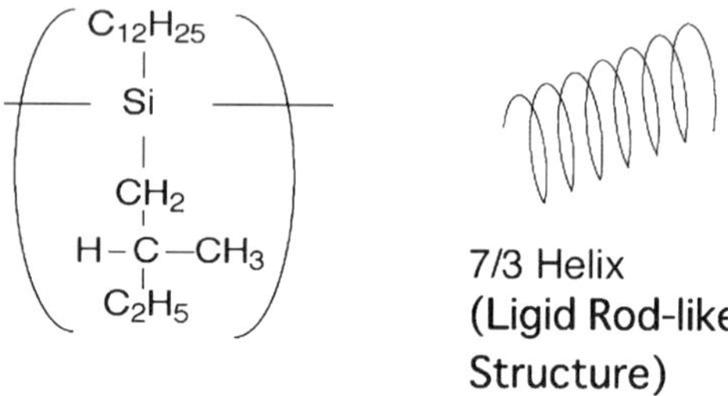

Fig. 9.15. Chemical structure of the chiral PS molecule

with large energy gaps. The UV-NSOM technique will also play a key role in operating molecular optoelectronic devices whose nanometer size is controlled by blue light. Here, we overview the UV-NSOM technique achieved by Mononobe et al. using a triple-tapered probe [52] (see Fig. 9.14) and an atomically flat sapphire plate [53].

Koshihara et al. demonstrated the UV-NSOM technique developed by observating the photoluminescent images of polysilane (PS) molecules. PS is an organic-inorganic hybrid material with a silicon quantum wire structure [54,55]. Among various PS molecules, they focused on the n-decyl-(S)-2-methylbutyl silane (hereafter abbreviated as chiral PS) molecule as a typical example in this study. The backbone structure of the chiral PS molecule is considered a rigid rod-like helix and is strongly fixed into a preferential single-screw sense by enantiomerically pure chiral alkyl pendents. The schematic structure of chiral PS is shown in Fig. 9.15. It has been also made clear by various spectroscopic studies that chiral PS molecule has a direct-type band structure with a large energy gap (about 4.5 eV) due to the one-dimensional confinement effect of σ-electrons [54]. The absorption and fluorescent spectra of chiral PS are plotted in Fig. 9.16. From these points of view, chiral PS can be classified as a polymer semiconductor with a large energy gap, so it is an important candidate for optoelectronic materials in the blue and UV

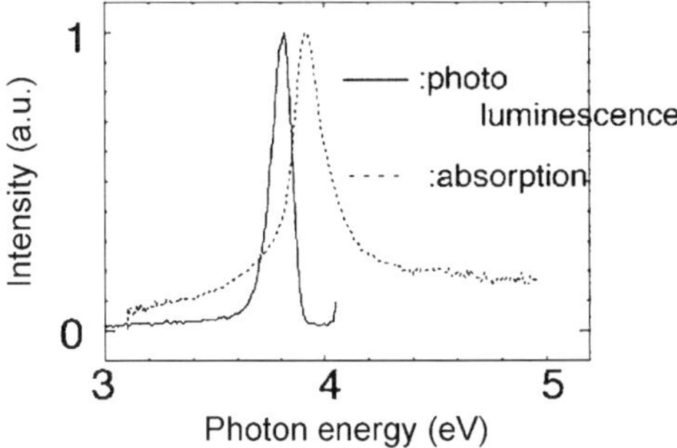

Fig. 9.16. Absorption (*dashed line*) and photoluminescent (*solid line*) spectra of a chiral PS thin film

regions. This is the reason that the chiral PS molecule is a suitable material for demonstrating the UV-NSOM technique.

The methods for preparing a triple-tapered probe and an atomically flat sapphire plate have been reported previously [54,55]. A multistep index fiber made of GeO_2 silica, nondoped silica, and fluorine-doped silica was used to increase the transmission efficiency of the fiber probe in the UV-region. The schematic structure of the fiber probe is shown in Fig. 9.14. The NSOM was operated in the illumination mode. The UV light was irradiated on the sample surface via a fiber probe, and scattered light and/or luminescence was collected by the quartz lens system. The photon energy for the observation was selected by monochromator and/or band-pass filters.

Koshihara et al. prepared highly purified chiral PS with a molecular weight of 1 000 000 by sodium-mediated polymerization of n-decyl-(S)-2-methylbutyldichlorosilane. Chiral PS was cast on the atomically flat sapphire substrate with a (1012) surface from a dilute solution in toluene of about 10^{-8} (Si unit)/L, and the sample was dried in a vacuum for several minutes. The sapphire plate is transparent to UV light, and its superflat surface with an atomic layer step is hydrophobic. This is the reason that the sapphire plate, rather than a hydrophilic mica substrate, is suitable for observing hydrophobic PS molecules by UV-light. All images were obtained in air at room temperature.

Figure 9.17a shows the AFM image of chiral PS dispersed on the atomically flat sapphire substrate [56]. As shown in this figure, chiral PS molecules dispersed on the plate form a fibril-like structure. The height of the observed fibril-like structure spread from 7 to 50 nm. The estimated molecular diameter of the chiral PS based on a space-filling model was about 1 nm and thus we guess that the observed fibril-like structures are constructed from 7–50 molecules. From the results of light scattering and viscosity measurement, it has been conjectured that chiral PS molecules take a rigid, rodlike heli-

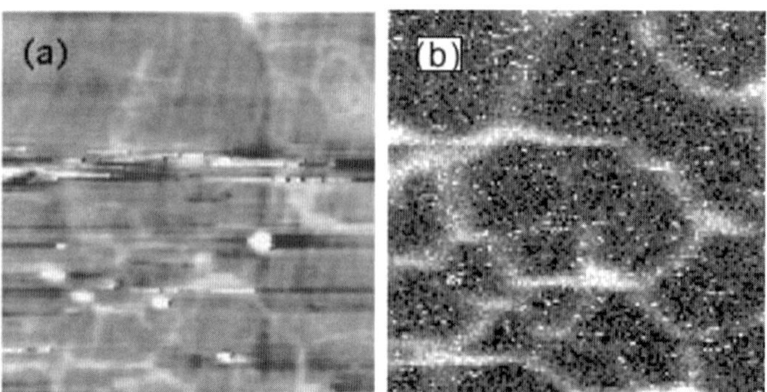

Fig. 9.17. Topographic (**a**) and NSOM (**b**) images of chiral PS molecules dispersed on an atomically flat sapphire substrate. The NSOM image was observed at 3.82 eV (325 nm) under illumination of 4.13-eV (300-nm) light via a fiber probe. The scanning area was 10 × 10 μm, and all observations were done at room temperature under ambient pressure [56]

cal structure, as introduced previously. The result obtained seems consistent with this expectation.

A UV near-field luminescent image of chiral PS in the same scanning area as Fig. 9.17a is shown in Fig. 9.17b. The 4.13-eV (300-nm) light was irradiated on the samples via the fiber probe. The photon energy of the irradiated light was sufficiently larger than the band-gap energy of the chiral PS molecule. The detection wavelength of the luminescent signal was tuned to 3.82 eV (325 nm) which is resonant with that of the luminescent band. The fibril-like structures have also been observed, even in fluorescent images. The spatial resolution of 60 nm was achieved by the present UV-NSOM technique. The fluorescent spectrum corresponded with that of chiral PS thin films, and thus it can be safely concluded that the observed structures are due to chiral PS molecules dispersed on the sapphire plate. Observed fibril-like images strongly support the idea that chiral PS molecules take a rigid rod-like structure even in thin films.

The fluorescent images of PS molecules dispersed on a sapphire plate with an atomically flat surface have been reported, using a UV-NSOM system. The images observed were identified as fibril-like structures of chiral PS molecules by fluorescent spectrum measurement. The results obtained clearly demonstrate that the UV-NSOM method is a really useful and powerful tool for promoting polymer science.

9.2.1 Morphologies and Quantum Size Effects of Single InAs Quantum Dots Studied by Scanning Tunneling Microscopy/Spectroscopy

The luminescent properties of InAs QDs on a submicron scale have been intensively investigated by near-field scanning optical microscopy [57], microphotoluminescence (PL) spectroscopy [58], and cathodoluminescence spectroscopy [59]. Scanning tunneling spectroscopy (STS) and scanning tunneling luminescence (STL) based on a STM technique are powerful tools for observing morphologies and local electronic structures simultaneously at a nanometer-scale resolution. In STS, the conductance spectrum obtained from current-voltage characteristics reveals the conduction- and valence-band edges of QDs.

Here, a correlation between dot size and band gap of single InAs quantum dots (QDs) was investigated by scanning tunneling microscopy (STM)/spectroscopy (STS). The height dependence of the gap energy is well explained by a quantum disk model.

A single InAs layer with a coverage of 2.5 monolayers (ML) was grown on an n-type ($n_{\mathrm{Si}} \sim 1 \times 10^{18}$ cm^{-3}) GaAs (001) substrate at 520°C by molecular beam epitaxy (MBE). The typical size of self-organized QDs is 3–7 nm high and 17–30 nm in lateral size. The As-capping layer was removed by thermal desorption at $\sim$ 380°C in an ultrahigh vacuum (UHV) chamber. The STS measurement was performed with a current imaging tunneling spectroscopy (CITS) mode in an UHV of 2×10^{-8} Pa at room temperature. To suppress tip-induced band-bending effects [60], the current-voltage characteristics were measured under the illumination of a laser light with a photon energy larger than the band gap of GaAs.

Figure 9.18 shows a topological image and normalized differential conductance spectra of InAs QDs on a GaAs (001) surface. Nine dots with heights of 3.5–5.9 nm and diameters of 18.0–26.7 nm were observed in the STM image. As shown in Fig. 9.19, the height distribution ranges from 1.7 to $\sim$ 7.7 nm indicating two distribution peaks. Assuming two Gaussian distributions, the peaks in the distribution are at 3.7 nm and 5.9 nm, and the standard deviations are 2.0 and 1.7 nm, respectively. The measured dot height h and diameter d are related by the following equation in units of nanometers:

$$d = 2.4h + 10.2 \text{ (nm)} \tag{9.4}$$

Figure 9.18 shows the normalized differential conductance spectra that were measured at the corresponding QDs. Because the spectrum is proportional to the densities of the states of the conduction and valence band, the gaps observed yield the band gaps of the QDs: 1.1 eV for $h = 4.2$ nm and 0.68 eV for 5.9 nm. The closed circles in Fig. 9.19 show the observed gap energy as a function of dot height. The gap energy increases with decreasing height. The photoluminescent (PL) spectrum, which was measured at 77 K in a UHV chamber, is also displayed in Fig. 9.19 by shifting the energy scale by 73 meV

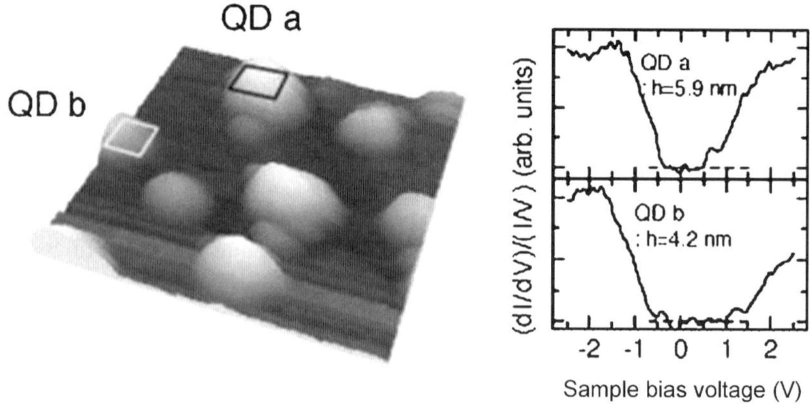

Fig. 9.18. STM image ($95 \times 95\ \mathrm{nm}^2$) of InAs QDs on GaAs (001) and normalized differential conductance spectra for $h = 4.2$ and 5.9 nm. The height variation in the STM image is 8.1 nm. The sample bias voltage and current are -3.0 V and 0.2 nA, respectively

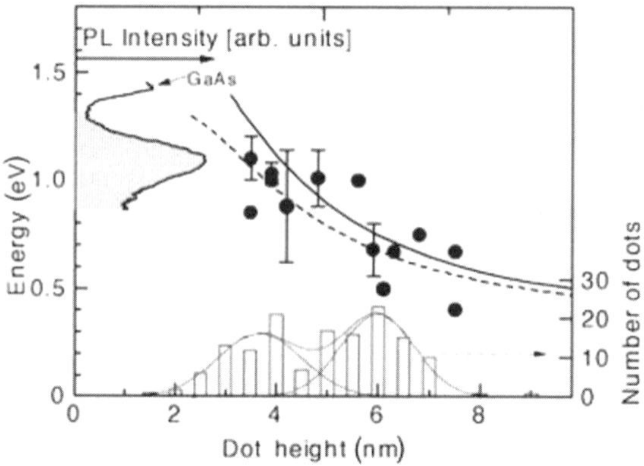

Fig. 9.19. Gap energy vs dot height, histogram of dot heights, and photoluminescent spectrum of InAs QDs

due to the temperature variation of the band gap. The emission peak at ~ 1.1 eV corresponds to dots with a diameter of ~ 4 nm, and PL corresponds to dots larger than 5 nm that cannot be detected by the sensitivity limit of the detector used in this study.

Quantum confinement energies of electrons and holes have been calculated as a function of dot height using a quantum disk model with finite barrier potentials within an effective mass approximation. The dashed and solid curves in Fig. 9.19 show the calculated energies for an infinite diameter and a finite diameter given by (9.4), respectively. The calculation based on the finite disk model explains the observed height dependence rather well. The gap energy of a single InAs QD is determined mainly by the quantum confinement effect in the growth direction.

9.2.2 Photonic Structures Consisting of Dielectric Spheres

History and Overview. In 1908, the first development in the theory of the diffraction of a plane electromagnetic wave by a sphere was done by Mie [61]. This theory has often been called the Mie theory and is widely adopted as the fundamental theory for understanding the various physical aspects of the interaction between a plane wave and a single sphere [62]. The Mie theory has been, furthermore, extended to investigate the coupling of an evanescent electromagnetic wave to a single sphere [63–68]. The effect of the multiple scattering that occurs between a sphere and a substrate has also been discussed [64–67]. Such a diffraction phenomenon has strong relevance to the photonic band effect, the properties of the point defect in photonic lattices, the characteristics of the optical sphere resonator that has an extremely high Q, and so forth.

The optical responses of a single sphere and spheres forming in line have been experimentally investigated. For a single sphere, for example, experiments have been carried out to obtain information about the decay length of the near-field of a Rhodamine-doped polystyrene sphere [69], the near-field coupling between a glass sphere and a glass substrate [70,71], the propagative characteristics of pulsed whispering gallery (WG) modes in glass spheres in attenuated total reflection (ATR) geometry [72], and the photothermal property of a polystyrene sphere [73].

A microsphere is a good candidate for studying the cavity effects on lasing. Spherical geometry has two advantages. First, the Mie theory allows rigorous quasi-analyses of cavity QED effects [74]. Second, high-Q spherical cavities can be readily self-assembled by strong surface tension. Efficient lasing has been observed with droplets of dye solution [75] and dye-doped polymer microspheres [76,77]. Systematic study of fluorescence and lasing spectra have been reported for dye droplets [78]. To reduce the number of high-Q WGMs, which give rise to multimode lasing and do not allow quantitative analysis, using microspheres with enhanced surface concentration of dye molecules has been proposed [77]. The reduction of the threshold and achievement of considerable lasing efficiency, however, requires reducing the sphere size to several microns. It has been found, in particular, that mode competition can be dramatically reduced for polystyrene dye-doped spheres of 5 µm diameter giving rise to an increase in the number of photons coupled with a given WGM [77,79].

For a dye-doped polystyrene bisphere, it was shown that the coherent resonant coupling of WG modes in the bispheres leads to splitting of the characteristic frequency of the WG modes in a single sphere [80]. Studies of a structure consisting of more than two spheres formed in line are in progress

[81]. These systems may be expected to be employed to manipulate the light path on a micrometer length scale; this has recently attracted considerable attention from both fundamental and application points of view.

Conventionally, the micromanipulation of light is based on the photonic crystal concept [82–84]. Photonic band effects are found in periodic dielectric materials, often called photonic crystals, that exhibit wavelength regions where light cannot propagate. These materials open a new world of photonic physics and photon technology and were widely exploited recently in optics and opoelectronics [83]. It is currently believed that the photonic band effect can be used to realize the localization of a photon, the small group velocity of a photon, the enhanced nonlinear optical response, and so on. Such features may, for example, lead to the realization of thresholdless lasers by controlling the spontaneous emission of light. In addition, photonic bands recently attracted growing attention because of the superprism [85], supercollimator [86], and superlens [86] which have been devised on the basis of features of photonic bands.

In photonic crystals, which have weak periodic modulation of the refractive index, propagation of a light wave can be described in terms of wave-packet propagation in a weak potential. Correspondingly, such an approach can be referred to as a nearly free photon approach, analogous to the nearly free electron approach in band theory. Alternatively, it is possible that light can be micromanipulated by confining it in a small unit of wavelength size. Light propagates through a system of such units due to the coupling between nearest neighbors. Such a tight binding photon approach [87] to the micromanipulation of light allows one to guide optical waves by connecting units, which may be referred to as photonic atoms, in arbitrarily shaped microstructures (photonic molecules). Microspheres are the most natural choice of the unit to be employed in a tight binding photon device. It is known that a dielectric sphere acts as a unique optical microcavity which has very long photon storage time within a small mode volume [76,77,88,89]. In particular, Q-factors of the order of 10^{10} have been observed for whispering gallery modes (WGMs) in quartz spheres with diameters of several tens of micrometers [90–94], and the mode structure of pairs of these large spheres in contact has been studied [95]. However, to explore the feasibility of micromanipulating light, one has to confirm the existence of coherent coupling between spheres whose size is a few times an optical wavelength. Lorenz–Mie theory predicts a long photon lifetime even for small spheres, giving, for example, nearly 30 picoseconds for a 4-µm sphere with a refractive index of 1.59. This has allowed one to propose using such relatively small spheres as photonic atoms [96] for a tight binding scheme.

The first theoretical prediction and formulations of the photonic band effect were introduced by Ohtaka [97] for a periodic array of polystyrene submicron spheres. The photonic band for such a structure has been analyzed by expanding it in terms of vector spherical waves for plane-wave incidence, based on the concept of sequential Mie diffraction. Subsequently, Ohtaka and his co-workers [98] extended their treatment to analyze many novel phenomena related to the photonic band effect and actually presented a number of beautiful results, including the near-field properties of photonic bands for ordered dielectric spheres [99].

The finite-difference time-domain (FDTD) method [100], which is a scheme first introduced by Yee [101], is a numerical modeling approach for solving Maxwell's equation and has been recently applied to investigate the photonic band effect [102–105]. This approach can easily give a solution which is not obtained with ease by analytical methods. For example, it makes possible the analysis of the response of ordered dielectric spheres to Gaussian light and provides information about the transient behavior of a wave front of light as it proceeds among spheres as time goes on [105].

The two-dimensional photonic band effect has been studied theoretically [98,99] and experimentally [98,106–108] in hexagonally ordered spheres of various diameters. Transmission spectra, for example, were measured as a function of frequency and angle of incidence for two dimensionally ordered Si_3N_4 spheres in the millimeter wave region [98]. In consequence, the dispersion curve of two-dimensional photonic bands was determined, and it was pointed out that the radiative decay of such bands is greatly suppressed.

SNOM images of a single layer of closely-packed polystyrene spheres, including a fluorescent sphere, were measured [108–112]. It was found that two types of light propagating over spheres exist. One is the optical mode which looks a guided wave in channel waveguides and thus goes straight through the center of the sphere. Another optical mode propagates along the sphere surface and enters the next sphere through a near-field coupling around the contact point between spheres.

The interaction between evanescent light and single layer of closely-packed polystyrene spheres in attenuated total reflection geometry was also investigated [113,114]. Besides the above two optical modes, the ATR signal due to the optical mode bound to one sphere was observed. Investigation of these three optical modes may give significant information about the properties of photonic bands.

A fully three-dimensional photonic crystal composed of dielectric spheres was fabricated [115–121]. The transmission and reflectance of the structure were measured, and a photonic stop gap was recognized [116–118,120,121].

The key point to realize promising photonic structures consisting of dielectric spheres is to find how to fabricate submicron spheres that have a high sphericity and how to control the size and the array of such spheres.

Although it is beyond the focus of this section, photonic structures consisting of metallic spheres have been theoretically and experimentally investigated [102,122,123]. It should be pointed out that electron relaxation time plays a crucial role in the photonic band effects of ordered metallic spheres in the visible region [123].

Resonant Coupling of Optical Whispering Galleries. In this section, we discuss the experimental observation of normal mode coupling in a system of two polymer spheres of diameters ranging from 2 to 5 μm in contact (bisphere) [79,80]. The sufficiently narrow line width and wide separation of WGMs in this size range allows us to avoid intricate band mixing. We use monodispersive polystyrene spheres (refractive index 1.59) that are soaked with a solution of dye (Nile Red, concentration is about 10^{-2} mol/L, fluorescent FWHM is about 70 nm). Dye doped spheres are spread on a glass plate under a microscope and can be manipulated with an optical fiber probe which is made by pulling a multimode optical fiber in a frame. The diameter of the fiber probe is about 3 μm. By using a microscope objective lens, an individual sphere can be excited with the second harmonics of a cw Q-switched Nd-YLF laser ($\lambda = 527$ nm) (see Fig. 9.20).

Figure 9.21 shows an example of the fluorescent spectrum from an individual sphere about 4.1 μm in diameter. Sharp lines, which originate from the strong modification of the vacuum field inside the sphere!!!! [76,77,89,125–128], appear at the resonant frequencies of WGMs with TE and TM polarization. Each WGM can be specified by the mode number n (indicates the number of light wavelengths around the circumference and indicates the order of spherical Bessel and Hankel functions, which gives the radial field distribution in the Lorentz–Mie theory [124]), the order number s (corresponds to the number of maxima in the radial dependence of the internal electric field), and the azimuthal mode number m (gives the orientation of the WGM's orbital plane). Since the frequency of the WGM does not depend on m, each resonance has $(2n+1)$-fold degeneracy.

For spheres from 2 to 5 μm in diameter, we can identify the mode indexes n and s of a high-Q WGM by comparing the positions of the lines in the fluorescent spectra with the predictions of the Lorenz–Mie theory. The width

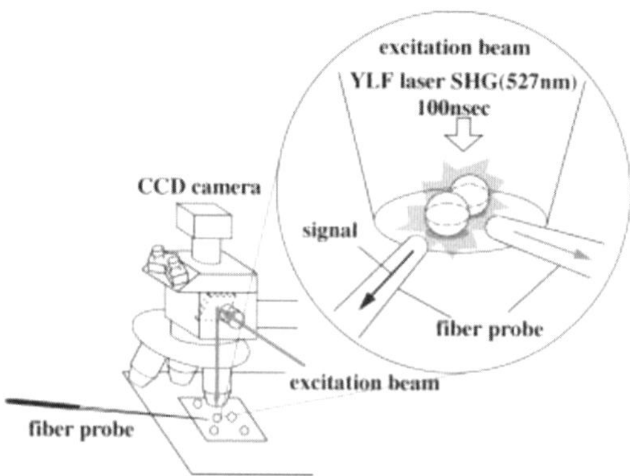

Fig. 9.20. The experimental setup for observing fluorescence from dye-doped microspheres

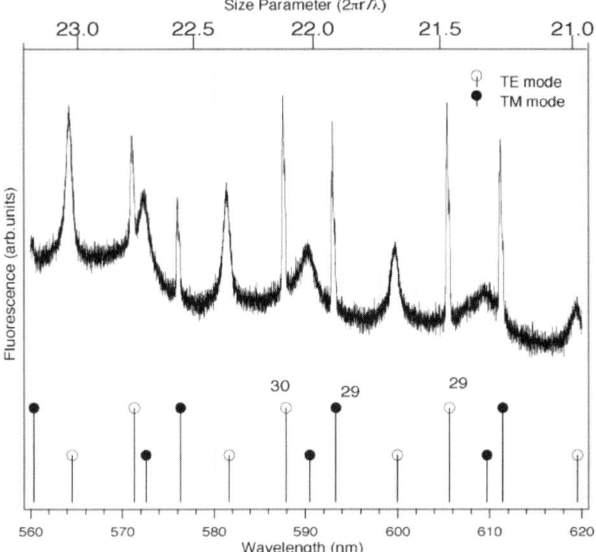

Fig. 9.21. Typical fluorescent spectrum from a 4.1-μm sphere. The mode assignment is shown at the bottom of the figure

of the resonances $\lambda_{n.s}$ gives the Q factor for a WGM of a given n and order number s, $Q_{n.s} = \lambda_{n.s}/\Delta\lambda_{n.s}$. Spheres 2 μm in diameter have a Q factor of the order of 10^2, which is comparable with that predicted by the Mie theory radiative leakage rate. Spheres more than 4 μm in diameter have Q factors of the order of 10^3. This is two orders less than the radiative leakage rate, and, therefore, the Q factor is determined mainly by the light absorption by dye molecules and light scattering by imperfect surfaces. Note that nonideal spherical shapes of spheres lift the degeneracy with respect to azimuthal mode indexes, giving rise to broadening and asymmetry of fluorescent lines. For example, the narrowest line width of 5-μm spheres is 0.4 nm, which corresponds to an eccentricity less than 0.08%.

Once we label the fluorescent lines with the mode indexes n and s, we can judge the relative size of the sphere within an error of $1/Q$ which is about 0.05% for a 5-μm sphere. To make a bisphere, we choose two spheres of the desired size by comparing the frequencies of the specific WGM resonance. We pick up one sphere and attach it to the other by handling the optical fiber probe attached to the three-axis mechanical stage.

The measurement configurations are presented in Fig. 9.22a. In parallel configuration (A), the fiber probe is set parallel to the axis of the bisphere, whereas in the perpendicular configuration (B), the fiber probe collects fluorescence in a direction perpendicular to the axis of the bisphere. Figure 9.22b,c shows the fluorescent spectra of the bisphere (A,B) and indi-

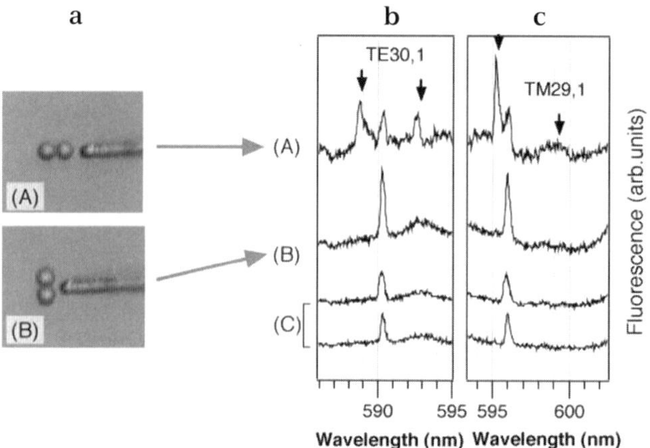

Fig. 9.22. **a** Microscope images of the bisphere and the fiber probe. The diameter of the probe is about 5–10 µm. We detected the emission in two different geometries; one is parallel configuration (A); the other is perpendicular configuration(B). **b, c** Spectra of resonant bisphere of TE30,1 mode and TM29,1 mode in parallel configuration (A) and perpendicular configuration (B). Spectra (C) show the fluorescence of individual spheres before contact. The two spheres have almost the same diameters. The arrows indicate the coupled modes

vidual spheres (C) in the vicinity of the TE mode of $n = 30$ and $s = 1$ and the TM mode of $n = 29$ and $s = 1$ resonances, respectively. For both polarizations, we observe new peaks due to intersphere coupling in the parallel configuration (A), but not in the perpendicular configuration (B). To explain this, one should recall that each WGM of a given n and s has $(2n + 1)$-fold degeneracy associated with the azimuthal mode number m. Hence $2n + 1$ orientations of the WGM orbital plane are permitted and, correspondingly, $(2n + 1)^2$ combinations of intersphere couplings are generally possible. However, intersphere coupling is maximum for the pair of modes whose orbitals include the contact point and lie in the same plane. Therefore, the signal from coupled intersphere modes is more pronounced in the parallel configuration than in the perpendicular configuration because the fluorescence of coupled modes is maximum in the direction parallel to the bisphere axis. The dependence of mode coupling on orbital plane orientation also breaks the degeneracy with respect to the azimuthal indexes. This manifests itself as asymmetry in the line shape of the coupled modes in Fig. 9.22b,c, which indicates that stronger the coupling between the WGM, the more the energy in the resulting bisphere mode.

To study the dependence of WGM coupling on sphere size (size detuning), we fix the diameter of one sphere and change the diameter of the second sphere by 0.3%. Figure 9.23 shows the detuning dependence of bisphere fluorescence in the vicinity of the TE29,1 mode for various pairs with size

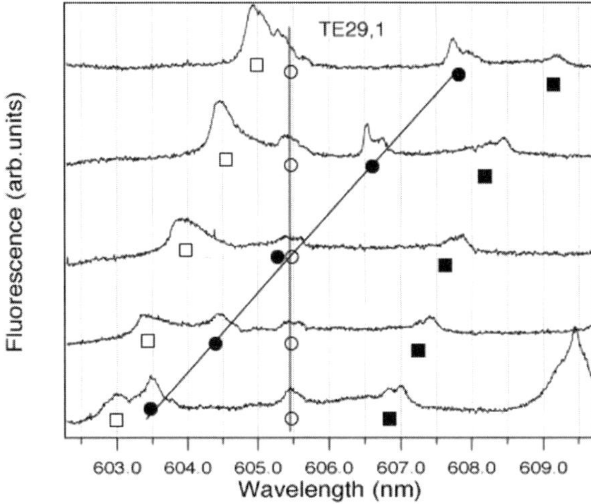

Fig. 9.23. Emission spectra of nine bispheres with slightly different sized spheres. The fiber probe is set parallel to the axis of the bisphere. The open and closed circles indicate uncoupled modes. The open and closed squares show the coupled modes of the bispheres

detuning. The open and closed circles show the uncoupled original resonance, and the open and closed triangles show coupled modes. We clearly see the anticrossing behavior.

Figure 9.24 shows the dependence of bisphere resonance on detuning for TE30,1 and TM29,1 modes. The positions of the modes before contact are shown by dashed lines. Both TE30,1 and TM29,1 modes show anticrossing with some asymmetry. Such asymmetry could be attributed to the influence of the broad second-order ($s = 2$) WGMs, and their positions are shown in the figure by thick shaded lines.

The features of intersphere coupling observed can be explained by the resonant enhancement of the internal field in bispheres. By expanding the internal and external fields of each sphere in terms of the vector spherical harmonics, we solve the coupled linear equation numerically by the matrix inversion method [129]. The energy of the internal field is obtained from the expansion coefficients as a function of incident plane wave parameters and sphere radii. Excellent numerical convergence is achieved with a summation of n up to 49. When the incident wave is almost parallel to the bisphere axis, the spectrum shows prominent peaks asymmetrically located on both sides of the TM29,1 and TE30,1 modes. Between these prominent peaks appears a series of less intense peaks converging to the original frequencies of WGMs. These substructures make the line shape asymmetrical, as shown in Fig. 9.22b,c. The open circles of Fig. 9.24 show the positions of the prominent peaks. One

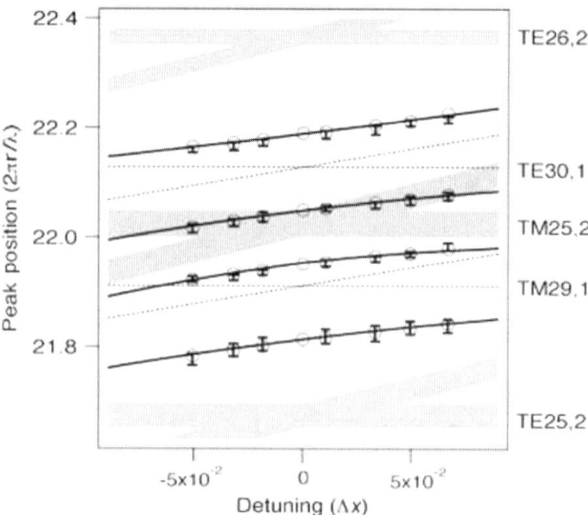

Fig. 9.24. Resonance frequencies of coupled TE30,1 and TM29,1 for various bispheres with different size detuning. Dashed lines indicate the resonances of uncoupled modes. Large open circles show the result of wave optics calculations. Solid lines show the calculated normal modes of bispheres as a function of size detuning using the coupled harmonic oscillator model

may observe excellent agreement between calculation and experiments from Fig. 9.24. Note that both the theory and experiment show asymmetry in mode splitting.

Bisphere fluorescence can be described in the standard terms of interacting harmonic systems by introducing phenomenological coupling parameters, which account for coupling of the relevant modes of different spheres. The corresponding detuning dependences, obtained from this model calculation, are shown in Fig. 9.24 as solid lines. One can see from Fig. 9.24 that the coupled oscillators model explains the overall features of the bispheres fluorescence well and, in particular, the asymmetrical behavior of the upper mode branch of the TM29,1 due to the strong mixing with the TM25,2 mode. The best fit values of the coupling parameters are almost constant for all of the modes involved in the narrow region of mode number variation from $n = 25$–30.

The results of the estimation of the coupling parameters for spheres with diameters of 2, 3, 4.1, and 5 µm and estimates of the coupling parameters are presented in Fig. 9.25 on the "size parameter – Q-factor" plane. The coupling parameters obtained normalized to the resonant frequencies are 7×10^{-3} for 2 µm ($n = 14$), 5×10^{-3} for 3 µm ($n = 21$), 4×10^{-3} for 4.1 µm ($n = 29$) and 3×10^{-3} for 5 µm ($n = 36$). The reduction obtained in coupling with an increase in sphere diameter is due to the fact that the WGM amplitude outside the sphere – and, therefore, the coupling – is for smaller spheres. Importantly, the

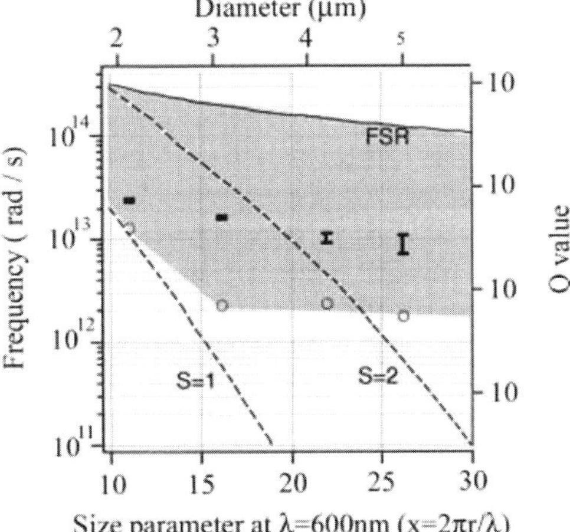

Fig. 9.25. The size dependence of the coupling constants and Q factors of microspheres. Open circles show the measured Q factors of the microspheres, and the inverse coupling constants are shown by solid squares and experimental bars. The dashed line represents the Q factor given by the Mie theory for the $s = 1$ and $s = 2$ modes. The coherent coupling regime is possible within the shaded area, whose lower boundary is given by the free spectral range (FSR)

coupling parameters obtained are much larger than the line widths, indicating the dominance of coherent resonant coupling. This ensures the feasibility of the tight binding manipulation of light waves in a structure composed of contacting spheres, which can be used in developing new optoelectronic devices.

9.2.3 Interaction of a Near-Field Light with Two-Dimensionally Ordered Spheres

The near-field light coupling technique may be one principal candidate for investigating the interaction of light and spheres because near-field light can couple to spheres in various fashions [69,70,72]. In near-field coupling, the manner of coupling can be changed by varying the penetration depth of the near-field light. An attenuated-total-reflection (ATR) method should be suitable for this type of experiment. The ATR phenomenon is caused by an absorbing coupling mechanism and is observed when the angle of incidence θ_i is varied above the critical angle θ_c or the wavelength λ of incident light is swept through an absorption band. It is known that an evanescent field is excited above θ_c and may be employed to record the spectra of absorbed materials.

The sample presented here is a hexagonal closely-packed, 2-D single layer of ordered polystyrene spheres on a BK 7 substrate. It was fabricated by

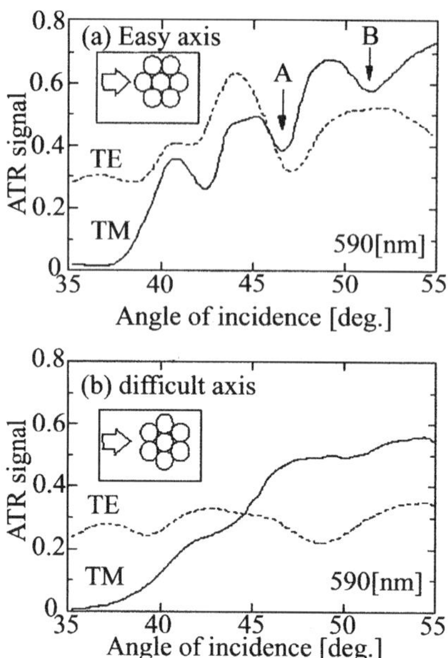

Fig. 9.26. Angle-scan ATR spectra for the TM (*solid line*) and TE (*dashed line*) polarizations at $\lambda = 590$ nm. **a** Easy axis configuration, and **b** difficult axis configuration

the self-organization method [106,113]. The angle- and frequency-scan ATR method is employed to investigate the interaction of a near-field light in an ATR configuration with 2-D spheres. The BK 7 glass substrate and a BK 7 prism were optically coupled to tailor one medium. A R6G dye laser pumped by an Ar^+-ion laser was employed as the light source. To obtain ATR signals due to a single domain of ordered spheres, the TM (or TE) polarized laser beam was focused by an objective lens to a diameter smaller than the typical area of a single domain. One must be careful to the configuration between the direction of propagation of the incident light and that of the ordered axis of the 2-D spheres because of the existence of an easy axis and a difficult axis, as described later, for light propagation in a 2-D array of spheres [104,108–113].

Figure 9.26a,b shows the angle-scan ATR spectra for the two directions of propagation of incident light at $\lambda = 590$ nm. The solid and dashed lines are for the TM and TE polarizations, respectively. The direction of propagation of the incident light in relation to the ordered axis of the spheres is shown in the inset in each figure. We call the configurations indicated in Figs. 9.26a–b the easy axis configuration and the difficult axis configuration, respectively.

Several prominent dips appear in Fig. 9.26a. These dips approach the position of θ_c ($\sim 41.3°$) and some of them cross it as λ becomes longer [113]. The angular positions of the corresponding dips, comparing both results for TM-

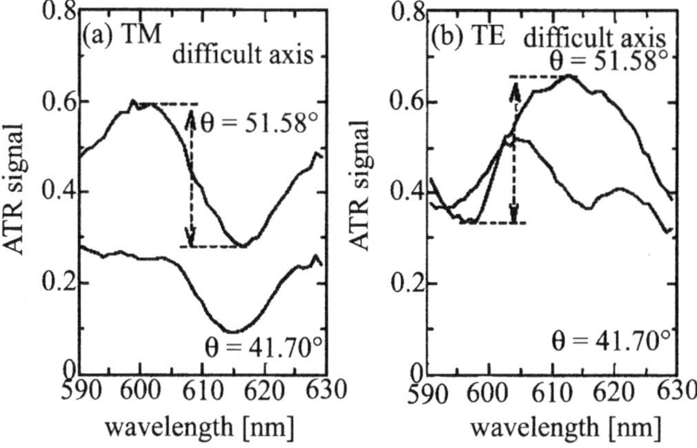

Fig. 9.27. Frequency-scan ATR spectra for $\theta_i = 41.70°$ and $\theta_i = 51.58°$ in the difficult axis geometry

and TE-polarized light, are slightly different. From the half minimum width of the dip at 42.5° in TM polarization, the propagative length of the resonant mode is estimated at about 2.4 μm. From the dispersion relation evaluated from the position of the dips in the ATR spectra at various λ, the resonant modes related to the dips observed are the propagating type [113,114]. The propagating velocities of the modes for the dips A and B in Fig. 9.26a are estimated at 1.87×10^8 m/s and 1.56×10^8 m/s, respectively. On the other hand, as shown in Fig. 9.26b, the dips for the difficult axis configuration are shallower compared with those for the easy axis configuration. The shallow dips indicate that it is difficult to excite the resonant modes observed in the difficult axis configuration in angle-scan ATR geometry.

The reflectance presented in Fig. 9.26 gradually decreases as a whole as θ_i decreases. This, it is considered, arises from an increase in energy loss due to light scattering, as the value of θ_i decreases. When $\theta_i > \theta_c$, evanescent light may be more strongly scattered at smaller θ_i because its evanescent decay length becomes longer. When $\theta_i < \theta_c$, the light scattering geometry is of Mie-diffraction [61], i.e., plane-wave diffraction by a sphere.

Figure 9.27 shows the frequency-scan ATR spectra in the wavelength range of 590 nm to 630 nm for $\theta_i = 41.70°$ and $\theta_i = 51.58°$ in the difficult axis geometry. Figure 9.27a,b shows the ATR spectra for the TM and TE polarizations, respectively. The wavelength positions of the dips and peaks shift when the polarization is switched. From the dispersion relation of the dips in the frequency-scan ATR spectra, the corresponding resonant modes are expected to be localized. It has been confirmed that these modes have the same properties as the WG mode in a single sphere [61,62].

The difference between the maximum and the minimum ATR signal intensities in the easy axis configuration is smaller than that for the difficult axis. In the easy axis configuration, the energy of the WG mode in a sphere can be transferred to neighboring spheres, and a considerable part of the light tends to be scattered. In consequence, it is apparent that the dips become shallower.

At $\theta_i = 51.58°$, the wavelength positions of the dips and peaks are almost independent of the axis configuration, as long as the polarization of the incident light is maintained. On the other hand, at $\theta_i = 41.70°$, the dips are shallower and shift in accordance with the axis configuration. These variations are induced by the change in the manner in which evanescent light spheres couple [130].

At a larger value of θ_i, the field associated with the evanescent light can strongly interact with the field of the WG modes whose orbits are in the incident plane. Then the TM-(TE-)polarized evanescent light mainly excites WG modes of the electric (magnetic) type. At a smaller value of θ_i, e.g., $\theta_i = 41.70°$, the field associated with the evanescent light can interact with the field of another WG mode whose orbit crosses the incident plane because of the large penetration depth of the evanescent field. Such additional WG modes are the modes of magnetic (electric) type excited by TM-(TE-) polarized light. The excitation of the additional WG modes leads to the change in ATR spectra.

By using numerical simulations, one can obtain the visual image of the light intensity distribution in 2-D spheres which helps us to interpret experimental results. Among them, the finite-difference time-domain (FDTD) method [100] can provide both spatial and temporal images from solutions of Maxwell's equations. Current high-end "personal" computers have potential enough to perform the FDTD analysis. The 2-D spheres are numerically described by the unit cells arranged consecutively along the y axis and the incidence plane is parallel to the x,z plane, taking into account the symmetry of the problem. Light intensity distributions are calculated only in a 2-D unit cell [113] to save time and memory space in the FDTD computation.

The unit cell consists of 20 polystyrene spheres with a diameter of 2.060 μm located on the BK 7 substrate. The absorbing boundary's perfectly matched layer [100] and the periodic boundary are adopted as appropriate truncation planes. The incidenct beam existing in the x,z plane was a box type whose width was nearly equal to the length of the unit cell in the x direction. The refractive indexes of the sphere and the substrate are set at 1.590 and 1.515, respectively. By applying the FDTD method to the above model, one can extract various information from the numerical data because both spatial and temporal electric and magnetic field images in the unit cell are calculated. Here, an example of FDTD analysis to interpret the origin of the dips in experimental ATR spectra is presented.

In Fig. 9.28, the calculated angle-scan ATR spectra of 2-D spheres are shown for the easy axis configuration at $\lambda = 600$ nm. The solid and dashed lines are for TM- and TE-polarized light, respectively. The calculated ATR

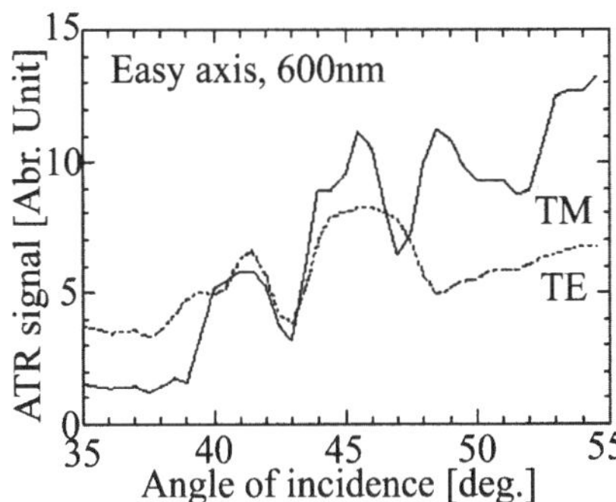

Fig. 9.28. Calculated ATR spectra for the easy axis configuration at $\lambda = 600$ nm

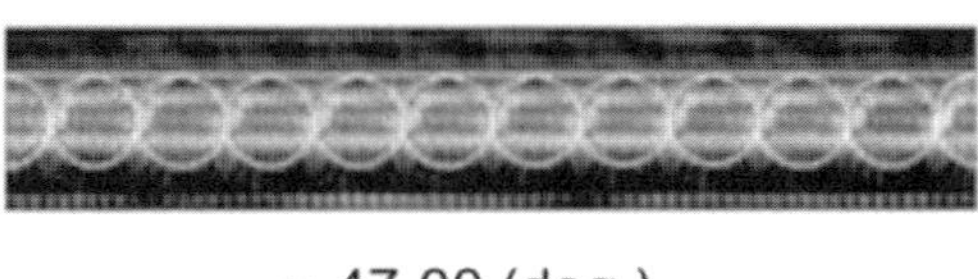

a 47.00 (deg.)

b 51.50 (deg.)

Fig. 9.29. The time-averaged light intensity distribution in the cross section for the TM-polarized light and the easy axis configuration at $\lambda = 600$ nm. **a** and **b** are for $\theta_i = 47.00°$ and $\theta_i = 51.50°$, respectively

spectra are in agreement with the experimental spectra. For the difficult axis configuration, the calculated ATR spectra do not show indicated any clear dip structures.

In Fig. 9.29, the time-averaged light intensity distribution in the cross section of 2-D spheres is shown for TM-polarized light and the easy axis configuration. Figure 9.29a,b is for $\theta_i = 47.00°$ and $\theta_i = 51.50°$ at $\lambda = 600$ nm, respectively. The white lines indicate the surfaces of the prism and spheres. Note that a white color indicates maximum light intensity.

In Fig. 9.29a, the light field coupled into the spheres from the substrate flows straightly toward the next spheres and moves to them. Light is scattered at the contact point with the neighboring sphere. In Fig. 9.29b, the existence of the WG mode along the sphere surface is confirmed at $\theta_i = 51.50°$. The WG modes are coupled to neighboring WG modes. Part of the energy of the WG mode flows into neighboring spheres. The manner of two types of propagating light, as shown in Fig. 9.29, throws light on the reason for the difference in the propagating speeds of the modes corresponding to dips A and B in Fig. 9.26.

To summarize, three types of optical modes in 2-D spheres can be excited by a near-field light of ATR configuration: the mode propagating straight, the mode propagating along the sphere's surface, and the mode localized in a sphere. These propagating modes are observed only in the easy axis configuration. These experimental results can be interpreted well by FDTD analysis.

9.2.4 Photonic-Band Effect on Near-Field Optical Images of 2-D Sphere Arrays

Self-assembled ordered layers of monodispersive polystyrene spherical particles (latex particles) whose diameter is an optical wavelength have a 2-D periodic structure for the dielectric constant on the scale of the optical wavelength and therefore, have characteristic electromagnetic eigenmodes, known as *photonic bands* [106,107].

The resonant phenomenon between the eigenmode of the 2-D photonic band and the incident plane wave appears as dips in the transmission spectra. The transmission spectrum of the latex monolayer for normal incidence [106] and the theoretical transmission spectrum [99] are shown in Fig. 9.30. In the calculation, a hexagonally close-packed array of dielectric spheres whose dielectric constant ϵ is 2.56 is assumed to be a model of the latex layer. The abscissa ω is the normalized frequency: $\omega = \sqrt{3}d/2\lambda$, where λ is the wavelength of incident light in free space and d is the diameter of the sphere. The experimental result is in qualitative agreement with the calculated result. The dip at $\omega = 0.71$ is quite obvious. The width of the experimentally

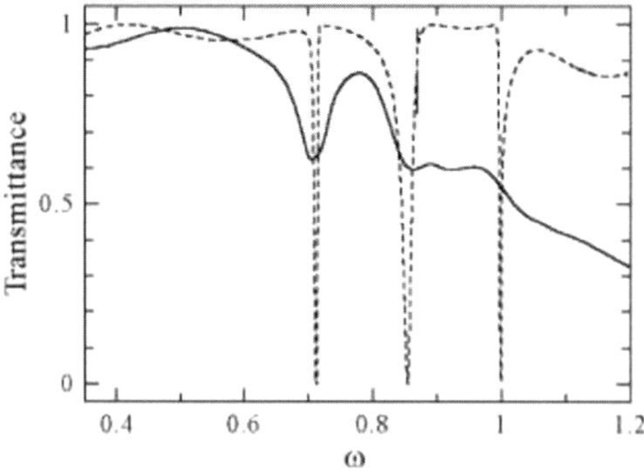

Fig. 9.30. Normal incidence transmission spectrum of a latex monolayer composed of 0.989 μm latex particles (*solid line*) and a theoretical transmission spectrum calculated by H. Miyazaki and K. Ohtaka (*dashed line*)

obtained dips is much wider than the theoretical dip that is caused mainly by glass substrate

The 2-D photonic bands of the latex monolayers are caused by intersphere light scattering in the array. When the layer is illuminated with local excitation from the probe of a scanning near-field optical microscope (SNOM), the light may hop among the latex particles mainly around the SNOM probe. Since the latex monolayer has a periodic surface roughness, the light easily leaks to free space while it hops. Recently Itoh et al. observed such light propagation and dissipation in a latex layer by fluorescent excitatory SNOM measurement [111,112]. This result has motivated H. Miyazaki and K. Ohtaka to plot *dissipative 2-D photonic bands* for a 2-D array of dielectric spheres [99]. When the combination of ω and $\mathbf{k}$ of the incident plane wave is close to one of the branches of the photonic bands, light scattering occurs resonantly in the array, the local field on the surface of the array is also expected to be strongly enhanced [99,131], and that may reflect on the SNOM image contrast patterns.

In the following experiment, we investigated how the photonic band effect in the latex layers is reflected on SNOM images and also to explain how the contrast patterns of SNOM images are formed.

Latex particles with diameters of 1.0, 0.38, and 0.23 µm were used to fabricate self-assembled ordered monolayers on glass substrates by two different self-organization methods: *the ring method* [132] and *the improved Dimitrov cell method* [133]. The SNOM instrument was equipped with an aluminum-coated, tapered optical fiber tip with a cantilever shape as a SNOM probe, controlled by the noncontact AFM mode. The light source was one of the three single lines of an Ar^+ laser: 457.9, 488.0, or 514.5 nm. The latex layer was illuminated with near-field light from the optical fiber tip, in the so-called illumination mode (I mode). The transmitted light coming out from the opposite side of the samples was collected with a 0.4 or 0.1 NA objective lens and detected by a photomultiplier. The "aperture size" of the probe tip end selected was so small that it did not modify the SNOM images.

Figure 9.31 shows 4.5×4.5 µm^2 images of a latex monolayer which is composed of 1.0-µm particles [111,112]. The incident wavelength is 488.0 nm. Here, the corresponding normalized frequency ω is 1.8, which is much larger than the lowest resonant frequency of 0.71. The topographic image in (a) indicates that the latex particles are well ordered in a hexagonal, closely-packed structure. The transmission SNOM image in (b) has a sharp, bright spot at the center of each latex particle, as indicated by white circles drawn on both images of the same particle. Although the contrast pattern of the transmission SNOM image seems to depend on the gap width between the tip aperture and the latex particles [134], yet the change in the coupling of the evanescent wave with the latex particles is, not found as the main cause of the SNOM image, as clearly demonstrated in the transmission SNOM image of a double layer [111].

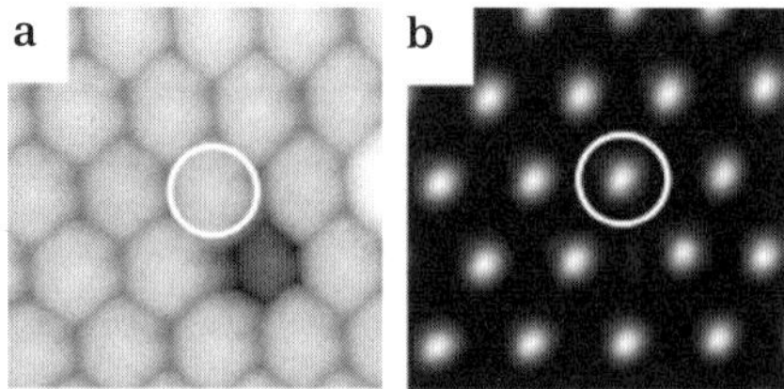

Fig. 9.31. $4.5 \times 4.5\ \mu m^2$ images of a 1.0 µm latex monolayer. **a** topographic image; **b** transmission SNOM image at the 488.0-nm wavelength. White circles indicate the position of the same particle

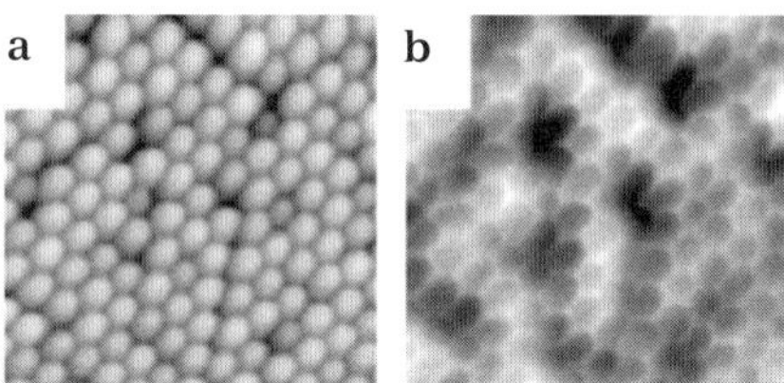

Fig. 9.32. $3 \times 3\ \mu m^2$ images of a 0.23 µm latex monolayer. **a** topographic image; **b** transmission SNOM image at the incident 488.0-nm wavelength

A dark spot is observed in (a) at the lower right-hand side of the white circle where the latex particle is slightly smaller than the others surrounding it. Nevertheless, the corresponding optical image in (b) differs little in size and intensity from the other spots. This suggests that the bright spot is predominantly due to the near-field pattern associated with individual latex particles and not much affected by the ordered structure of the particles.

When the diameter is reduced, the drastic change in transmission SNOM images occurs. Figure 9.32 shows the transmission SNOM image of a monolayer composed of 0.23-µm latex particles [108]. Here, the normalized frequency ω corresponds to 0.41 which is much smaller than the lowest resonant frequency. In the transmission SNOM image in (b), the latex particles appear as a reversed image with the darkness at the central part of each particle superimposed on a background. The domain boundary is reflected on the background as a dark stripe that always appears when the diameter of the particle is smaller than the incident wavelength.

Itoh et al. showed the drastic dependence of transmission SNOM images of latex layers on the wavelength of incident light caused by the resonant effect and discussed the relation between SNOM image contrast patterns and photonic bands [108].

They chose a latex layer composed of 0.38-µm particles. The resonance for the lowest eigenmodes of the photonic band was observed as a dip in the transmission spectrum at the 467-nm wavelength that corresponded to $\omega = 0.71$ in Fig. 9.30. Three Ar^+ laser lines, 457.9, 488.0, and 514.5 nm,

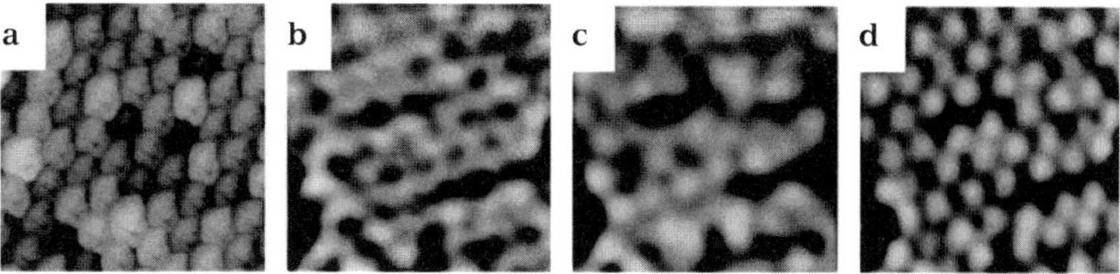

Fig. 9.33. $3 \times 3\ \mu m^2$ images of a 0.38-μm latex monolayer; **a** topographic image, **b–d** transmission SNOM images at 457.9, 488.0, and 514.5 nm, respectively

are close to the observed resonant wavelength. In the transmission SNOM measurement, an objective lens with a small numerical aperture ($NA = 0.1$) was used so as to restrict the light collection angle almost normally to the layer, that is, to get the resonance at the zone center of the 2-D photonic band.

Figure 9.33 shows $3 \times 3\ \mu m^2$ images of the 0.37-μm latex monolayer. The topographic image is shown in (a), and the transmission SNOM images are shown from (b) to (d) for the three different Ar^+ laser lines, respectively. In (b) and (d), the SNOM image patterns mainly reflect the arrangement of the latex particles, although the contrast of the patterns is quite different. In (c), the pattern does not clearly reflect the regular arrangement of the particles, but rather reflects the domain boundaries. Thus, the transmission SNOM images change drastically depending on the incident wavelength closely resonant with the eigenmode of the photonic band.

To make a theoretical approach to the experimental results, H. Miyazaki calculated the wavelength dependence of the electric near-field intensity on an ordered dielectric particle layer. Details of the calculation are reported in [99]. The geometry of the system is shown in Fig. 9.34, where a 2-D periodic array of dielectric particles at diameter d located on the x,y plane is illuminated by an x-polarized plane-wave light coming along the z axis from the negative side. Four sampling points on the $z = 0.5d$ plane are chosen, as illustrated in Fig. 9.34, and the x-component electric near-field intensities at these points, namely, on top of each particle $|E_c^x|^2$ and at three different points ($i = 1$–3) between adjacent particles $|E_i^x|^2$, respectively. This configuration corresponds to the so-called collection mode (C mode). Since the I mode and the C mode give almost the same transmission SNOM images for latex layers [99,135], one can directly compare the calculated results with the experimental results.

The normalized values $|E_i^x|^2/|E_c^x|^2$ as a function of the normalized frequency ω are shown in Fig. 9.35. Note that a dispersive structure appears around the resonant frequency $\omega = 0.71$ for all three points and the center of each particle becomes brighter or darker than the surrounding three points ($i = 1$–3) in a complicated manner, especially near the resonant frequency. Therefore, it is considered that the drastic change in the SNOM images shown in Fig. 9.33 is attributable to the photonic band effect in the latex layer.

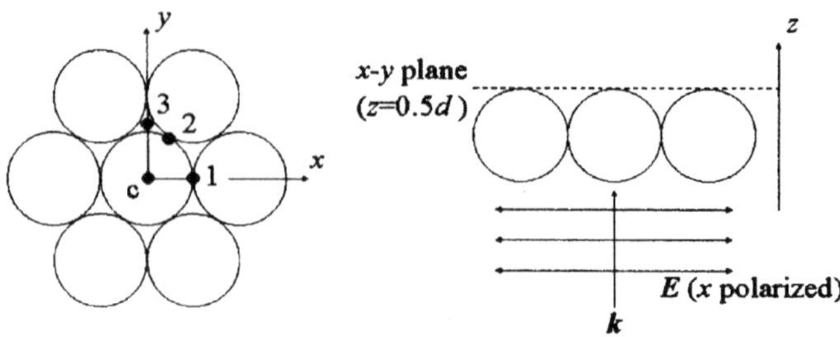

Fig. 9.34. System geometry for calculation. A 2-D dielectric particle array is illuminated by a monochromatic plane wave, and near-field intensities are calculated at four points on the layer surface

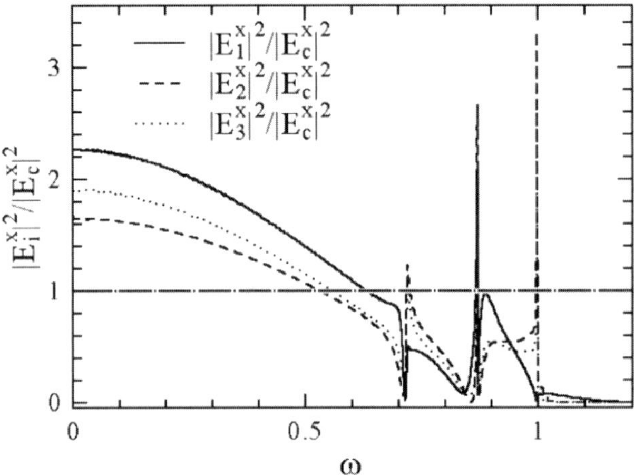

Fig. 9.35. The value $|E^x_{i=1-3}|^2/|E^x_c|^2$ as a function of the normalized frequency ω. A dashed line is drawn at unity

Figure 9.35 also shows that the point at the center of each particle is expected to be much brighter than the other points when the normalized frequency is larger than unity, except for some regions of higher resonance. This is so for the image in Fig. 9.31 for $\omega \sim 1.8$. On the contrary, when the frequency ω is much smaller than the resonant frequency, the center of each particle is expected to be darker than the surrounding points. This is so for the 0.23-μm latex particles in Fig. 9.32 for $\omega = 0.41$.

To summarize, SNOM images represent the electromagnetic near-field structure of the latex layers. Resonant light scattering occurs especially when the incident light emitted from the SNOM probe excites electromagnetic eigenmodes of the latex layers and it is reflected on the SNOM images as a drastic change in the contrast patterns, depending on the incident wavelength. SNOM image patterns for arbitrary dielectric objects, should correspond to electromagnetic intensity patterns in the near-field regime, which are specified by the contrast patterns of the refractive indexes and also by the relative ratio of the incident light wavelength and the dimensions of the objects.

9.3 Near-Field Photon Tunneling

The notion of photonic bands in photonic crystals is based on the analogy of light propagation in periodic structures to that of electrons in periodic potentials of crystals. Similarly, photon tunneling analogous to electron tunneling through barrier structures is possible. In photon tunneling, evanescent waves or near-fields play a crucial role as in electron tunneling. In this section, after a brief introduction, photon tunneling through a photonic double barrier structure and a photonic dot are discussed with particular emphasis on resonant tunneling.

9.3.1 What is Photon Tunneling?

The term *photon tunneling* is used very often in research in near-field optics, in particular, in scanning near-field optical microscopy (SNOM). However, what is meant by photon tunneling in the literature is not always clear, and the physical origin of photon tunneling has not been discussed in detail. The intention of this section is to give a brief introduction into photon tunneling as a fundamental physical process and present some experimental results for resonant photon tunneling, which might be useful for further application of photon tunneling phenomena to near-field optics.

In his "Opticks" in 1704 [136], Sir Isaac Newton described photon-tunneling experiments done by using two glass prisms. His experiment is schematically shown in Fig. 9.36. He compressed two prisms A and B hard together so that their sides might touch one another somewhere. The basal planes of the prisms were a little convex. When light was incident on prism A so obliquely that total reflection took place, light transmission to prism B occurred through the contact point. He found that the transmission spot was larger than the size of the contact point. This means that the light could be transmitted through the thin air gap between the prisms. This phenomenon is a consequence of photon tunneling.

The equivalence of photon tunneling to electron tunneling through a potential barrier can be established from the following consideration [137]. For simplicity, let us consider electron tunneling through a one-dimensional potential barrier, as depicted in Fig. 9.37a. The Schrödinger equation can be written as

$$\left\{\frac{\mathrm{d}^2}{\mathrm{d}z^2}+\frac{2m}{\hbar^2}[E-V(z)]\right\}\Psi(z)=0\,. \tag{9.5}$$

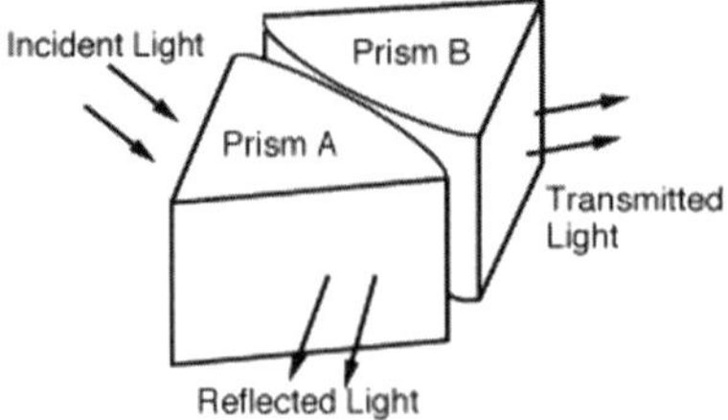

Fig. 9.36. Newton's experiment in photon tunneling

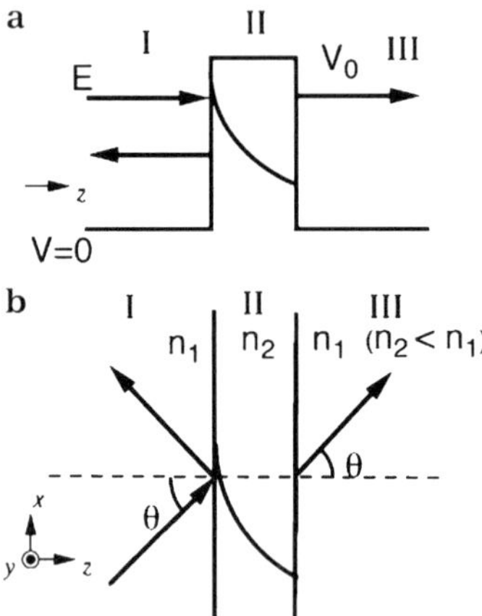

Fig. 9.37. a Tunneling of an electron through a one-dimensional potential barrier and **b** barrier structure for photon and photon tunneling

Letting $V(z) = V_0$ in the barrier region (region II) and $E < V_0$, the wave function in the barrier region can be written as $\Psi(z) \propto \exp(-\kappa z)$, with $\kappa = [2m(V_0 - E)/\hbar^2]^{1/2}$. The wave function is evanescent in this region. When the barrier region is sufficiently thin, the probability of finding an electron in the region III is nonzero, even when $E < V_0$. This is a simple description of electron tunneling.

Simplifying Newton's experiment, an optical analog of the single barrier structure can be constructed as shown in Fig. 9.37b. The structure consists of two dielectric materials with refractive index n_1 separated by a thin gap layer with refractive index n_2 ($n_2 < n_1$). The light wave propagating with a wave vector $\mathbf{k} = (k_x, 0, k_z)$ and a radial frequency ω must satisfy the wave equation derived from the Maxwell's equations,

$$\left\{\frac{\mathrm{d}^2}{\mathrm{d}z^2} + [\frac{\omega^2}{c^2}\epsilon(z) - k_x^2]\right\} A(z) = 0\,. \tag{9.6}$$

In this equation, $\epsilon(z)$ is the dielectric constant and $A(z)$ is the field amplitude of the light wave. When the light is incident at an angle θ, $k_x = (\omega/c)n_1 \sin\theta$.

For s-polarized light, $A(z)$ is taken as the y component of the electric field Ey, and for p-polarized light, $A(z)$ is taken as the y component of the magnetic field Hy. Note that (9.5) is equivalent to (9.6) when the following correspondence is established:

$$\frac{2m}{\hbar^2}[E - V(z)] \Leftrightarrow \frac{\omega^2}{c^2}\epsilon(z) - k_x^2. \tag{9.7}$$

This demonstrates that photon tunneling and electron tunneling can be described by the same equation.

The transmittance of light incident on an optical barrier structure at an incident angle larger than the critical angle of total reflection can be obtained from a standard electromagnetic calculation by applying boundary conditions. In region II, the field amplitude is evanescent as for an electron and can be written as $A(z) \propto \exp(-\kappa' z)$ with $\kappa' = (\omega/c)n_1[\sin^2\theta - (n_2/n_1)^2]^{1/2}$. When the gap layer is thin enough, the transmittance is nonzero, resulting in photon tunneling analagously to electron tunneling. It should be noted that (9.7) does not give one to one correspondence, but one to many correspondence. In fact, the variation of electron energy can correspond to either that of radial frequency (photon energy) or that of the incident angle. Therefore, experiments in photon tunneling can be done in two different ways: by varying the wavelength while fixing the incident angle (larger than the critical angle) and by varying the incident angle while fixing the wavelength.

9.3.2 Resonant Photon Tunneling Through a Photonic Double-Barrier Structure

It is well known that an electron can tunnel through a double-barrier structure, as shown in Fig. 9.38a. The tunneling probability (transmittance) plotted as a function of the energy of an incident electron exhibits a peak when the energy coincides with that of a quasi-bound state inside the potential well (Fig. 9.38b). This leads to so-called resonant electron tunneling that has been realized in semiconductor double-barrier structures [138]. A question then arises as to whether or not resonant tunneling is possible for photons. The answer is yes. An extension of the discussion of photon tunneling given in the preceding subsection allows us to construct a photon analog of the double-barrier structure (photonic double-barrier structure, PDBS), shown in Fig. 9.38c–d. This is a stack of thin layers with the distribution of the refractive index presented in Fig. 9.38c.

Hayashi et al. realized the PDBS and experimentally observed resonant photon tunneling [139,140]. The PDBS prepared is shown schematically in Fig. 9.39. Multilayer samples consisting of an active layer sandwiched by SiO_2 layers were prepared on glass substrates (SF10 glass). Al and Al_2O_3 layers were used as active layers. The SiO_2 and Al_2O_3 layers were deposited by *rf* magnetron sputtering. The Al layer was deposited by vacuum evaporation.

For photon tunneling experiments, the glass substrate with the multilayer on it was pasted to the face of a 60° high-index prism (SF10) with an index matching fluid. Since the glass substrate and the prism are made of the same material, the substrate can be regarded as part of the prism in the present structure. Another prism, also made of SF10, was brought into optical contact with the SiO_2 surface of the multilayer sample to complete the PDBS. The PDBS was mounted on a rotating table driven by a stepping motor. The 632.8-nm line of a He–Ne laser was incident on the structure through the first prism. The intensity of the reflected light was measured as a function of incident angle θ using a Si photodiode. The intensity of light transmitted through the second prism (tunneling photons) was also measured using the Si photodiode. The polarization of the incident light was set to the p or s polarization.

Figure 9.40 shows reflectance and transmittance spectra obtained for a PDBS containing an Al_2O_3 active layer. The incident light was p-polarized. Rough estimates of the thickness of the active Al_2O_3 and the coupling and de-

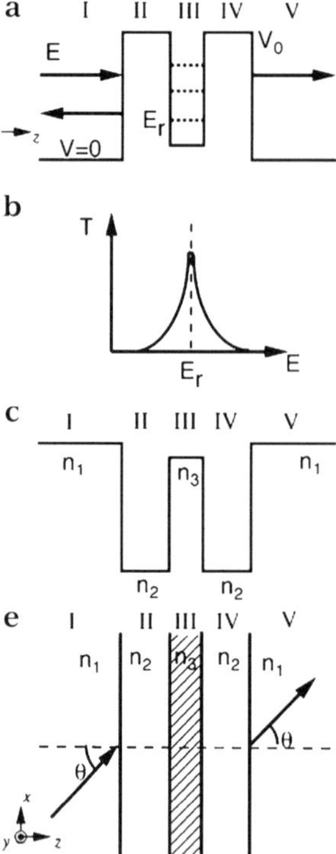

Fig. 9.38. **a** Electron tunneling through a double-barrier structure; **b** transmission probability as a function of the incident energy; and **c** variation of refractive index in the photonic double-barrier structure shown in **d**

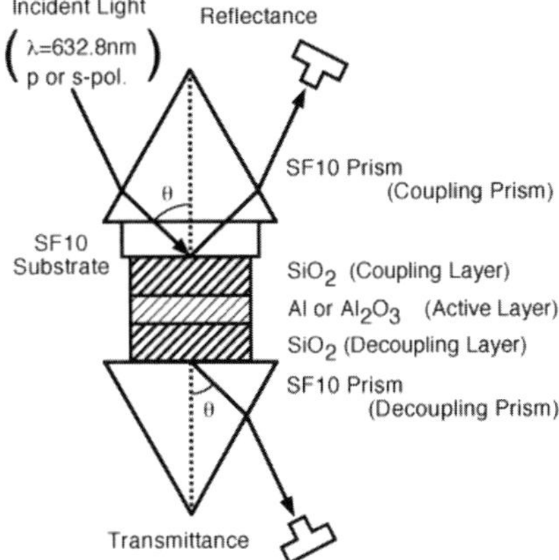

Fig. 9.39. Prepared sample and experimental arrangement

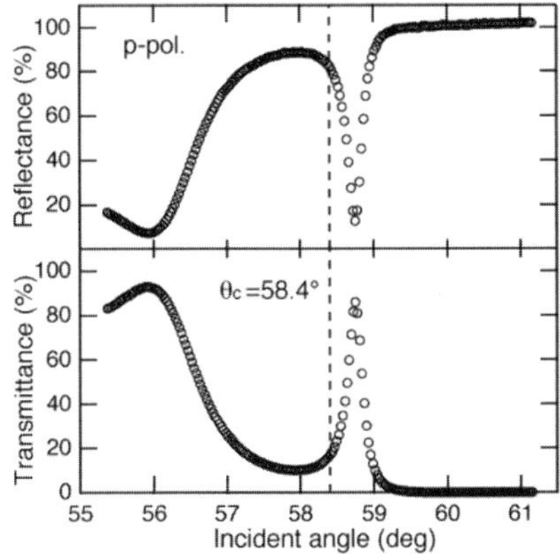

Fig. 9.40. Reflectance and transmittance of a PDBS containing an Al_2O_3 active layer obtained for p-polarized incidence

coupling SiO_2 layers were 35 nm, 800 nm, and 1 µm, respectively. In Fig. 9.40, a sharp dip in the reflectance spectrum is clearly seen at an angle of incidence larger than the critical angle of total reflection for the prism–SiO_2 system ($\theta_c = 58.4°$). At the same angle, a sharp peak appears in the transmittance spectrum. This is a manifestation of resonant photon tunneling. A similar reflection dip and transmission peak can also be observed for s-polarized incident light.

A detailed analysis of the data based on electromagnetic calculations of the reflectivity and transmittance as well as the field distributions in the multilayer systems suggests that the tunneling observed is mediated by the excitation of TM or TE guided modes inside the Al_2O_3 active layer. When

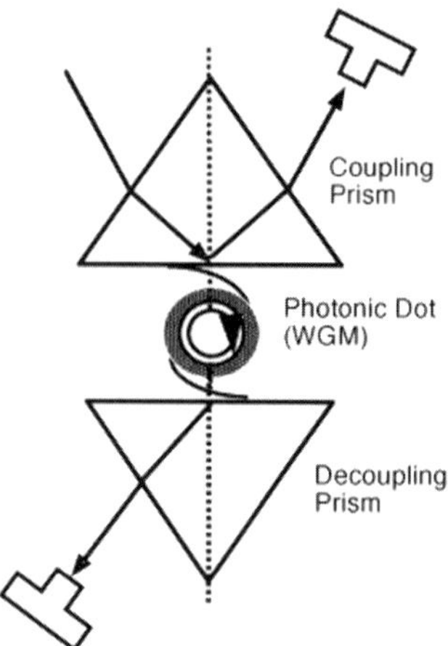

Fig. 9.41. Schematic presentation of photon tunneling mediated by a photonic dot

the light is incident on the PDBS at an angle larger than the critical angle, an evanescent wave is generated in the coupling SiO_2 layer. At a certain incident angle, the phase velocity of the evanescent wave matches that of a guided mode supported by the $SiO_2/Al_2O_3/SiO_2$ system, and the guided mode is excited. Upon excitation of the guided mode, the evanescent wave associated with the mode is induced in the decoupling layer, which is decoupled into the second prism giving rise to transmitted light. Surface plasmon modes, which can be regarded as a special type of guided mode, are excited in the Al active layer. Since the surface plasmon modes are TM modes, resonant photon tunneling is observed only for p-polarized incident light. It should be stressed that resonant photon tunneling can clearly be observed in the PDBS and the intermediate states for the resonance are electromagnetic normal modes of the well, i.e., TM or TE guided modes for the dielectric well and surface plasmon modes for the metallic well.

9.3.3 Resonant Photon Tunneling Mediated by a Photonic Dot

In recent years, a lot of work has been done on electron tunneling mediated by quantum dots made of a semiconductor or a metal. For a variety of tunneling structures containing dots, a Coulomb blockade and staircases, which are the consequence of single electron transport, were generally observed [141]. Resonant tunneling mediated by confined electronic states in the dots was also reported. Now, a question arises, "is it possible to realize photon tunneling mediated by a photonic dot?" The answer is believed to be "yes," as described in the following.

An optical analog of electron tunneling through a dot can be constructed, as shown schematically in Fig. 9.41. The structure is similar to the PDBS presented in Fig. 9.40, but now the active thin layer is replaced by a photonic dot. A dielectric or metallic sphere can be used as a photonic dot. It is well known that a dielectric or metallic sphere can support electromagnetic normal modes, often called whispering gallery modes (WGM). Resonant tunneling of photons mediated by WGM may be possible. The whole tunneling process is as follows. When light is incident on the first prism at an angle larger than the critical angle, an evanescent wave is generated in the vicinity of the basal plane of the prism. When the wavelength of incident light is varied, while fixing the incident angle, a WGM can be excited at a certain wavelength by the evanescent wave. Upon excitation of the WGM, the near-field associated with the mode is induced in the vicinity of the basal plane of the second prism. The near-field is then decoupled into the second prism giving rise to transmitted light. This is resonant photon tunneling mediated by a photonic dot, and the intermediate states for the resonance are WGMs in the dot. Experiments on resonant photon tunneling mediated by a polystyrene sphere are now under way in the author's laboratory, and results will be reported elsewhere [142].

9.3.4 Concluding Remarks

Recently, photonic crystals and photonic bands have been the subject of intensive research. As discussed, the potential $V(z)$ for electrons corresponds to the dielectric function $\epsilon(z)$ for photons. On the basis of the equivalence of wave equations (9.5) and (9.6) together with the correspondence given by (9.7), it is easy to argue that in a periodic structure that has a periodic dielectric function $\epsilon(z)$ (photonic crystals), there are energy bands and band gaps for photons analogous to electrons in a periodic potential in a crystal. Photon tunneling and related phenomena seem to be less familiar compared with photonic crystals and related phenomena. However, it should be noted that they both have the same footing, i.e., the analogy between the electron wave propagation and light wave propagation. It should be stressed that photon tunneling is essentially governed by evanescent waves and near-fields. Although the application of photon tunneling has not been attempted, further studies on photon tunneling in the framework of near-field optics may open the way to the fabrication of novel photonic materials and devices.

References

1. R. D. Grober et al.: Appl. Phys. Lett. **64**, 1421 (1994); R. D. Grober et al.: Rev. Sci. Instrum. **65**, 626 (1994)
2. H. F. Hess et al.: Science **264**, 1740 (1994)
3. E. Betzig and R. Chichester: Science **262**, 1422 (1993)
4. J. K. Trautman et al.: Natúre **369**, 40 (1994)
5. X. S. Xie and R. C. Dunn: Science **265**, 361 (1994)
6. W. P. Ambrose et al.: Science **265**, 364 (1994)

7. N. H. Bonadeo et al.: Phys. Rev. Lett. **81**, 2759 (1998)
8. B. A. Nechay et al.: Appl. Phys. Lett. **74**, 61 (1999)
9. M. Mansfield and G. S. Kino: Appl. Phys. Lett. **57**, 2615 (1990)
10. S. M. Mansfield, W. R. Studenmund, G. S. Kino, and K. Osato: Opt. Lett. **18**, 305 (1993)
11. B. D. Terris, H. J. Mamin, D. Ruger, W. R. Studenmund, and G. S. Kino: Appl. Phys. Lett. **65**, 388 (1994)
12. B. D. Terris, H. J. Mamin, and D. Ruger: Appl. Phys. Lett. **68**, 141 (1996)
13. J. A. H. Stotz and M. R. Freeman: Rev. Sci. Instrum. **68**, 4468 (1997)
14. A. Chekanov, M. Birukawa, Y. Itoh, and T. Suzuki: J. Appl. Phys. **85**, 5324 (1999)
15. G. S. Kino: Proc. SPIE **3609**, 56 (1999)
16. L. P. Ghislain, and V. B. Elings: Appl. Phys. Lett. **72**, 2779 (1998)
17. L. P. Ghislain, V. B. Elings, K. B. Cozier, S. R. Manalis, S. C. Minne, K. Wilder, G. S. Kino, and C. F. Quante: Appl. Phys. Lett. **74**, 501 (1999)
18. T. Sasaki, M. Baba, M. Yoshita, and H. Akiyama: Jpn. J. Appl. Phys. **36**, L962 (1997)
19. C. D. Poweleit, A. Gunther, S. Goodnick, and J. Menendez: Appl. Phys. Lett. **73**, 2275 (1998)
20. M. Yoshita, T. Sasaki, M. Baba, and H. Akiyama: Appl. Phys. Lett. **73**, 635 (1998)
21. M. Yoshita, M. Baba, S. Koshiba, H. Sakaki, and H. Akiyama: Appl. Phys. Lett. **73**, 2965 (1998)
22. M. Yoshita, N. Kondo, H. Sakaki, M. Baba, and H. Akiyama: unpublished
23. Q. Wu, R. D. Grober, D. Gammon, and D. S. Katzer: Phys. Rev. Lett. **83**, 2652 (1999)
24. M. Vollmer, H. Giessen, W. Stolz, W. W. Ruehle, L. P. Ghislain, and V. B. Elings: Appl. Phys. Lett. **74**, 1791 (1999)
25. Q. Wu, G. D. Fuke, R. D. Grober, and L. P. Ghislein: Appl. Phys. Lett. **75**, 4064 (1999)
26. M. Baba, T. Sasaki, M. Yoshita, and H. Akiyama: J. Appl. Phys. **85**, 6923 (1999)
27. K. Koyama, M. Yoshita, M. Baba, T. Suemoto, and H. Akiyama: Appl. Phys. Lett. **75**, 1667 (1999)
28. K. Koyama, M. Yoshita, M. Baba, T. Suemoto, and H. Akiyama: unpublished
29. K. H. Drexhage: Prog. Opt. **12**, 163 (1974)
30. E. H. Hellen and D. Axelrod: J. Opt. Soc. Am. B **4**, 337 (1987)
31. J. J. Macklin, J. K. Trautman, T. D. Harris, and L. E. Brus, Science **272**, 255 (1996).
32. Y. Arakawa and H. Sakaki: Appl. Phys. Lett. **40**, 939 (1982)
33. D. Leonard, M. Krishnamurthy, C. M. Reaves, S. P. Denbaars, and P. M. Petroff: Appl. Phys. Lett. **63**, 3203 (1993)
34. K. Nishi, H. Saito, S. Sugou, and J. S. Lee: Appl. Phys. Lett. **74**, 1111 (1999)
35. H. Bensity, C. M. Sotomayor-Torres, and C. Weisbuch: Phys. Rev. B **44**, 10945 (1991)
36. U. Bockelmann and G. Bastard: Phys. Rev. B **42**, 8497 (1990)
37. F. Adler, M. Geiger, A. Bauknecht, F. Scholz, H. Schweizer, M.H. Pilkuhn, B. Ohnesorge, and A. Forchel: J. Appl. Phys. **80**, 4019 (1996)

38. S. Raymond, S. Fafard, P. J. Poole, A. Wojs, P. Hawrylak, S. Charbonneau, D. Leonard, R. Leon, P. M. Petroff, and J. L. Merz: Phys. Rev. B **54**, 11548 (1996)
39. J. H. H. Sandmann, S. Grosse, G. von Plessen, J. Feldmann, G. Hayes, R. Phillips, H. Lipsanen, M. Sopanen, and J. Ahopelto: Phys. Status Solidi (b) **204**, 251 (1997)
40. S. Grosse, J. H. H. Sandmann, G. Von Plessen, J. Feldmann, H. Lipsanen, M. Sopanen, J. Tulkki, and J. Ahopelto: Phys. Rev. B **55**, 4473 (1997)
41. V. Zwiller, M. Pistol, D. Hessman, R. Coderstrom, W. Seifert, and L. Samuelson: Phys. Rev. B **59**, 5021 (1999)
42. Y. Toda, O. Moriwaki, M. Nishioka, and Y. Arakawa: Phys. Rev. Lett. **82**, 4114 (1999)
43. M. Ono, K. Matsuda, T. Saiki, K. Nishi, T. Mukaiyama, and M. Kuwata-Gonokami: Jpn. J. Appl. Phys. **38**, L1460 (1990)
44. K. Nishi, R. Mirin, D. Leonard, G. Medeiros-Ribeiro, P. M. Petroff, and A. C. Gossard: J. Appl. Phys. **80**, 3446 (1996)
45. T. Saiki, and K. Matsuda: Appl. Phys. Lett. **74**, 2773 (1999)
46. H. Yu, S. Lycett, C. Roberts, and R. Murray: Appl. Phys. Lett. **69**, 4087 (1996)
47. L.-Q. Li and L. M. Davis: Rev. Sci. Instrum. **64**, 1524 (1993)
48. K. Mukai, N. Ohtsuka, H. Shoji, and M. Sugawara: Appl. Phys. Lett. **68**, 3013 (1996)
49. A. V. Uskov, J. Mclnerney, F. Adler, H. Schweizer, and M. H. Pilkuhn: Appl. Phys. Lett. **72**, 58 (1998)
50. T. S. Sosnowski, T. B. Norris, H. Jiang, J. Singh, K. Kamath, and P. Battacharya: Phys. Rev. B **57**, R9423 (1998)
51. G. N. Patel et al.: J. Chem, Phys. **70**, 4387 (1979); F. Kajzar et al.: J. Appl. Phys. **60**, 3040 (1986); J.-C. Baumert et al.: Appl. Phys. Lett. **53**, 1147 (1988); K. Yokoyama and M. Yokoyama: Philos. Mag. **B61**, 59 (1990); M. Stolka et al.: J. Polym. Sci., Poly. Chem. Educ. **25**, 823 (1987)
52. S. Mononobe et al.: Opt. Commun. **146**, 45 (1998)
53. M. Yoshimoto et al.: Appl. Phys. Lett. **67**, 2615 (1995)
54. H. Tachibana et al.: Phys. Rev. **B47**, 4363 (1993)
55. K. Ebihara et al.: Jpn. J. Appl. Phys. **36**, L1211 (1997)
56. M. Arai et al.: J. Luminescence (in press), (2000)
57. T. Saiki, K. Nishi, and M. Ohtsu: Jpn. J. Appl. Phys **37**, 1638 (1998)
58. J.-Y. Marzin, J.-M. Gérard, A. Izraël, and D. Barrier: Phys. Rev. Lett. **73**, 716 (1994)
59. M. Grundmann, J. Christen, N. N. Ledentsov, J. Bohrer, and D. Bimberg: Phys. Rev. Lett. **74**, 4043 (1995)
60. T. Takahashi, M. Yoshita, I. Kamiya, and H. Sakaki: Appl. Phys. A **66**, S1055 (1998)
61. G. Mie: Ann. Phys. **25**, 377 (1908)
62. P. W. Barber and S. C. Hill: *Light Scattering by Particles: Computational Methods* (World Scientific, Singapore 1990)
63. H. Chew, D. S. Wang, and M. Kerker: Appl. Opt. **18**, 2679 (1979)
64. D. C. Prieve and J. Y. Walz: Appl. Opt. **32**, 1629 (1993)
65. E. Almaas and I. Brevik: J. Opt. Soc. Am. **B12**, 2429 (1995)
66. S. Chang and S. S. Lee: Opt. Commun. **151**, 286 (1998) and the references therein

67. A. V. Zvyagin and K. Goto: J. Opt. Soc. Am. **A15**, 3003 (1998)
68. M. Quinten and P. R. Wannemacher: Appl. Phys. **B68**, 87 (1999)
69. T. Saiki, M. Ohtsu, K. Jang, and W. Jhe: Opt. Lett. **21**, 674 (1996)
70. H. Ishikawa, H. Tamaru, and K. Miyano: Opt. Lett. **24**, 643 (1999)
71. H. Ishikawa, H. Tamaru, and K. Miyano: J. Opt. Soc. Am. **A17**, 802 (1999)
72. R. W. Shaw, W. B. Whitten, M. D. Barnes, and J. M. Rasey: Opt. Lett. **23**, 1301 (1998)
73. K. Mawatari, T. Kitamori, and T. Sawada: Anal. Chem. **70**, 5037 (1998)
74. M. O. Scully and M. S. Zubairy: *Quantum Optics* (Cambridge University Press, Cambridge 1997)
75. H.-M. Tzeng, K. F. Wall, M. B. Long, and R. K. Chang: Opt. Lett. **9**, 499 (1984)
76. M. Kuwata-Gonokami, K. Takeda, H. Yasuda, and K. Ema: Jpn. J. Appl. Phys. **31**, L99 (1992)
77. M. Kuwata-Gonokami, and K. Takeda: Opt. Mater. **9**, 12 (1998)
78. H.-B. Lin, J. D. Eversole, and A. J. Campillo: J. Opt. Soc. Am. B. **9**, 43 (1992)
79. M. Kuwata-Gonokami: *Proc. SPIE, Laser Resonators III*, San Jose, CA, 2000, **3930**, pp. 170–185
80. T. Mukaiyama, K. Takeda, H. Miyazaki, Y. Jimba, and M. Kuwata-Gonokami: Phys. Rev. Lett. **82**, 4623 (1999) and the references therein
81. M. Kuwata-Gonokami: private communication (1999)
82. A. Mekis, J. C. Chen, I. Kurland, S. H. Fan, P. R. Villeneuve, and J. D. Joannopoulos: Phys. Rev. Lett **77**, 3787 (1996)
83. J. D. Joannopoulos, P. R. Villeneuve and S. Fan: Nature **386**, 143 (1997)
84. J. D. Joannopoulos, P. R. Villeneuve, and S. Fan: Solid State Commun. **102**, 165 (1997)
85. H. Kosaka, T. Kawashima, A. Tomita, M. Notomi, T. Tamamura, T. Sato, and S. Kawakami: Phys. Rev. **B58**, R10096 (1998)
86. H. Kosaka, T. Kawashima, A. Tomita, M. Notomi, T. Tamamura, T. Sato, and S. Kawakami: Appl. Phys. Lett. **74**, 1212 (1999)
87. E. Lidorikis, M. M. Sigalas, E. N. Economou, and C. M. Soukoulis: Phys. Rev. Lett. **81**, 1405 (1998)
88. P. R. Conwell, P. W. Barber, and C. K. Rushforth: J. Opt. Soc Amer.**B1**, 2 (1984)
89. R. E. Benner, P. W. Barber, J. F. Owen and R. K. Chang: Phys. Rev. Lett. **44**, 475 (1980)
90. V. B. Braginsky, M. L. Gorodetsky, and V. S. Ilchenko: Phys. Lett. **A137**, 393 (1989)
91. L. Collot, V. Lefevreseguin, M. Brune, J. Raimond, and S. Haroche: Europhys. Lett. **23**, 327 (1993)
92. V. Sandoghdar, F. Treussart, J. Hare, V. LefevreSeguin, J. M. Raimond, and S. Haroche: Phys. Rev. **A54**, R1777-R1780 (1996)
93. M. L. Gorodetsky, A. A. Savchenkov, and V. S. Ilchenko: Opt. Lett. **21**, 453 (1996)
94. D. W. Vernooy, V. S. Ilchenko, H. Mabuchi, E. W. Streed, and H. J. Kimble: Opt. Lett. **23**, 247 (1998)
95. V. S. Ilchenko, M. L. Gorodetsky, and S. P. Vyatchanin: Opt. Communs. **107**, 41 (1994)

96. S. Arnold, J. Comunale, W. B. Whitten, J. M. Ramsey, and K. A. Fuller: J. Opt. Soc. Amer. **B9**, 819 (1992)
97. K. Ohtaka: Phys. Rev. **B19**, 5057 (1979)
98. K. Ohtaka, Y. Suda, S. Nagano, T. Ueta, A. Imada, T. Koda, J. S. Bae, K. Mizuno, S. Yano, and Y. Segawa: Phys. Rev. B **61**, 5267 (2000) and the references therein. This paper contains information about the papers related to ordered spheres published by various groups
99. H. Miyazaki and K. Ohtaka: Phys. Rev. B **58**, 6920 (1998)
100. A. Taflove: *Computational Electrodynamics-The Funite-Difference Time-Domain Method* (Artech House, Boston 1995)
101. K. S. Yee: IEEE Trans. Antennas and Propagation **14**, 302 (1966)
102. S. Fan, P. R. Villeneuve, and J. D. Joannopoulos: Phys. Rev. B **54**, 11245 (1996)
103. A. J. Ward and J. B. Pendry: Phys. Rev. B **58**, 7252 (1998)
104. A. Shinya and M. Fukui: Opt. Rev. **6**, 215 (1999)
105. A. Shinya, M. Haraguchi, and M. Fukui: J. Opt. Soc. Am. A (submitted)
106. R. Shimada, A. Imada, T. Koda, T. Fujimura, K. Edamatsu, T. Itoh, K. Ohtaka, and K. Takeda: Mol. Cryst. Liq. Cryst. **327**, 95 (1999)
107. R. Shimada, Y. Komori, T. Koda, T. Fujimura, T. Itoh, and K. Ohtaka: Mol. Cryst. Liq. Crystl. (in press)
108. T. Fujimura, T. Itoh, A. Imada, R. Shimada, T. Koda, N. Chiba, H. Muramatsu, H. Miyazaki, and K. Ohtaka : J. Lumin. **87-89**, 954 (2000)
109. T. Fujimura, T. Itoh, T. Koda, H. Miyazaki, and K. Ohtaka: in *Atomic-Scale Surface and Interface Dynamics, Proc. 3rd Symp.*, Fukuoka, Japan, 1999, pp. 47–51
110. S. I. Matsushita, Y. Yagi, T. Miwa, D. A. Tryk, T. Koda, and A. Fujishima: Langmuir **16**, 636 (2000)
111. T. Fujimura, K. Edamatsu, T. Itoh, R. Shimada, A. Imada, T. Koda, N. Chiba, H. Muramatsu, and T. Ataka: Opt. Lett. **22**, 489 (1997)
112. T. Fujumura, T. Itoh, K. Hatashibe, K. Edamatsu, K. Shimoyama, R. Shimada, A. Imada, T. Koda, Y. Segawa, N. Chiba, H. Muramatsu, and T. Ataka: Mater. Sci. Eng. **B48**, 94 (1997)
113. M. Haraguchi, T. Nakai, A. Shinya, T. Okamoto, M. Fukui, T. Koda, R. Shimada, and K. Takeda: Opt. Rev. **6**, 261 (1999)
114. M. Haraguchi, T. Nakai, A. Shinya, T. Okamoto, M. Fukui, T. Koda, R. Shimada, K. Ohtaka, and K. Takeda: Jpn. J. Appl. Phys. **39**, 1471 (2000)
115. A. van Blaaderen, R. Ruel, and P. Wiltzius: Nature **385**, 321 (1997)
116. H. Míguez, C. López, F. Meseguer, A. Blanco, L. Vázquez, R. Mayoral, M. Ocaña, V. Formés, and A. Mifsud: Appl. Phys. Lett. **71**, 1148 (1997)
117. K. Fukuda, H. Sun, S. Matsuno, and H. Misawa: Jpn. J. Appl. Phys. **37**, L508 (1998)
118. J. E. G. Wijinhoven and W. L. Vos: Science 802 (1998)
119. A. A. Zakhidov, R. H. Baughman, Z. Iqbal, C. Cui, I. Khayrullin, S. O. Dantas, J. Marti, and V. G. Ralchenko: Science **282** 897 (1998)
120. T. Yamasaki and T. Tsutsui: Appl. Phys. Lett. **72**, 1957 (1998)
121. G. Subramania, K. Constant, R. Biswas, M. M. Sigalas, and K.-M. Ho: Appl. Phys. Lett. **74**, 3933 (1999)
122. E. R. Brown and O. B. McMahon: Appl. Phys. Lett. **67**, 2138 (1995)
123. V. Yannopapas, A. Modinos, and N. Stefanou: Phys. Rev. B **60**, 5359 (1999)

124. M. Born and E. Wolf. *Principles of Optics* (Pergamon Press, Oxford 1975)
125. K. Ujihara, M. Osige, and M. Takagu: **32**, L1808 (1993)
126. A. J. Campillo, J. D. Eversole, and H.-B. Lin: Phys. Rev. Lett. **67**, 437 (1991)
127. H.-B. Lin and A. J. Campillo: Phys. Rev. Lett. **73**, 2440 (1994)
128. S. Arnold: J. Chem. Phys. **106**, 8280 (1997)
129. M. Inoue and K. Ohtaka: J. Phys. Soc. Jpn. **52**, 1457 (1983)
130. M. Haraguchi, T. Nakai, A. Shinya, T. Okamoto, M. Fukui, T. Koda R. Shimada, K. Ohtaka, and K. Takeda: *Proc. 2nd Workshop on Near-Field Optics*, Beijing, Oct. 20–23, 1999 (World Scientific, 2000)
131. M. Inoue and K. Ohtaka, Phys. Rev. B **26**, 3487 (1982)
132. C. D. Dushkin, H. Yoshimura, and K. Nagayama, Chem. Phys. Lett. **204**, 455 (1993)
133. S. Matsushita, T. Miwa, and A. Fujishima, Langmuir **13**, 2582 (1997)
134. L. Novotny, B. Hecht and D. W. Pohl, Ultramicroscopy **71**, 341 (1998)
135. T. Fujimura, T. Itoh, A. Imada, R. Shimada, T. Koda, S. Takabayashi, H. Miyazaki, and K. Ohtaka, Proceedings of The Second Asia-Pacific Workshop on Near-field Optics, World Scientific Publishing Co. Ltd. (2000) to be published
136. Sir I. Newton: *Opticks or A Treatise of the Reflections, Refractions, Inflections and Colours of Light* (Dover Publications, New York 1952)
137. R.Y. Chiao, P. G. Kwiat, and A. M. Steinberg: Physica B, **175**, 257 (1991)
138. H. Mizuta and T. Tanoue: *The Physics and Applications of Resonant Tunneling Diodes* (Cambridge University Press, Cambridge 1995)
139. S. Hayashi, H. Kurokawa, and H. Ohga: *5th Int. Conf. Near Field Opt. Relat. Tech.*, Shirahama, 1998, pp. 457–458
140. S. Hayashi, H. Kurokawa, and H. Ohga: Opt. Rev. **6**, 204 (1999)
141. H. Grabert and M. H. Devoret, eds.: *Single Charge Tunneling, Coulomb Blockade Phenomena in Nanostructures*, NATO ASI Series, Series B: Physics **294** (Plenum Press, New York 1992)
142. S. Hayashi: NFNO News. **11**, 46 (1999)

10 Other Imaging and Applications

N. Umeda, A. Yamamoto, R. Nishitani, J. Bae, T. Tanaka, and S. Yamamoto

In this chapter, the following five topics are selected within various applications using near field optics. First, we describe the principle and equipment of the birefringent-contrast near-field optical microscope and observations of samples in Sect. 10.1 The observation of birefringent distribution is important in view of the evaluation of liquid crystal or polymer materials by molecular orientation. In Sect. 10.2, the development of a simple low-temperature near-field optical microscope is explained. Low-temperature near-field optical microscopes are necessary for semiconductor spectral analysis, e.g., quantum dot, because conventional microscopes are difficult to handle and change samples. In Sect. 10.3, a near-field microscope system is developed that can detect tunneling light emission from the probe tip of the scanning tunneling microscope. The feature of this microscope is observing the optical emission characteristic distribution of a sample with lateral resolution on the nanometer scale. In Sect. 10.4, a new method for detecting an evanecent wave is proposed that uses an electron beam as a probe; it has succeeded in observing an evanescent wave in the infrared region experimentally. In Sect. 10.5, we describe a method and features for driving Mie particles by an evanescent photon force formed in an optical channeled waveguide. The driving performance of a multimode waveguide has been especially clarified.

10.1 Birefringent Imaging with an Illumination-Mode Near-Field Scanning Optical Microscope

The advent of near-field scanning optical microscopy (NSOM) allows for the development of research with a resolution beyond the diffractve limit. In the early stage of near-field optics study, localized light intensity near the surface of a sample was used as a detection quantity in most NSOMs developed. In the further progress of near-field optics, various NSOM instruments that can detect an optical phase or spectroscopic quantity have been developed. Some polarization contrast NSOMs have been proposed by researchers, since polarization is necessary in measuring the orientation of polymer material, in cell structure analysis, in phase transition of liquid crystals, and in the stress analysis of solid materials. Betzig and et al., who first realized polarization NSOM with cross Nicol prisms [1], achieved near-field imaging for magneto-optic recording bits [2]. However, quantitative analysis of polarization in sample is difficult; therefore, novel polarization NSOMs that can measure dichroism and birefringence have been discussed and studied. Vaez-Iravani et al. developed the polarization NSOM using an electro-optic phase modulator [3], but the azimuthal angle of birefringence cannot be measured. On the other hand, a polarization modulation NSOM which allows for the simultaneous imaging of the magnitude and direction of optical anisotropy, was investigated [4], and a dichroic single crystal of rhodamine 110 was imaged. Optical anisotropy, in particular, the birefringent contrast of NSOM has received a great deal of attention in many fields, e.g., optical memory, biology, polymers and liquid crystals.

In this section, we report on a technique that can provide the magnitude and orientation of the birefringent images of a sample.

10.1.1 Principle

Figure 10.1 depicts the basic principle of birefringent measurement [5]. A He–Ne laser to which an axial magnetic field is applied is used as a light source. A frequency-stabilized axial Zeeman laser (SAZL) has two oscillating components that consist of right- and left-hand circularly polarized lights that are slightly separated at a frequency of 70 kHz. The laser beam goes through a quarter-wave plate (QWP) with a fast axis of 45° and a linear polarizer (LP) with a transmission axis parallel to the horizontal plane, after passing through a sample that is to be measured. The emergent beam from the LP is incident on the photodetector (PD). The output signal from the photodetector consists of ac components and a dc component.

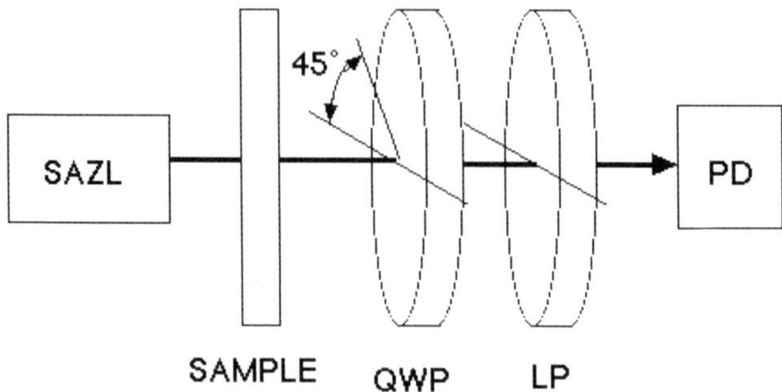

Fig. 10.1. Basic principle of birefringent measurement

The polarization state at each position shown in Fig. 10.1 can be described by the Stokes vector. The matrix formula corresponding to this configuration is given by

$$S = L \cdot Q \cdot X \cdot Sz\,, \tag{10.1}$$

where L, Q and X are the Muller matrices of LP, QWP, and the sample, respectively, and Sz is the Stokes vector of SAZL. The Sz is also written as

$$Sz = \begin{vmatrix} a_{\mathrm{r}}^2 + a_{\mathrm{l}}^2 \\ -2a_{\mathrm{r}}a_{\mathrm{l}} \sin \omega_{\mathrm{b}} t \\ 2a_{\mathrm{r}}a_{l} \cos \omega_{\mathrm{b}} t \\ a_{\mathrm{r}}^2 - a_{\mathrm{l}}^2 \end{vmatrix} \tag{10.2}$$

where a_{l} and a_{r} are the amplitudes of the left- and right-hand circularly polarized components and ω_{b} is the angular Zeeman beat frequency. For the sake of simplicity, we have used an approximation of the measured retardation that is smaller than unity. Therefore, the detected component which is proportional to the irradiance component of the Stokes vector is given by

$$\begin{aligned} I &= \frac{a_{\mathrm{r}}^2 + a_{\mathrm{l}}^2}{2} + a_{\mathrm{r}}a_{\mathrm{l}} \sin \Delta \cos 2\phi \cos \omega_{\mathrm{b}} t + a_{\mathrm{r}}a_{\mathrm{l}} \sin \Delta \sin 2\phi \sin \omega_{\mathrm{b}} t \\ &= I_{\mathrm{dc}} + I_{\mathrm{x}} \cos \omega_{\mathrm{b}} t + I_{\mathrm{y}} \sin \omega_{\mathrm{b}} t\,, \end{aligned} \tag{10.3}$$

where Δ and ϕ are the retardation and fast axis of the sample, respectively. In addition, I_{dc} is the dc component and I_{x} and I_{y} are the quadrature ac components of the beat signal. Since we can adjust the equal amplitudes of the left and right circular polarization components in (10.3), i.e. $a_{\mathrm{l}} = a_{\mathrm{r}}$, the retardation D and fast axis f of the sample are calculated as follows:

$$\Delta = \sin^{-1} \left\{ \frac{\sqrt{I_x^2 + I_y^2}}{I_{\mathrm{dc}}} \right\}\,, \tag{10.4}$$

and

$$\phi = \frac{1}{2} \tan^{-1} \left\{ \frac{I_y}{I_x} \right\}\,. \tag{10.5}$$

In Fig. 10.1, I_{dc} is filtered from the detected signal by using a low-pass filter (LPF), and I_{x} and I_{y} are synchronously detected by a two-phase lock-in amplifier whose reference frequency is supplied from the controlled beat signal of the axial Zeeman laser. These data are acquired by a personal computer to calculate D and f by using (10.4) and (10.5).

10.1.2 Apparatus

A birefringent-contrast NSOM (B-NSOM) based on the above principle was constructed with a shear-force regulated technique, as shown in Fig. 10.2.

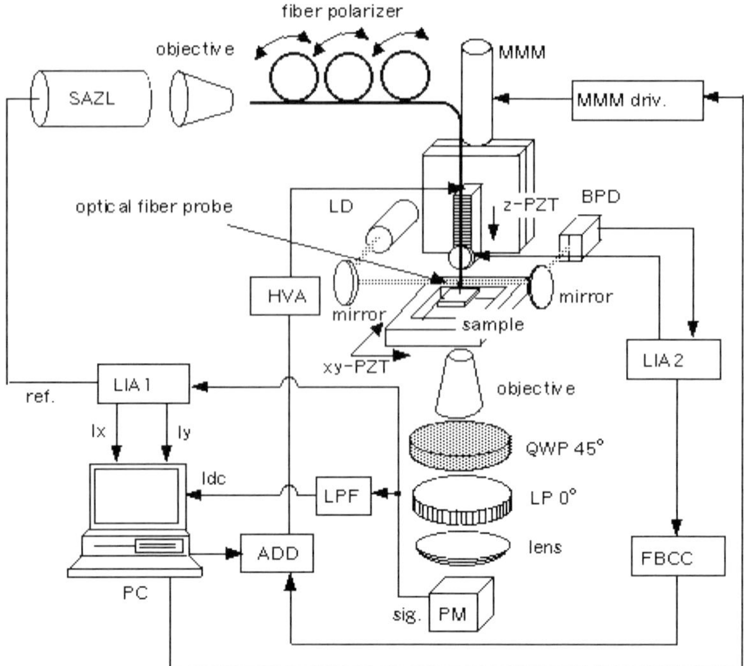

Fig. 10.2. Diagram of the B-NSOM using a right- and left-hand circularly polarized laser

The light from an axial Zeeman laser (SAZL) was launched into a single-mode optical fiber with an in-line polarization compensator which consists of three fiber loops. This is used to compensate for the polarization state change due to the stress-induced birefringence in the fiber. The other end of the fiber is pulled to form a tip by the melting and pulling method. The emergent light from the fiber tip is recovered into right- and left-hand circular polarization with the fiber polarizer. The transmitted light through a sample goes to an objective lens, a quarter-wave plate and a linear polarizer, and is incident on a photomultiplier (PM). The detected beat signal via the PM is synchronously rectified by a two-phase lock-in amplifier (LIA1) with a reference signal of the beat frequency from the SAZL. The quadrature signals from LIA1 and the dc component of the beat signal are acquired by a computer, and the retardation and azimuthal angle of birefringence are calculated by using (10.4) and (10.5). The gap between the tip and sample is regulated by a shear-force technique, which is commonly used in usual NSOMs.

10.1.3 System Performance

Polarization Property. The polarization property of the light emerging from the gold-coated optical-fiber probe fabricated by the melting and drawing method was examined. The beat signal strength for the azimuthal angle of the LP in Fig. 10.2 was measured. If the polarization state of the SAZL is maintained to the end of fiber, the right- and left-hand circular polarization states of the SAZL are converted to orthgonal linear polarizations through a quarter-wave plate with an azimuth of 45°. Therefore, the peak of the beat signal intensity should appear at four points, when the LP is rotated by 360°. The measurements are shown in Fig. 10.3. It is found that the four peaks of beat signal intensity have appeared with the rotation of the LP, though the rotation of the LP has been limited within 330° due to the configuration of the equipment. From this figure, we can confirm that the light from a sharpened optical fiber preserves the right- and left-hand circularly polarized light from the SAZL.

Measurement Accuracy. The measurements in Fig. 10.4a show when the retardation of birefringence of a Babinet–Soleil compensator (BSC) changes at a constant azimuthal angle of the fast axis. The straight line in the figure is the calibrating line of the Babinet–Soleil compensator. From this, it is proven that the measured value differs from the calibrating line of the BSC over about 30°. The following are considered as the cause: the effect of birefringence on the optical fiber is not perfectly cancelled by the fiber polarizer and the intensities of the right- and left-hand circularly polarized light from the SAZL are not equal. Therefore, the measured value was corrected by using a correction function obtained from both the calibrating line of the BSC and the measured data shown in Fig. 10.4a. The result of the corrected value

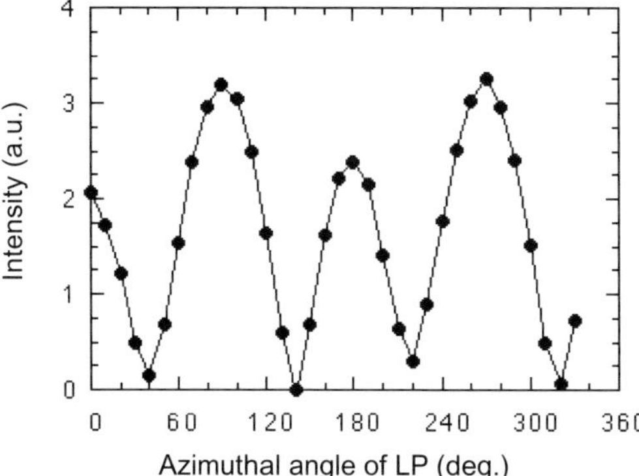

Fig. 10.3. Measurements of the polarization of a beam emerging from a sharpened fiber probe

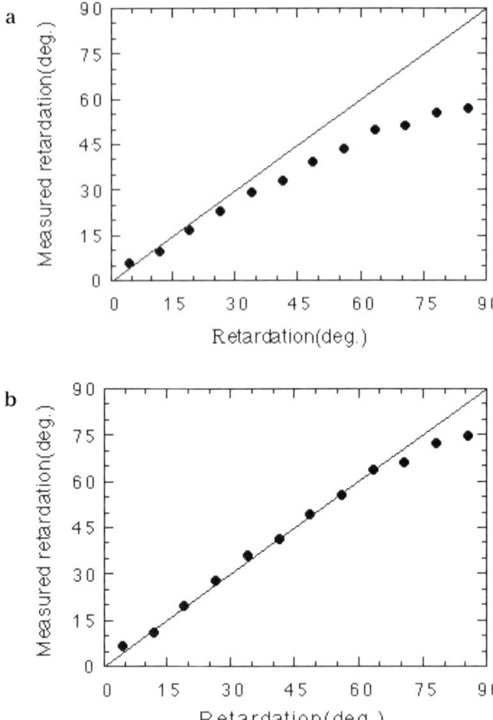

Fig. 10.4. Measured retardation response when changing the settings of the Babinet–Soleil compensator **(a)** before correction and **(b)** after correction with the calibration line

is shown in Fig. 10.4b. This figure shows that the retardation of birefringence agreed well with the calibrating line within about 60°.

The azimuthal angle of birefringence was also measured by rotating the fast axis of the BSC at the fixed retardation to confirm the effectiveness of the measurement method. The result is shown in Fig. 10.5. The straight line in the figure is the calibrating line of the BSC. There exists an error of 0.4°, on average, and 3.04° maximum in measuring the azimuthal angle of birefringence by this method.

10.1.4 Observation of Sample

We prepared a highly oriented sample to evaluate the imaging capability of the B-NSOM. The sample was fabricated by the frictional transfer technique, wherein the surface of a polymer rod is slid on a clean glass slide [6]. The polymer rod (ethylene vinyl acetate) came into contact with the heated coverslip at 180°C and was moved by hand at a constant rate in one direction.

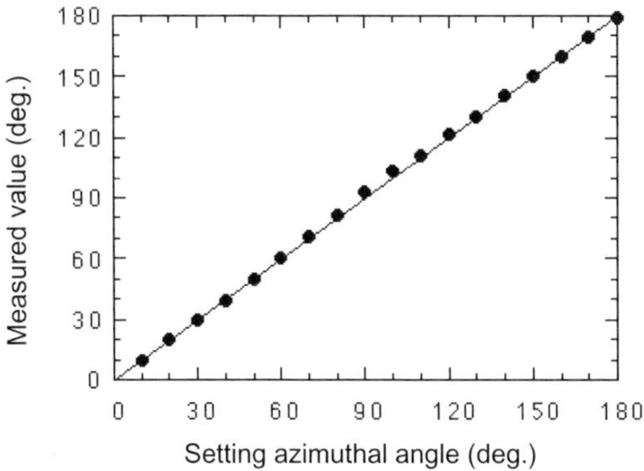

Fig. 10.5. Measured azimuthal angle of birefringence when changing the azimuthal of the Babinet–Soleil compensator

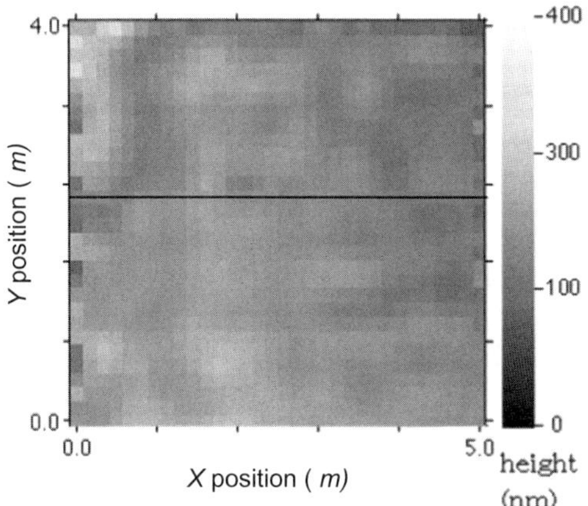

Fig. 10.6. Topographic image of a polymer film

The topographic image of the polymer film fabricated by frictional transfer is shown in Fig. 10.6. Many corrugated structures with a height difference of 50 to 100 nm can be seen. Figure 10.7a,b shows images for the retardation and azimuthal angle of birefringence in the sample. The sliding direction of the polymer rod is vertical in these figures. Figure 10.7a shows an inhomogeneous structure in the retardation of birefringence, which appears as bright and dark portions in the vertical direction. The structure is most likely due to the different degree of orientation in the polymer sample.

On the other hand, Fig. 10.7b shows the mapping of the azimuthal angle of birefringence in a sample. In this figure, the direction of each small bar represents an azimuthal angle of birefringence at each pixel. The direction

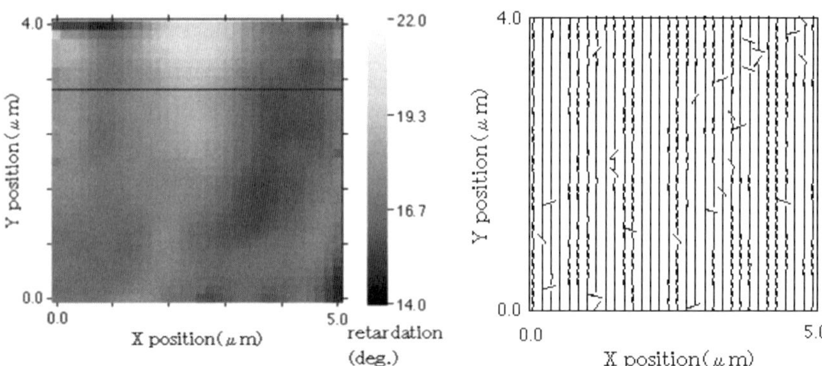

Fig. 10.7. a Retardation and **b** azimuthal angle of birefringent images of a polymer film

bars at each pixel are almost parallel to the sliding direction of the polymer rod during sample formation.

10.1.5 Conclusion

We constructed a new polarization contrast NSOM capable of obtaining retardation and azimuthal angle mapping of a sample. A highly oriented polymer film fabricated by the frictional transfer technique was imaged to demonstrate the utility of this technique. The B-NSOM will be valuable for evaluating anisotropic microcrystalline samples and microstress in materials.

10.2 Plain-Type Low-Temperature NSOM System

Recently, great interest has been shown in nanometer-sized semiconductors because they have potential for application in various kinds of optical devices. Microcrystallites of nanometer size, i.e., quantum dots, are one of the attractive material systems and have been studied very extensively. Early discussions of their optical properties have been based on large inhomogeneous broadening due to size distribution. To extract the effect of size distribution, the spectroscopy of a single dot has been strongly desired. Recent progress on near-field scanning optical microscopy (NSOM) makes it possible to overcome this barrier.

To study the optical properties of a single dot, it is essential to work at low temperature. Several groups have already constructed NSOM systems for low-temperature experiments. However, the systems become larger and complicated. The whole scanning setup is immersed in liquid He [7] or installed in a helium gas-flow cryostat [8], and it is not easy to access and prepare an experiment. Furthermore, the sample cannot be changed after it is cooled down. To improve accessibility, we constructed a plain-type low-temperature NSOM system.

10.2.1 Experimental Setup

Figure 10.8 shows a schematic view of the NSOM system we constructed. This system was expanded from a STM/AFM system (JEOL Ltd.; JSTM-4200D/A/S). Since our system used lots of know-how from the STM measurement technique, there are many advantages. The stage driven by a piezoelectric transducer is drift-free, and it allows us to work within a drift of 0.05 nm/s. The feedback system for controlling the height of the fiber tip can be monitored by either a tunneling current or a shear force. Hence, we can choose the feedback mode according to the measurement condition and/or the electrical conductivity of the sample. It is easy to install various kinds of excitatory lasers by using an objective lens to couple the fiber. The excitatory laser light that passes through the fiber reaches the tapered tip provided by Nanonics Imaging Ltd., and an evanescent wave from the tip excites the sample. The aperture size of the fiber tip is around 100 nm. The scattered light or emitted light from the sample is collected by an objective lens, and the intensity is detected by a photomultiplier tube (HAMAMATSU; R647). The light reflected by the mirror in front of the detector is caught by a TV monitor. Since we can look at the position of the fiber tip and sample surface with this TV monitor, it is easy to approach the sample surface with the fiber tip.

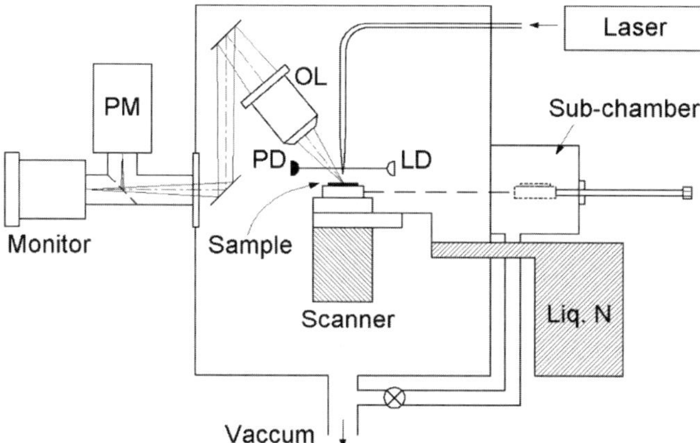

Fig. 10.8. The experimental setup of the NSOM system. Shear force is monitored by a pair of LDs (laser diode) and PDs (photo diode). The light scattered or emitted from the sample is collected by an objective lens (OL), and the intensity is detected by a photomultiplier tube (PM)

The sample stage contacts a cold finger, which can be cooled down by liquid nitrogen. There are heat and temperature sensors at the stage, which allow us to control the sample temperature from 130 K to RT. Since the previously reported low-temperature NSOM systems [7,8] are very large and complicated, it is not so easy to prepare experiments. On the contrary, we can operate the present experimental system at low temperatures as easily as at room temperature. When the sample is cooled down, the sample surface may be contaminated. To avoid contamination, we set a sample in a sub-chamber, and after cooling the sample stage, we install it in the main chamber. Furthermore, using this subchamber, we can change the sample without going back to room temperature.

10.2.2 Results and Discussion

We prepared Au microcrystals as a test sample. Figure 10.9 shows an STM image of Au microcrystals, many of which can be clearly seen. Figure 10.10a shows an AFM image of another portion of the same sample at room temperature. Although the spatial resolution is a little poorer than that of the STM, many Au particles are observable. Figure 10.10b shows the NSOM image of the same portion. A 514.5-nm Ar-ion laser line was used as an excitatory source, and light scattered from the sample was detected. High intensity regions seem to be the same as those imaged by AFM. However, there is no reason that a larger portion in the AFM image must be at a higher intensity in the NSOM image. Actually, the NSOM image is more complicated than the AFM image. For example, the region of the circled area indicated

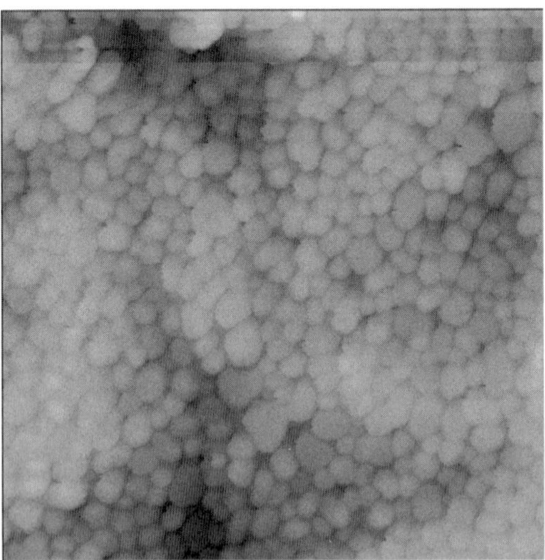

Fig. 10.9. STM image of Au microcrystals (6×6 mm)

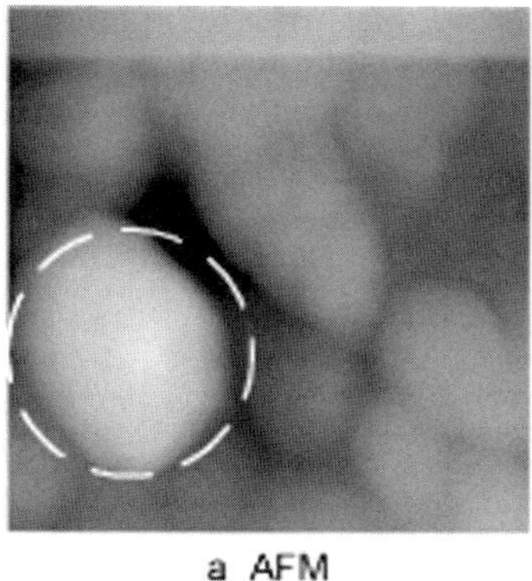

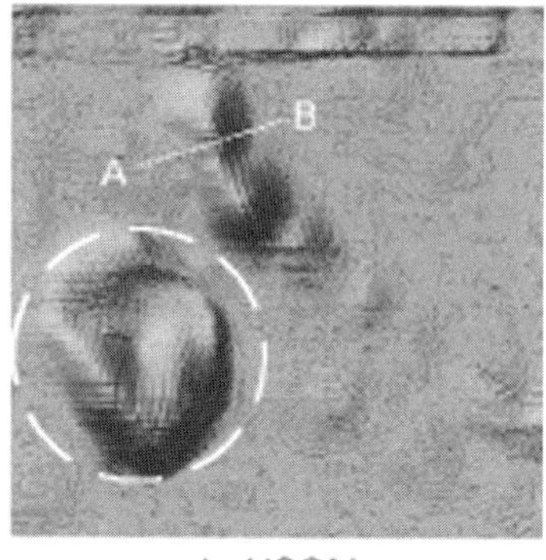

Fig. 10.10. AFM (**a**) and NSOM (**b**) images of Au microcrystals (4×4 mm). The NSOM image (**b**) shows a view of a bird's eye with the light from the direction of the upper left side of this paper

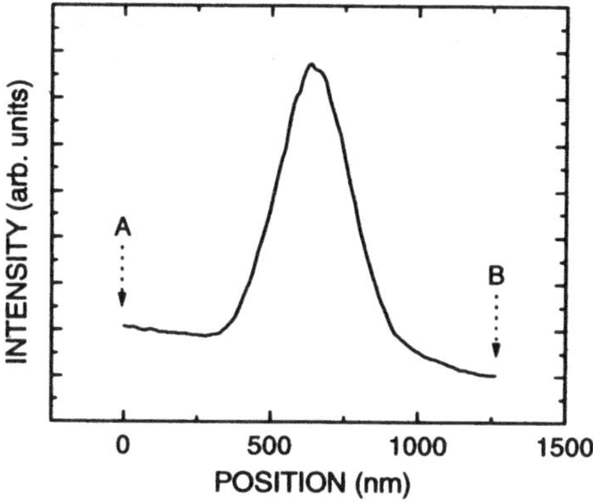

Fig. 10.11. Cross-sectional intensity along the line from A to B in Fig. 10.10b

in Fig. 10.10 seems to be one particle in the AFM image, whereas the NSOM image of the same region is not uniform and not so simple. This may be due to electromagnetic-field enhancement of the surface plasmon in Au microcrystals [9].

The cross-sectional intensity along the line from A to B in Fig. 10.10b is shown in Fig. 10.11. From this figure, the spatial resolution of the NSOM is estimated at about 200 nm by measuring the distance between the bottom and the peak position. The aperture size of the fiber tip, it is thought, becomes larger during scanning, and as a result, the spatial resolution is reduced a little.

10.2.3 Conclusion

We constructed a plain-type low temperature NSOM system. The tip height is controllable by monitoring either tunneling current or shear force. The experimental accessibility is improved, and we can change the sample easily even at low temperatures. We measured Au microcrystals as a test sample, and observed clear images by STM, AFM, and NSOM.

10.3 STM-Induced Luminescence

Optical properties have also been detected locally by electron tunneling in scanning tunneling microscopy (STM) as well as by scanning near-field optical microscopy(SNOM). The former gives higher spatial resolution than SNOM. Photon emission stimulated by electron tunneling was originally discovered in 1976 by Lambe and McCarthy from a metal–oxide–metal junction [10]. Spatially-resolved measurement of electron-tunneling-induced photon emission was achieved at a STM tip–vacuum–sample junction in 1988 by Gimzewski et al. [11]. Since this pioneering work, photon emission induced by STM has attracted considerable interest for surface analysis on a nanometer scale because the intensity map reflects various surface properties at high resolution in STM [12–17,19,22]. Photon emission by STM is induced by electron-impinging on a sample, but it is not a simple extension of cathode luminescence (CL) in SEM to the STM version. In STM-induced photon emission, the STM tip is so close to the sample that the resulting electromagnetic coupling of tip and sample is strong enough to cause different emission characteristics from CL in SEM, i.e., high emission efficiencies and different emission spectra [12]. Some researchers have succeeded in obtaining STM-induced photon maps with nanometer-scale or atomic-scale resolution in ultrahigh vacuum [13,14,16,19] as well as in air [15] and liquid [23]. The emission spectra at selected points on an STM image can be obtained for comparison with geometric structures [17]. The spectrally resolved image from the mapping of emission spectra can also be made at the same time as the STM measurements for a detailed comparison of optical properties with the geometric structure [22]. Recently, the effect of magnetic circular dichroism of emitted light was used to study magnetic properties on a nanometer scale [24,33,34].

10.3.1 Theoretical Model

The photon emission from a STM junction is considered the radiation from a local plasmon which is excited by inelastic electron tunneling near the local region between the STM tip and sample metals [12,25–27]. The nature of the local plasmon is determined by the boundary conditions of potentials of the STM tip and the metal surface [25–27]. Then, the spectra of the local plasmon should be closely related to the geometric structure of the STM tip and the sample metals. In addition, the current flow of tunneling electrons in the range of 1 nA is a discrete event on the timescale of plasmon oscillation, so that the fluctuation of tunneling current has influences on tunneling-induced dipole oscillation [28].

Theoretical analyses of emission spectra are carried out by solving Maxwell's equation in which the response of materials, tip, and sample is defined as a complex dielectric function, based on the following assumption about the tunneling system: the STM tip is modeled by a metallic sphere, and the sample is assumed to be a flat surface or a metallic sphere. In the calculation, the effective dipole induced by electron tunneling is calculated in the nonretarded limit for the above model following Rendell and Scalapino [29], and the radiation intensity from the dipole obtained is calculated by Green's function method [30]. The effect of retardation on the emission spectra studied by Johansson is rather small for the sample material of Au or Cu and a W or Ir tip, but is pronounced for Ag samples [31].

10.3.2 Experimental Method

To measure STM-induced luminescence, an STM system is combined with an optical detection system, as shown in Fig. 10.12 [18,19]. The optical measurements were made at the same time as the STM images by using an in-house made UHV-STM combined with the photon detection system.

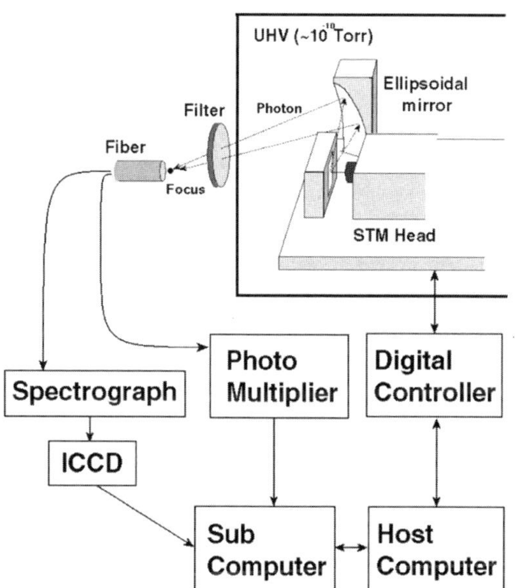

Fig. 10.12. System for measureing STM-induced luminescence

For STM control of the optical measurements, a digital computer system with microcomputers and a digital feed back controller with a digital signal processor (DSP; TMS320C30, Texas Instruments) is used because variable and complex data acquisition is easily realized by changing the software without a significant change in the hardware.

Photons emitted from the tip–sample junction are collected by using an ellipsoidal mirror. The photon focused by the mirror is guided through optical fibers to photomultipliers (Hamamatsu, R943-02) or a spectrograph (Kaiser Optical Systems, Inc., Holo Spec f/2.2). A fiber cannot be used for polarization measurements of emitted light for the study of magnetic properties because of the depolarization of light through it. The spectra were recorded with an ICCD (intensified charge coupled device, ICCD-1024E, Princeton Instruments, Inc.) detector with 1024 × 256 pixels. Using this spectrograph and ICCD, this system can cover a wave length range of 385–800 nm.

By acquiring emission spectra while scanning the tip position, a constant current topographic image and spatially resolved emission spectra or photon intensity can be obtained simultaneously. Based on the spatially resolved emission spectra, a spectrally-resolved photon image with a given spectral range can be displayed after the measurement.

10.3.3 Results

Correlation between Photon Intensity Map and Topography [21]. Many researchers have reported a close correlation between photon intensity and topographic height for metal particles in STM [13,15,19], as shown in Fig. 10.13. Figure 10.13 shows (a) an STM topographic image of Au grains and (b) a simultaneously recorded photon map of the corresponding scan area. Figure 10.13c compares the cross section of the STM topographic image with the photon intensity map on a line A across the particles. We can see that the photon intensity has a linear correlation with the height of the Au

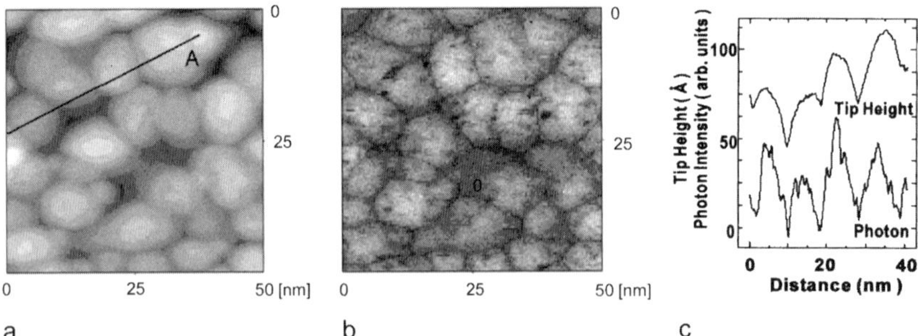

Fig. 10.13. **a** STM image of Au grains. **b** Photon map. **c** Cross-sectional profile of figures **a** and **b**

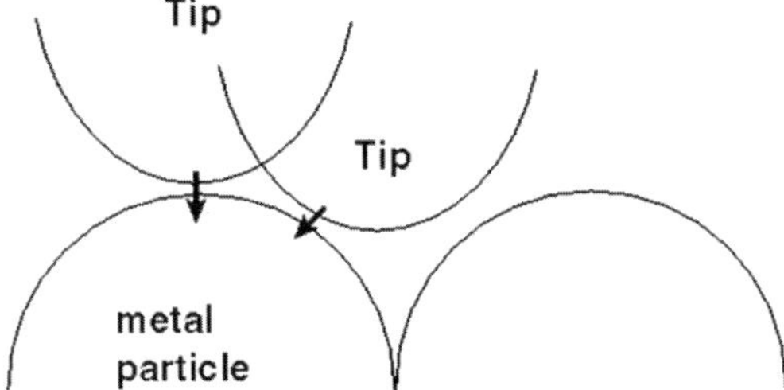

Fig. 10.14. Orientation change of an STM-induced dipole accompanying the movement of the STM tip across the particles

grains in the STM image. In this section, a simple model to explain the linear correlation between photon intensity and the topographic height of sample is described.

STM-induced luminescence is interpreted in terms of the dipole radiation induced during the tunneling process at the tip apex. The oscillating dipole in STM-induced luminescence can be created at a tunneling site, connecting the nearest points on the surfaces of tip and sample, in the direction shown in Fig. 10.14. As the tip scans along the sample surface, the position of the two points in the tunneling gap keeps changing, depending on the topography of the sample surface. The orientation of the dipole also changes accordingly and causes the total intensity of light emission to vary.

The total intensity of light from an STM-induced dipole as a function of the direction of the dipole can be calculated on the basis of the theory by Greenler [32] who calculated the polarization dependence of the intensity of dipole radiation from a molecule adsorbed on a metal surface for the following three cases of dipole orientation: case 1 is a dipole axis perpendicular to the surface, case 2 is a dipole axis parallel to the surface and parallel to the plane of the emitted light and the surface, and case 3 is a dipole axis parallel to the surface and perpendicular to the plane of the emitted light and the surface normal.

The tunneling-induced dipole with an arbitrary orientation in STM may be described as a linear combination of two independent polarizations, case 1 and case 2. When the induced dipole D is oriented in a direction α, measured from the normal to the surface, the dipole D is given as

$$D = D_{\mathrm{n}} \cos\alpha + D_{\mathrm{p}} \sin\alpha \,, \tag{10.6}$$

where $D_{\mathrm{p}}, D_{\mathrm{n}}$ are the dipoles of case 1 and case 2, respectively. On the basis of this model, Fig. 10.15 shows the total intensity of light from an STM-induced dipole as a function of $\cos\alpha$ for gold with optical constants $n = 0.205$, $k = 3.305$ at a wavelength of 600 nm. If we assume that metal particles are spherical, $\cos\alpha$ is proportional to the movement of the STM tip along the normal to the surface, i.e., the topographic height of the metal particles. Fig-

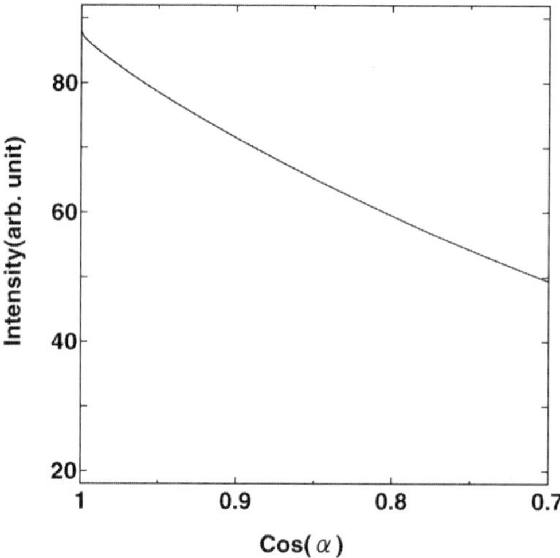

Fig. 10.15. The dipole-angle dependence of light intensity on the STM-induced dipole

ure 10.15 indicates that the intensity is nearly proportional to $\cos\alpha$ at small angles. This fact explains the experimental observation in which the STM induced photon intensity from spherical metal particles is closely correlated with the topographic height, as shown in Fig. 10.13.

Spectrally Resolved Mapping of STM-Induced Luminescence. To study the optical properties of various materials on a nanometer scale in correlation with the nanometer scale structure, we need to carry out spectral measurements at every pixel in an STM image [20,22].

The authors have made a spectral mapping measurement of STM-induced light from metal films which is recorded simultaneously with an STM image using an ICCD detector and a spectrograph [22]. The spectrally resolved photon maps are obtained from the map of the emission spectra. We have observed that the correlation between the photon map and the STM topography depends on the spectral region in which the emission intensity is used for the spectrally-resolved photon map.

In Fig. 10.16a–f, we show a topographic STM image of Au grains (a) and the spectrally resolved photon maps (b–f) which are obtained from a spectral mapping measurement. Each photon image is the map of the emission intensity which is integrated for a given spectral band 40 nm wide in wavelength for the spectral range from 600 through 800 nm, as indicated in the figure caption.

Nearly spherical Au particles (grains) ranging from 5 to 20 nm in diameter are observed in the STM image, and the photon maps reveal some character-

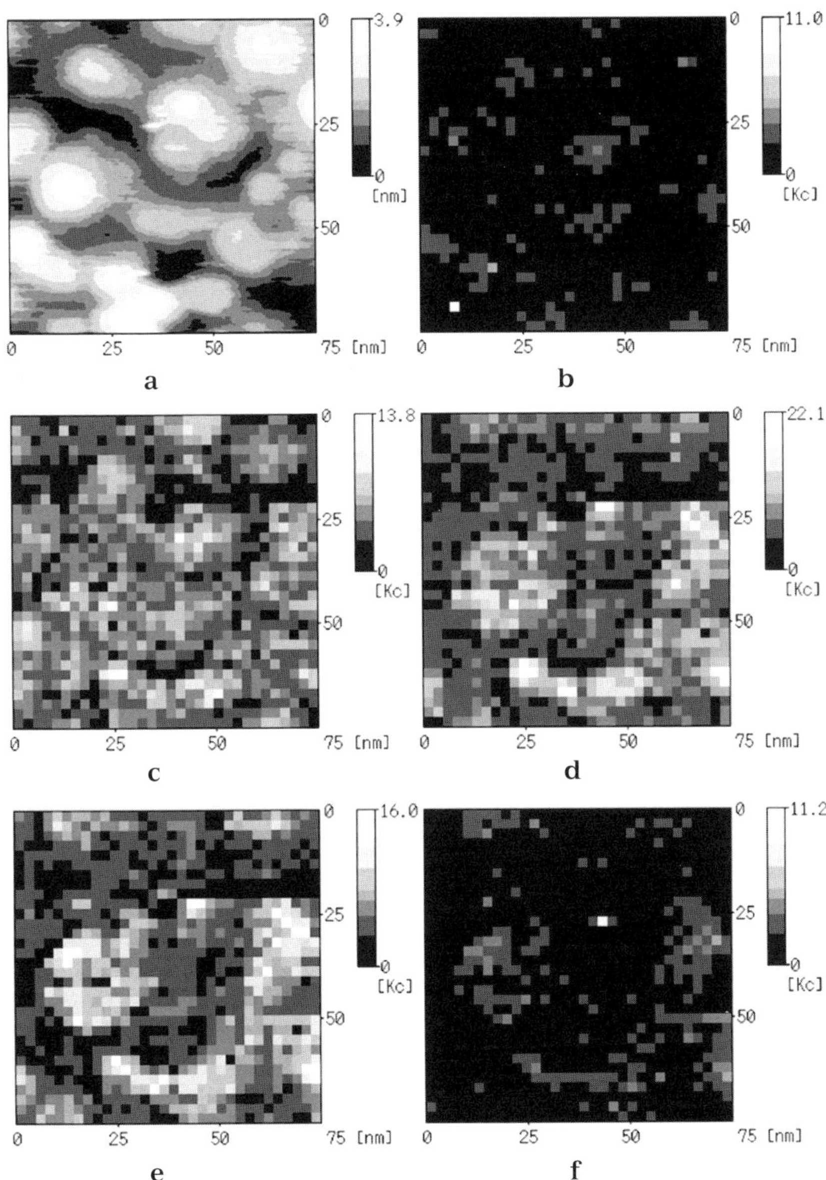

Fig. 10.16. **a** STM image of gold particles (scan area 75×75 nm^2, current 10 nA, sample bias +2 V) and spectrally resolved photon maps in the spectral range of **b** 600–640 nm, **c** 640–680 nm, **d** 680–720 nm, **e** 720–760 nm, and **f** 760–800 nm

istic correlations with the topography. The comparison between isochromatic photon maps reveals different correlations with STM topography. Almost all grains in the photon map at shorter wavelengths (Fig. 10.16c) show contrast similar in the photon map to the topography of the grains. On the other hand, at longer wavelength (Fig. 10.16e), only specific grains show high intensity. This result shows that there are two kinds of grains: one is bright at longer wavelengths and the other is not.

10.3.4 Conclusion

Spectrally and spatially resolved luminescence on a nanometer scale can be measured by STM combined with an optical detection system. The theoretical and experimental methods were described. The correlation of the emitted intensity with topography was interpreted by a simple model. The spectrally resolved photon map can be correlated with the topographic STM image. STM-induced luminescence gives us various information on the electronic and magnetic properties of fine particles and surface layers on a nanometer scale if we analyze the relation between the emission spectra and geometrical structures and the polarization of the emitted light on the basis of the magneto-optical effect.

10.4 Energy Modulation of Electrons with Evanescent Waves

10.4.1 Sensing an Optical Near-Field with Electrons

A recent advance in scanning near-field optical microscopy (SNOM) provides a useful means for observing small objects at high spatial resolution beyond the diffractive limits of about a half wavelength in conventional optics [35,36]. Since the spatial resolution for imaging in SNOM is determined primarily by the tip size of the near-field probe [37], much effort has been made to develop fine probes with a submicroscopic dimension at the apex. So far, various kinds of near-field probes with a tip size less than several tens of nanometers have been developed and used, as described in Sects. 10.2 and 10.3.

From this point of view, an electron beam would be a potential near-field probe because electrons are smaller than an atom [38,39]. In this section, the feasibility of probing optical near-fields by using an electron beam is discussed.

10.4.2 Metal Microslit

An optical near field is induced in the proximity of a small object by light illumination (refer to Fig. 10.17). The near-field contains an evanescent wave which has a wave number k_{ev} much greater than that of the light in free space

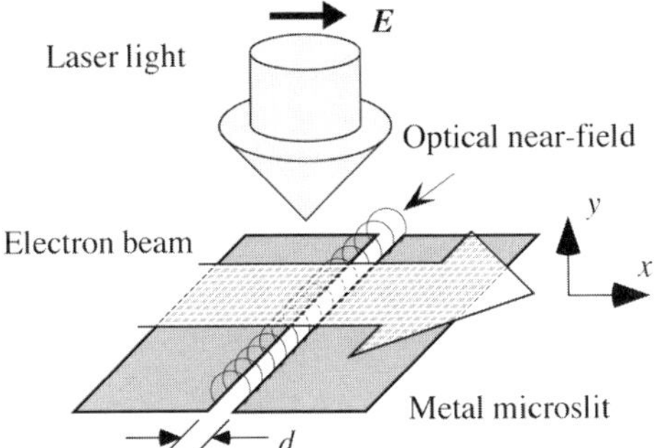

Fig. 10.17. Theoretical model for energy modulation of an electron beam with an optical near-field induced on the metal microslit by laser illumination

[40]. When an electron beam with a velocity v passes through the near-field region, it is expected that the electrons will interact with the evanescent wave when the phase velocity, $v_{\mathrm{p}} = \omega/k_{\mathrm{ev}}$, is equal to v, where ω is the angular frequency of the light, because energy and momentum conservation are satisfied for the interaction [41,42]. So far, the interaction between electrons and an evanescent wave has been experimentally demonstrated by several authors using a diffraction grating as an interaction circuit, but their experimental frequencies still remain in the far-infrared regon [43–45].

To evaluate the electron–light interaction quantitatively in the optical region, a metal microslit has been proposed as an optical near-field generator [46]. The simple structure of the slit enables us to precisely analyze the field distributions and the interaction with electrons theoretically and also experimentally. Energy changes in electrons passing through the microslit have been estimated using the theoretical model shown in Fig. 10.17 [39]. The electrons with an initial velocity v pass through the slit of width d in the x direction at a distance y from the slit surface. A laser light polarized in the x direction is normally incident on the surface of the slit. First, the near-fields, including electric and magnetic field components on the slit, are determined using the method of moments [47]; second, energy changes in electrons with the near-field are estimated by computer simulation using Lorentz force equations; finally, the optimum slit structures for various experimental parameters such as the laser wavelength and the electron velocity are determined from the calculated results.

Figure 10.18 shows calculated wave-number spectra for the near-field distributions of the electric vector E_x on three different slits at $y = 0.01\lambda$. The abscissa is the wave number k_{ev} and the corresponding electron velocity v that is determined by the relation $v = \omega/k_{\mathrm{ev}}$. These are normalized to $k_0 = 2\pi/\lambda$ and the speed of light c, respectively. The ordinate is the amplitude of the wave component. The thick solid curve represents the calculated maximum energy changes in electrons with v for $d = 0.62\lambda$. In the calculation, a CO_2

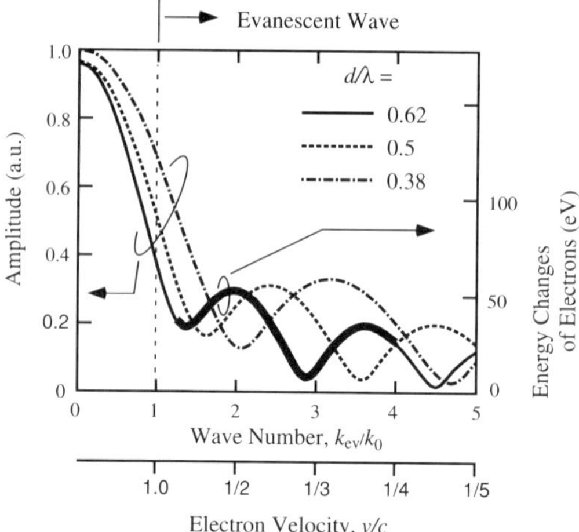

Fig. 10.18. Calculated wave-number spectra of near-field distributions (x component) on the three slits with $d/\lambda = 0.38, 0.5$, and 0.62. The thick curve represents the maximum electron-energy changes for different initial velocities between $v/c = 0.25$ and 0.8 which correspond to electron energy between 17 and 340 keV

laser with a wavelength of 10.6 μm and a power density of $10^8 \omega$ cm^2 was assumed as an incident wave. This power density corresponds to a 10-kW output power focused on a 100-mm diameter area.

As seen from Fig. 10.18, the variation in electron-energy changes with v is proportional to the amplitude of the evanescent wave with corresponding k_{ev}. This calculated result indicates that the electrons interact only with the evanescent wave whose phase velocity is equal to the electron velocity. Therefore, using electron beams with different initial velocities, we can measure the wave-number distribution of a near-field as well as its spatial intensity distribution, with minimum disturbance to the field distributions. This is the distinct feature of the electron-beam probe.

10.4.3 Experiment

First, experimental demonstration for energy modulation of an electron beam with the microslit was carried out using a CO_2 laser at $\lambda = 10.6$ μm [48]. Figure 10.19 is a schematic of the experimental setup. The microslit consists of two polished copper blocks about 12 μm wide. The electron beam had an initial energy of 80 keV ($v \sim 0.5c$) and a diameter of 240 μm. The electron energy was measured by using a retarding field analyzer. This analyzer has a resolution better than 0.8 eV (full width at half maximum). The electromechanical Q-switched (EMQ) CO_2 laser [49] used in the experiments oscillates in the TEM00 mode and generates output pulses with a

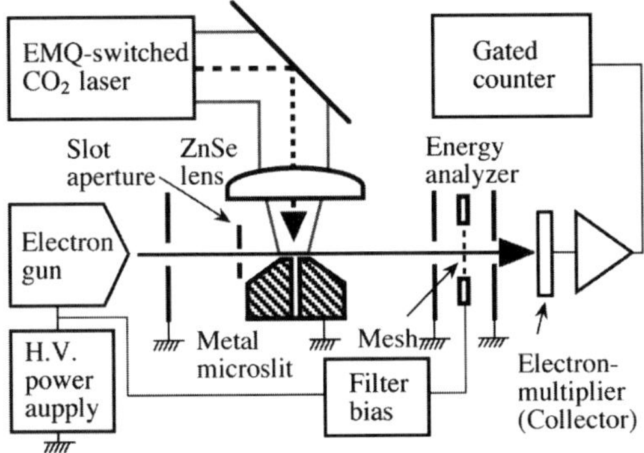

Fig. 10.19. Experimental setup

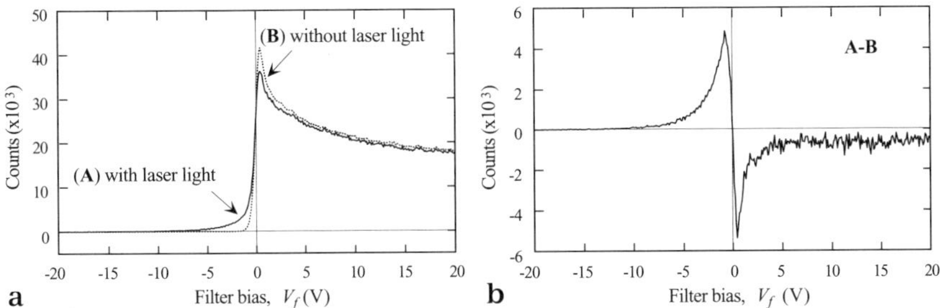

Fig. 10.20. **a** Measured electron-energy spectra (A) with and (B) without laser illumination, and **b** the difference, A–B, between the two spectra

maximum peak power of 9 kW and 140 nsec wide. The laser beam was focused onto the slit surface, whose a diameter was about 200 μm, through a ZnSe lens.

The pulsed laser output modulates the energy of the electron beam, so that the electron current through the analyzer changes during the pulse. The electrons that passed through the energy analyzer were detected by a secondary electron-multiplier (collector) connected to a gated counter which is triggered by the laser pulse.

Figure 10.20 shows (a) the measured energy spectra of electrons with and without laser illumination and (b) the difference between the two energy spectra. The abscissa is the filter bias voltage V_f which is the retarding potential of the energy analyzer. The ordinate is the number of electrons counted by the gated counter with an integration time of 1 s. When the laser beam irradiates the slit, energy spectrum B changes to spectrum A with a wider energy spread. The difference, A–B, between the two energy spectra in Fig. 10.20b represents the number of electrons that interacted with the laser light at the modulated energy. The measured maximum energy spread is about ± 18 eV which compares to the theoretical value of ± 29 eV calculated by computer simulation. These experimental results demonstrate that an optical near-field can be detected by using an electron beam at an infrared wavelength.

10.4.4 Conclusion

Electron–light energy exchanges in an optical near-field region on a metal microslit have been investigated. The theory has shown that an electron beam interacts with an evanescent wave in the near field, when the phase velocity of the wave is equal to the electron velocity. The experiments at the wavelength of 10.6 μm indicate that the energy of an 80-keV electron beam can be modulated more than 10 eV by a 9-kW laser beam. In the measurements, a high signal-to-noise ratio of more than 500 was achieved. Those results imply that detecting of an optical near-field with an electron beam at a shorter wavelength may be feasible.

10.5 Manipulation of Particles by Photon Force

Since Ashkin demonstrated the laser trapping of Mie particles in 1970 by using photon pressure, the laser trapping technique has been applied to various fields such as optical acceleration, levitation, optical tweezers, and so on [50]. As a technique of manipulating small particles on a surface, Kawata et al. demonstrated using an evanescent wave produced in the near-field of a high-refractive-index prism under conditions of total internal reflection [51,52]. This method allows the easy separation of the light propagating region from the driven particles and allows driving the particles located near the boundary discretely because an evanescent field is localized only the near region of the surface of a prism. In 1996, Kawata also demonstrated a technique for driving both dielectric and metallic small particles by using the evanescent field generated in the vicinity of a single-mode channeled waveguide [53]. In this section, we report experimental results of manipulating small particles in an evanescent field generated by a multimode channeled waveguide.

10.5.1 Method

Figure 10.21 shows the schematic diagram of the experimental setup. We used polystyrene beads 4.0 μm in diameter for the driven particle. The refractive index of the polystyrene bead is $n = 1.59$, and the specific gravity is 1.05 g/cm^3. The particles were dispersed randomly in purified water of index $n = 1.33$, dripped on the channeled waveguide with spacers (made of microscope cover glass 170 μm thick), and covered by another piece of cover glass (170 μm). Laser light is fed onto the end face of the waveguide using the end-fire method with a microscope objective of 20× (NA = 0.40). An evanescent field is generated in the aqueous region above the waveguides. Particles are observed using a microscope (40× magnification and N.A. = 0.65) and a CCD camera.

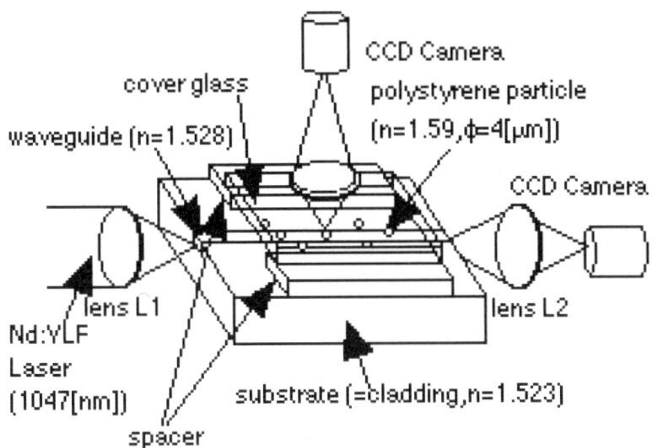

Fig. 10.21. Experimental setup for the optical propulsion of particles in a multimode channeled waveguide

10.5.2 Experiments

We made straight channeled waveguides on the glass surface by using the thermal ion-exchange method [54]. The widths of the waveguides were from 4.0–11.0 μm. The cladding glass substrate was immersed in molten KNO_3 to exchange the glass Na^+ ions with K^+ ions. The temperature of the KNO_3 was 370°C during ion exchange, and the exchange time was 120 min. After ion exchange, the waveguides were annealed at 370°C for 180 min. In the constructed K^+ ion-exchanged waveguides, the refractive index of the core was almost $n = 1.528$ and that of glass substrate (cladding) was $n = 1.523$.

Manipulation Using an IR-Laser (Nd:LYF laser). We used a cw Nd:YLF laser ($l = 1047$ nm) as the light source. When the laser beam propagated inside the waveguide, scattered light was observed only from the particles located on or in the neighborhood of the core, as shown in Fig. 10.22. This scattered light was produced due to the conversion of evanescent light to propagating light. From this picture, we confirmed that the evanescent field was localized in the vicinity of the core and that only the particles close to the waveguide region interact with the evanescent field and scatter it.

Figure 10.23 shows an experimental result of the propulsion of small particles. In Fig. 10.23, the width of the waveguide core is 3.8 μm. This waveguide can propagate only the fundamental mode (E^x_{00} mode), whereas higher order modes are cut off. (The mode number corresponds to the number of nodes (dark points) in the near-field patterns). Figure 10.23a shows the near-field pattern of the guided mode propagating in the waveguide and demonstrates that only the fundamental mode is excited and propagated. The power of the laser light coming out of the end facet of the waveguide was 6 mW. Figure 10.23b is a series of photographs of the particles taken at 3-s intervals

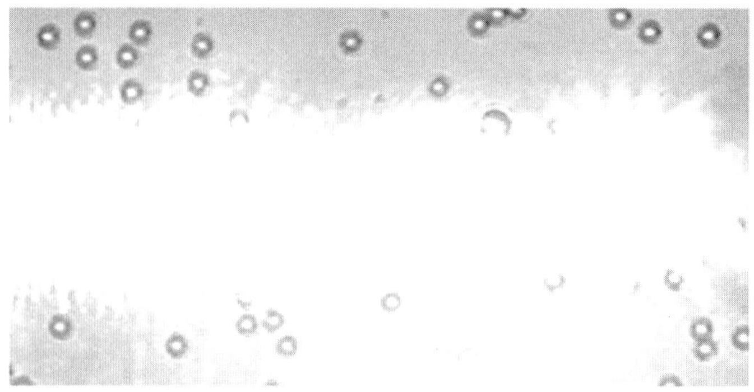

Fig. 10.22. Experimental scattering of the evanescent field by small particles

a

b

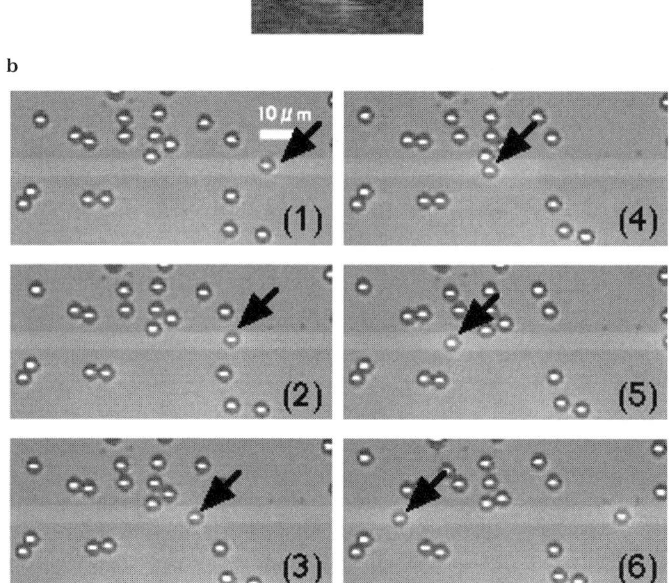

Fig. 10.23. **a** Near-field pattern of the E_{00}^{x} guided mode, and **b** photograph sequence taken at 3-s intervals, illustrating the movement of polystyrene spheres 4 μm in diameter along the waveguide channel

with an IR cut filter for cutting off the scattered light of the Nd:YLF laser. From Fig. 10.23b, we found that only the particles localized in the waveguide channel are trapped at the center of the waveguide and are driven longitudinally in the direction of the waveguide channel and particles outside the

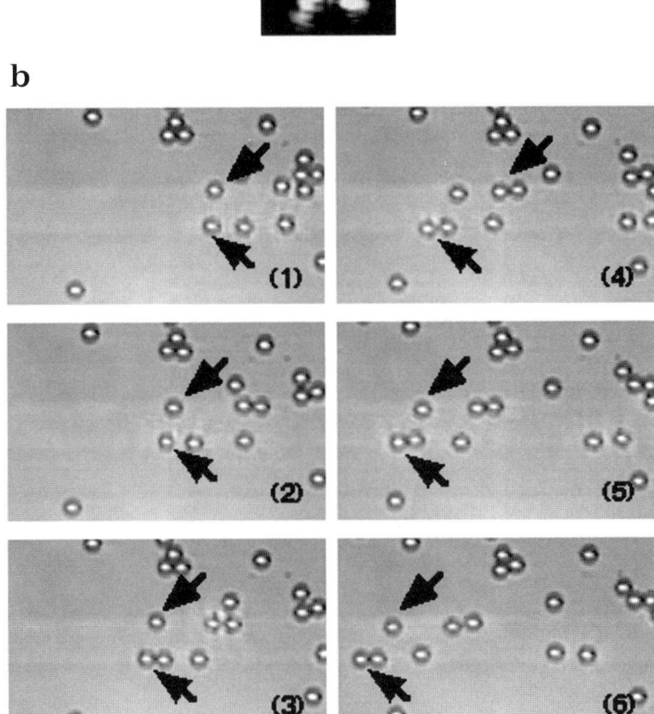

Fig. 10.24. **a** Near-field pattern of the E_{01}^x mode, and **b** photographic sequence taken at 5-s intervals displaying the motion of spheres propelled by radiative force from the E_{02}^x guided mode

waveguide remained unmoved. The microspheres moved at a constant speed of 4.7 µm/s across a total distance of more than 1.0 cm.

Figure 10.24 is also an experimental result. The width of the waveguide was 11.0 µm and this waveguide behaved as a multimode waveguide. In the multimode waveguide, each guided mode can be excited independently by slightly changing the position of the focused beam spot at the input waveguide end. We excited the E_{01}^x propagative mode in the waveguide. Figure 10.24a shows the near-field pattern of the guided modes: it demonstrates the two characteristic high-intensity spots of the E_{01}^x mode. Figure 10.24b shows a series of photographs taken at 5-s intervals that illustrates the movement of polystyrene spheres along the waveguide. We found that the microbeads are transversely trapped at two positions which correspond to the high field-intensity region of the E_{01}^x mode, and they moved in a straight line, longitudinally. From this result, we confirmed that radiative force attracts the

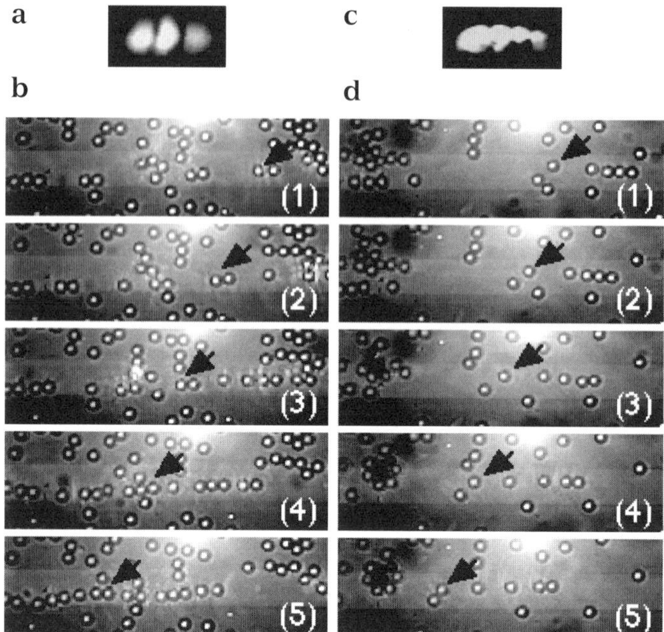

Fig. 10.25. **a** Near-field pattern of the E^x_{02} mode, **b** movement of the particles taken at 6-s intervals, **c** near-field pattern of the E^x_{03} mode, and **d** movement of the particles

particles laterally toward the high-intensity region, and even if the E^x_{01} mode was propagating inside the waveguide, the radiative force propels the particles in the light propagating direction along the waveguide channel as well as in the E^x_{00} mode.

We measured the velocity of the particles against the guided modes under the same output power of 14mW at the end facet of the waveguide. Figure 10.26 shows the relationship of the driven velocity to the propagating mode. This result indicates that the driven velocity decreases with the excitement of higher modes. This can be explained as follows: The field distribution of the lower mode represented by the E^x_{00} mode is concentrated at the center of the waveguide, whereas the field distribution of the higher mode extends the entire length of the waveguide. Therefore, the field density of the higher mode is weaker than that of the fundamental mode. Low field density leads to weak radiative pressure.

The large dispersion of the velocity of the E^x_{01} mode seen in Fig. 10.26 is due to the strong dependence of the radiative pressure on the distance between the driven particle and the waveguide. The intensity of the evanescent field decreases rapidly from the surface of the waveguide, and the penetration depth of the field is almost equal to the wavelength. Therefore, this leads to

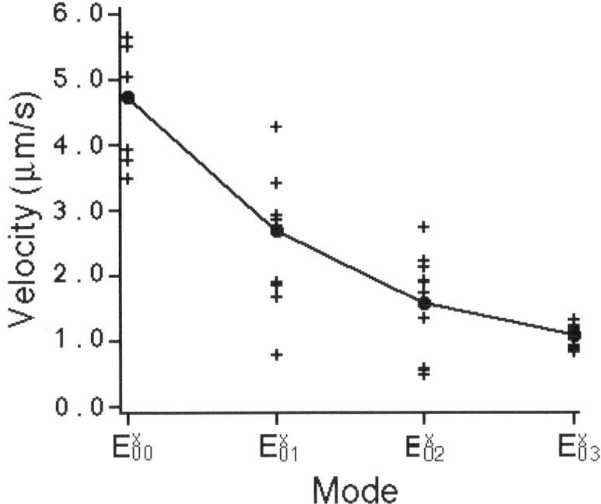

Fig. 10.26. Plot of the velocity of the driven particles versus the excited mode of the waveguide

strong radiative pressure on the particles on the waveguide which drives them quickly. However, if the particles are floating slightly away from the surface, the photon pressure is drastically weakened, thus decreasing the velocity.

Experimental Result Using a Green Laser (Ar^+ Laser). Figure 10.27 shows another experimental result. The same optical setup was used, except that a cw Ar^+ laser (514.5 nm) was substituted for the light source. The power of the light coming out of the end facet of the waveguide was 180 mW. Figure 10.27a shows the near-field pattern of the guided mode propagating in the waveguide, demonstrating that only the fundamental mode is excited. Figure 10.27b is a series of photographs of small particles taken at 20-s intervals with the Ar^+ laser cutoff filter. When we used the Ar^+ laser as the light source, the particles were swept out of the core region and were not forced along the waveguides. As shown in Figs. 10.23–10.25, when we used the Nd:YLF laser as the light source, particles dispersed on the waveguide were trapped above the core and accelerated along the waveguide channel, whereas in this result, all of the particles on the waveguide were removed from the core region. These experimental results do not agree with classical analysis, which predicts an attractive radiative force acting on the high-index particles, trapping them in high light-intensity regions (above the waveguide) due to the gradient force [55,56]. The radiative force exerted on a particle is peculiar and strongly dependent on several parameters such as particle diameter, a complex-valued refractive index, the distance between the particle and the waveguide, the penetration depth of the evanescent field, and the wavelength of the light. Therefore, it is quite difficult to estimate the radiative force given to the particular movement. The reason for the differ-

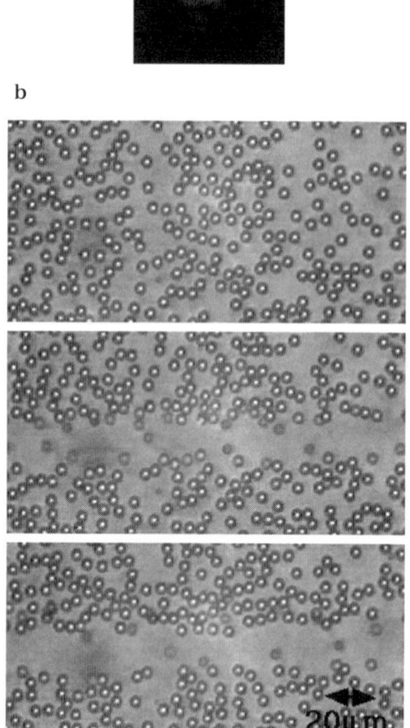

Fig. 10.27. **a** Near-field pattern of the guided mode, and **b** series of photographs of the particles taken at 20-s intervals

ence between the experimental results with the Nd:YLF laser and those with the Ar^+ laser remains to be elucidated. We may also consider the thermal effect due to light absorbance by the particle because the imaginary part of the refractive index of the polystyrene for the Ar^+ laser is much higher than the imaginary part of the refractive index of the Nd:YLF, and particles that absorb the Ar^+ laser become heated.

10.5.3 Conclusion

We have demonstrated the experimental results of a technique for manipulating small particles optically along a channel waveguide using radiative pressure caused by the higher guided mode of a multimode channeled waveguide. Even if we used the multimode channeled waveguide and excited the higher guided modes, we could drive the particles along the waveguide by trapping them in the waveguide channel. A combination of this technique with the techniques employed in optical integrated circuits such as branching waveguides, optical switching devices, and optical modulators will be applicable in cell sorting, transporting, and mixing devices. We also demonstrated that changing the wavelength of the light source leads to the intriguing result in which high-index small particles were repulsed from the waveguide.

References

1. E. Betzig, J. K. Tratman, L. S. Weiner, T. D. Harris, and R. Wolfe: Appl. Opt. **31**, 4563 (1992)
2. E. Betzig, J. K. Tautman, R. Wolfe, E. M. Gyorgy, P. L. Finn, M. H. Kryder, and C.-H. Chang: Appl. Phys. Lett. **61**, 142 (1992)
3. M. Vaez-Iravani and R. Toledo-Crow: Appl. Phys. Lett. **62**, 138 (1993)
4. D. A. Higgins, D. A. Vanden Bout, J. Kerimo, and P. F. Barbara: J. Phys. Chem. **100**, 13794 (1996)
5. N. Umeda, S. Wakayama, S. Arakawa, A. Takayanagi, and H. Kohwa: Proc. SPIE **2873**, 119–122 (1996)
6. J. C. Wittmann and P.Smith: Nature **352**, 414 (1991)
7. R. D. Grober, T. D. Harris, J. K. Trautman, and E. Betzig: Rev. Sci. Instrum. **65**, 626 (1994)
8. Y. Toda, S. Shinomori, K. Suzuki, and Y. Arakawa: Solid State Electron. **42**, 1083 (1998)
9. V. A. Markel, V. M. Shalaev, P. Zhang, W. Huynh, L. Tay, T. L. Haslett, and M. Moskovits: Phys. Rev. B **59**, 10903 (1999)
10. J. Lambe and S. L. McCarthy: Phys. Rev. Lett. **37**, 923 (1976)
11. J. K. Gimzewski, B. Reihl, J. H. Coombs, and R. R. Schlittler: Z. Phys. **B72**, 497 (1988)
12. R. Berndt and J. K. Gimzewski: Phys. Rev. Lett. **67**, 3796 (1991)
13. R. Berndt and J. Gimzewski: Phys. Rev. B **48**, 4746 (1993)
14. R. Berndt, R. Gaisch, W. D. Schneider, J. K. Gimzewski, B. Reihl, R. R. Schlitter, and M. Tschudy: Phys. Rev. Lett. **74**, 102 (1995)
15. V. Sivel, R. Coratger, F. Ajustron; and J. Beauvillain: Phys. Rev. B **45**, 8634 (1992)
16. A. W. McKinnon, M. E. Welland, and T. M. H. Wong: Phys. Rev. B **48**, 15250 (1993)
17. K. Ito, S. Ohyama, Y. Uehara, and S. Ushioda: Surf. Sci. **324**, 282 (1995)
18. R. Nishitani, K. Suga, T. Umeno, A. Kasuya, and Y. Nishina: Mater. Sci. Eng. **A217/218**, 99 (1996)
19. T. Umeno, R. Nishitani, A. Kasuya, and Y. Nishina: Phys. Rev. B **54**, 13499 (1996)
20. R. Nishitani, T. Umeno, A. Kasuya, and Y. Nishina: Surf. Rev. Lett. **4**, 1009 (1997)
21. R. Nishitani, T. Umeno, and A. Kasuya: Jpn. J. Appl. Phys. **36**, L1545 (1997)
22. R. Nishitani, T. Umeno, and A. Kasuya: Appl. Phys. A **66**, S139 (1998)
23. R. Nishitani and A. Kasuya: Surf. Sci. **433/435**, 283 (1999)

24. A. L. Vazquez de Parga and S. F. Alvarado: Phys. Rev. Lett. **72**, 3726 (1994)
25. P. Johansson, R. Monreal, and P. Apell: Phys. Rev. B **42**, 9210 (1990)
26. B. N. J. Persson and A. Baratoff: Phys. Rev. Lett. **68**, 3224 (1992)
27. Y. Uehara, Y. Kimura, S. Ushioda, and K. Takeuchi: Jpn. J. Appl. Phys. **31**, 2465 (1992)
28. D. Hone, B. Muhlschlegel, and D. J. Scalapino: Appl. Phys. Lett. **33**, 203 (1978)
29. R. W. Rendell and D. J. Scalapino: Phys. Rev. B **24**, 3267 (1981)
30. A. A. Maradudin and D. L. Mills: Phys. Rev. B **11**, 1392 (1975)
31. P. Johansson: Phys. Rev. B **58**, 10823 (1998)
32. R. G. Greenler: Surf. Sci. **69**, 647 (1977)
33. N. Majlis, A. Levy Yeyati, F. Flores, and R. Monreal: Phys. Rev. **52**, 12505 (1995)
34. A. L. Vazquez de Parga and S. F. Alvarado: Europhysics Lett. **36**, 577 (1996)
35. D. W. Pohl, W. Denk, and M. Lanz: Appl. Phys. Lett. **44**, 651 (1984)
36. A. Harootunian, E. Betzig, M. Isaacson, and A. Lewis: Appl. Phys. Lett. **49**, 674 (1986)
37. U. Duing, D. W. Pohl, and F. Rohner: J. Appl. Phys. **59**, 3318 (1986)
38. H. Cohen, T. Maniv, R. Tenne, Hacohen Y. Rosenfeld, O. Stephan, and C. Coliex: Phys. Rev. Lett. **80**, 782 (1998)
39. R. Ishikawa, J. Bae, and K. Mizuno: Electron Energy Modulation with Near-Fields. 5th Int. Conf. Near Field Opt. Relat. Tech., Shirahama, Japan, 1998, pp. 173–174
40. G. A. Massey: Appl. Opt. **23**, 658 (1984)
41. R. H. Pantell: *Physics of High Energy Particle Accelerators, AIP Conf. Proc.* No. 87, 1981, American Institute of Physics, pp. 863–918
42. J. Bae and K. Mizuno: Jpn. J. Opt. **28**, 502 (1999) (in Japanese)
43. K. Mizuno, J. Bae, T. Nozokido, and K. Furuya: Nature **328**, 45 (1987)
44. J. Bae, H. Shirai, T. Nishida, T. Nozokido, K. Furuya, and K. Mizuno: Appl. Phys. Lett. **61**, 870 (1992)
45. J. Urata, M. Goldstein, M. F. Kimmitt, A. Naumov, C. Platt, and J. E. Walsh: Phys. Rev. Lett. **80**, 516 (1998)
46. J. Bae, S. Okuyama, T. Akizuki, and K. Mizuno: Nucl. Instrum. Methods Phys. Res. A **331**, 509 (1993)
47. T. Y. Chou and A. T. Adams: IEEE Trans. Electromagnetic Compatibility EMC-**19**, 65 (1977)
48. J. Bae, R. Ishikawa, S. Okuyama, T. Miyajima, T. Akizuki, T. Okamoto, and K. Mizuno: Appl. Phys. Lett. **76**, 2292 (2000)
49. J. Bae, T. Nozokido, H. Shirai, H. Kondo, and K. Mizuno: IEEE J. Quantum Electron. **30**, 887 (1994)
50. A. Ashkin: Phys. Rev. Lett. **24**, 156 (1970)
51. S. Kawata and T. Sugiura: Opt. Lett. **17**, 772 (1992)
52. T. Sugiura and S. Kawata: Bioimaging **1**, 1 (1993)
53. S. Kawata and T. Tani: Opt. Lett. **21**, 1768 (1996)
54. H. Nishimura, M. Haruna, and T. Suhara: *OpticaL Integrated Circuits* (McGraw-Hill, New York 1989)
55. A. Ashkin, J. M. Dziedzic, J. E. Bjorkholm, and S. Chu: Opt. Lett. **11**, 288 (1986)
56. A. Ashkin: Phys. Rev. Lett. **40**, 729 (1978)

Index

Springer Series in
OPTICAL SCIENCES

New editions of volumes prior to volume 60

1 **Solid-State Laser Engineering**
By W. Koechner, 5th revised and updated ed. 1999, 472 figs., 55 tabs., XII, 746 pages

14 **Laser Crystals**
Their Physics and Properties
By A. A. Kaminskii, 2nd ed. 1990, 89 figs., 56 tabs., XVI, 456 pages

15 **X-Ray Spectroscopy**
An Introduction
By B. K. Agarwal, 2nd ed. 1991, 239 figs., XV, 419 pages

36 **Transmission Electron Microscopy**
Physics of Image Formation and Microanalysis
By L. Reimer, 4th ed. 1997, 273 figs. XVI, 584 pages

45 **Scanning Electron Microscopy**
Physics of Image Formation and Microanalysis
By L. Reimer, 2nd completely revised and updated ed. 1998,
260 figs., XIV, 527 pages

Published titles since volume 60

60 **Holographic Interferometry in Experimental Mechanics**
By Yu. I. Ostrovsky, V. P. Shchepinov, V. V. Yakovlev, 1991, 167 figs., IX, 248 pages

61 **Millimetre and Submillimetre Wavelength Lasers**
A Handbook of cw Measurements
By N. G. Douglas, 1989, 15 figs., IX, 278 pages

62 **Photoacoustic and Photothermal Phenomena II**
Proceedings of the 6th International Topical Meeting, Baltimore, Maryland,
July 31 - August 3, 1989
By J. C. Murphy, J. W. Maclachlan Spicer, L. C. Aamodt, B. S. H. Royce (Eds.),
1990, 389 figs., 23 tabs., XXI, 545 pages

63 **Electron Energy Loss Spectrometers**
The Technology of High Performance
By H. Ibach, 1991, 103 figs., VIII, 178 pages

64 **Handbook of Nonlinear Optical Crystals**
By V. G. Dmitriev, G. G. Gurzadyan, D. N. Nikogosyan,
3rd revised ed. 1999, 39 figs., XVIII, 413 pages

65 **High-Power Dye Lasers**
By F. J. Duarte (Ed.), 1991, 93 figs., XIII, 252 pages

66 **Silver-Halide Recording Materials**
for Holography and Their Processing
By H. I. Bjelkhagen, 2nd ed. 1995, 64 figs., XX, 440 pages

67 **X-Ray Microscopy III**
Proceedings of the Third International Conference, London, September 3-7, 1990
By A. G. Michette, G. R. Morrison, C. J. Buckley (Eds.), 1992, 359 figs., XVI, 491 pages

68 **Holographic Interferometry**
Principles and Methods
By P. K. Rastogi (Ed.), 1994, 178 figs., 3 in color, XIII, 328 pages

69 **Photoacoustic and Photothermal Phenomena III**
Proceedings of the 7th International Topical Meeting, Doorwerth, The Netherlands,
August 26-30, 1991
By D. Bicanic (Ed.), 1992, 501 figs., XXVIII, 731 pages

Springer Series in

OPTICAL SCIENCES

---

70 **Electron Holography**
By A. Tonomura, 2nd, enlarged ed. 1999, 127 figs., XII, 162 pages

71 **Energy-Filtering Transmission Electron Microscopy**
By L. Reimer (Ed.), 1995, 199 figs., XIV, 424 pages

72 **Nonlinear Optical Effects and Materials**
By P. Günter (Ed.), 2000, 174 figs., 43 tabs., XIV, 540 pages

73 **Evanescent Waves**
From Newtonian Optics to Atomic Optics
By F. de Fornel, 2001, 277 figs., XVIII, 268 pages

74 **International Trends in Optics and Photonics**
ICO IV
By T. Asakura (Ed.), 1999, 190 figs., 14 tabs., XX, 426 pages

75 **Advanced Optical Imaging Theory**
By M. Gu, 2000, 93 figs., XII, 214 pages

76 **Holographic Data Storage**
By H.J. Coufal, D. Psaltis, G.T. Sincerbox (Eds.), 2000
228 figs., 64 in color, 12 tabs., XXVI, 486 pages

77 **Solid-State Lasers for Materials Processing**
Fundamental Relations and Technical Realizations
By R. Iffländer, 2001, 230 figs., 73 tabs., XVIII, 350 pages

78 **Holography**
The First 50 Years
By J.-M. Fournier (Ed.), 2001, 266 figs., XII, 460 pages

79 **Mathematical Methods of Quantum Optics**
By R.R. Puri, 2001, 13 figs., XIV, 285 pages

80 **Optical Properties of Photonic Crystals**
By K. Sakoda, 2001, 95 figs., 28 tabs., XII, 223 pages

81 **Photonic Analog-to-Digital Conversion**
By B.L. Shoop, 2001, 259 figs., 11 tabs., XIV, 330 pages

82 **Spatial Solitons**
By S. Trillo, W.E. Torruellas (Eds), 2001, 194 figs., 7 tabs., XX, 454 pages

83 **Nonimaging Fresnel Lenses**
Design and Performance of Solar Concentrators
By R. Leutz, A. Suzuki, 2001, 139 figs., 44 tabs., XII, 272 pages

84 **Nano-Optics**
By S. Kawata, M. Ohtsu, M. Irie (Eds.), 2002, 258 figs., 2 tabs., XIII, 313 pages